Recent Advances in Differential Equations

edited by

ROBERTO CONTI
Universita Degli Studi
Istituto Matematico "Ulisse Dini"
Firenze, Italy

ACADEMIC PRESS 1981
A Subsidiary of Harcourt Brace Jovanovich, Publishers
New York London Toronto Sydney San Francisco

ACADEMIC PRESS, INC.
111 Fifth Avenue, New York, New York 10003

United Kingdom Edition published by
ACADEMIC PRESS, INC. (LONDON) LTD.
24/28 Oval Road, London NW1 7DX

Library of Congress Cataloging in Publication Data
Main entry under title:

Recent advances in differential equations.

Discussions from a meeting held at the International Center for Theoretical Physics, Aug. 24-28, 1978 and sponsored by the U.S. Army Research Office.
Includes index.
1. Differential equations--Addresses, essays, lectures.
I. Conti, Roberto. II. United States. Army Research Office.
QA371.R34 515.3'5 81-15042
ISBN 0-12-186280-1 AACR2

PRINTED IN THE UNITED STATES OF AMERICA

81 82 83 84 9 8 7 6 5 4 3 2 1

Contents

Contributors

Numbers in parentheses indicate the pages on which the authors' contributions begin.

Jean-Pierre Aubin (1), *CEREMADE, CNRS Université Paris XI Dauphine, Paris, France*

Andrea Bacciotti (23), *Istituto Matematico, Università di Siena, Siena, Italy*

Karl Wilhelm Bauer (37), *Technische Universität Graz, Institut für Mathematik I, Graz, Austria*

S. R. Bernfeld (45), *Department of Mathematics, University of Texas at Arlington, Arlington, Texas*

I. Bihari (59), *Mathematical Institute, Hungarian Academy of Sciences, Budapest, Hungary*

Arigo Cellina (65), *Istituto di Matematica Applicata, Università di Padova, Padova, Italy*

Jagdish Chandra (71), *U. S. Army Research Office, Research Triangle Park, North Carolina*

Earl A. Coddington (81), *Department of Mathematics, University of California, Los Angeles, California*

Paul Wm. Davis (71), *Department of Mathematics, Worcester Polytechnic Institute, Worcester, Massachusetts*

Djairo G. de Figueiredo (89), *Departmento de Matematica, Universidade de Brasília, Brasília, Brazil*

Klaus Deimling (101), *Gesamthochschule Paderborn, Paderborn*

M. C. Delfour (111), *C. R. M., Université de Montréal, Montréal, Montréal, Québec, Canada*

J. G. Dos Reis (209), *Universidade de São Paulo, Ribeirão Preto, São Paulo, Brazil*

Angelo Favini (135), *Istituto di Matematica Generale e Finanziaria, Università di Bologna, Bologna, Italy*

A. A. Freiria (209), *Universidade de São Paulo, Ribeirão Preto, São Paulo, Brazil*

Wolfgang Hahn (143), *Technische Universität Graz, Institut für Mathematik I, Graz, Austria*

A. Halanay (155), *Faculty of Mathematics, Bucharest 1, Romania*

Chaim Samuel Hönig (199), *Instituto de Matematica e Estatistica, Universidade de São Paulo, São Paulo, Brazil*

A. F. Izé (209), *Universidade de São Paulo, São Carlos—São Paulo, Brazil*

R. Kannan (231), *Department of Mathematics, University of Texas at Arlington, Arlington, Texas*

V. Lakshmikantham (243), *Department of Mathematics, University of Texas at Arlington, Arlington, Texas*

J. J. Levin (261), *University of Wisconsin, Madison, Wisconsin*

K. Magnusson (271), *Control Theory Center, University of Warwick, Warwick, England*

Patrizia Marocco (281), *Istituto di Matematica, Università degli Studi, Trieste, Italy*

Jean Mawhin (287, 295), *Institut Mathématique, Université de Louvain, Louvain-La-Neuve, Belgium*

A. N. Michel (309), *Electrical Engineering Department and Engineering Research Institute, Iowa State University, Ames, Iowa*

R. K. Miller (309), *Mathematics Department, Iowa State University, Ames, Iowa*

John A. Nohel (317), *University of Wisconsin–Madison, Madison, Wisconsin*

Livio Clemente Piccinini (337), *Istituto di Matematica, Facoltà di Science Statistiche, Università di Padova, Padova, Italy*

A. Plís (373), *Mathematical Institute, PAN, Kraków, Poland*

A. J. Pritchard (271), *Control Theory Centre, University of Warwick, Warwick, England*

Paul H. Rabinowitz (379), *Department of Mathematics, University of Wisconsin–Madison, Madison, Wisconsin*

D. R. K. Rao (387), *Faculty of Sciences, Razi University, Kermanshah, Iran*

Vl. Răsvan (155), *I. C. P. E. T., Bucharest, Romania*

L. Salvadori (45), *Istituto di Matematica, Universitá di Trento, Trento, Italy*

George R. Sell (393), *School of Mathematics, University of Minnesota, Minneapolis, Minnesota*

Yasutaka Sibuya (405), *School of Mathematics, University of Minnesota, Minneapolis, Minnesota*

Steven Sperber (405), *School of Mathematics, University of Minnesota, Minneapolis, Minnesota*

Ulrich Staude (421), *Mathematisches Institut, Universität Mainz, Mainz, West Germany*

Roberto Triggiani (431), *Mathematics Department, Iowa State University, Ames, Iowa*

Michel Willem (287, 295), *Institut Mathématique, Université de Louvain, Louvain-La-Neuve, Belgium*

Preface

An international conference on recent advances in differential equations was held at the International Center of Theoretical Physics, Miramare (Trieste, Italy), August 24–26, 1978.

The purpose of the conference was that of reviewing the present status of research in the field of differential equations (ordinary, partial, and functional). Seven general lectures and a number of shorter communications provided a picture of a very wide area covering both theoretical aspects (differential operators, periodic solutions, stability and bifurcation, asymptotic behavior of solutions, etc.) and problems arising from applications (reaction–diffusion equations, control problems, heat flow, etc.).

The total number of participants was over 50, among whom were representatives of 15 countries.

We are most grateful to all participants in the conference and especially to the speakers for their contributions to the meeting. In addition we want to thank the U.S. Army European Research Office for financial support, the International Center of Theoretical Physics whose facilities and services were made available at no cost for the organization, and Academic Press for the assistance offered during the preparation of the proceedings.

Firenze, August 1981

R. Conti, Firenze
F. Kappel, Graz
A. Pasquali, Firenze
G. Vidossich, Trieste

Academic Press Rapid Manuscript Reproduction

NONCOOPERATIVE TRAJECTORIES OF n-PERSON DYNAMICAL GAMES AND STABLE NONCOOPERATIVE EQUILIBRIA

Jean-Pierre Aubin[1,2]

CEREMADE, CNRS
Université Paris IX Dauphine
Paris, France

I. INTRODUCTION

We consider an n-person game. We say that a multistrategy x "*improves*" y if each player i prefers the strategy x^i to the strategy y^i when all the other players $j \neq i$ implement the strategies x^j. We say that a multistrategy $\bar{y}$ is a "*stable noncooperative equilibrium*" if there is no multistrategy x that strictly improves $\bar{y}$. We check indeed that any stable noncooperative equilibrium is a *noncooperative equilibrium* (in the sense of Nash) and stable in some sense.

Let us consider a sequence of multistrategies x_m (i.e., a trajectory). We say that a *trajectory is noncooperative* if each state x_m improves the previous state x_{m-1}.

In this paper, we consider both an n-person game and a

---

[1]Visiting Professor and Honorary Research Associate, Department of Economics and Modelling Research Group, University of Southern California, Los Angeles, California 90007.

[2]Present address: CEREMADE, CNRS, Université Paris IX Dauphine, Place du Maréchal De Lattre de Tassigny, 75775 Paris.

ISBN 0-12-186280-1

discrete dynamical system; we give reasonable assumptions relating the game and the dynamical system under which noncooperative trajectories do exist.

We shall also consider the case of continuous dynamical systems and show that analogous assumptions imply the existence of continuous noncooperative trajectories.

We define discrete and continuous noncooperative trajectories in the second section, state the existence theorems in the third, prove existence of discrete noncooperative trajectories in the fourth and of continuous noncooperative trajectories in the fifth. We end this paper with the definition of stable noncooperative equilibria and a study of their properties.

II. NONCOOPERATIVE TRAJECTORIES

Let us consider an n-player game defined by the strategy sets X^i and the loss functions f_i of the n players i. We assume that $X^i \subset V^i$, where V^i is a Banach space. We set

$$X = \prod_{i=1}^{n} X^i \text{ is the set of multistrategies } x = (x^1,\ldots,x^n) \tag{1}$$

and

$$X^{\hat{i}} = \prod_{j\neq i} X^j \text{ is the set of multistrategies } x^{\hat{i}} = (x^j)_{j\neq i} \text{ controlled by the adverse coalition } \hat{i} = \{j=1,\ldots,n,\ j\neq i\}. \tag{2}$$

From the point of view of player i, a multistrategy x can be written $x = (x^{\hat{i}},x^i) \in X = X^{\hat{i}} \times X^i$. So, we write loss functions in the following form:

$$x = (x^{\hat{i}},x^i) \mapsto f_i(x) = f_i(x^{\hat{i}},x^i) . \tag{3}$$

We assume that *they are* at least *continuous*. Let us consider now a multivalued dynamical system defined by a multifunction $S : X \to X$ associating to any multistrategy x a set $S(x) = \prod_{i=1}^{n} S_i(x)$ of feasible velocities.

Definition 1.1. *Let us consider a multistrategy* $y = (y^1, \ldots, y^n) \in X$. *We say that another multistrategy* $x \in X$ *"improves" (resp. "strictly improves")* y *if*

$$f_j(x^{\hat{j}}, x^j) \leq f_j(x^{\hat{j}}, y^j) \quad \text{for all} \quad j = 1, \ldots, n. \tag{4}$$

(resp. if (4) *holds and if* $f_j(x^{\hat{j}}, x^j) < f_j(x^{\hat{j}}, y^j)$ *for at least one player* j *).*

In other words, to say that "x improves y" amounts to saying that each player i prefers the strategy x^i to the strategy y^i when all the other players $j \neq i$ implement the strategies x^j. We denote by $\mathcal{N}(y)$ the set of multistrategies x improving y. The binary relation $x \in \mathcal{N}(y)$ (x improves y) is *reflexive*, but is not in general transitive.

In this paper, we are looking for trajectories x_m whose state x_m improves the previous state x_{m-1}: Namely, we shall solve the following:

Problem 1. *Find* n *sequences of strategies* $x_m^i \in X^i$ $(m = 0, 1, \ldots,)$ *satisfying for all* $i = 1, \ldots, n$

$$\begin{aligned} &\text{i)} \quad x_m^i - x_{m-1}^i \in S_i(x_m) \quad (m = 1, 2, \ldots) \\ &\text{ii)} \quad x_o^i \in X^i \quad \text{is given} \\ &\text{iii)} \quad f_i(x_m^{\hat{i}}, x_m^i) \leq f_i(x_m^{\hat{i}}, x_{m-1}^i) \quad (m = 1, 2, \ldots). \end{aligned} \tag{5}$$

Relation (5)i) are called "implicit" discrete dynamical systems. We may also study "explicit" discrete dynamical systems.

Problem 1 bis. *Same problem as problem* 1 *where* (5)i) *is replaced by*

$$x_m^i - x_{m-1}^i \in S_i(x_{m-1}) \quad (m = 1, 2, \ldots) .$$

We shall mention that the assumptions needed for solving problem 1 bis are more demanding.

Now, we will describe what is the "continuous version" of this problem. For that purpose, we introduce the Clarke derivative with respect to i of the functions f_i, defined by

$$D_i f_i(\hat{x}^i, x^i)(v^i) = \limsup_{\substack{y^i \to x^i \\ \sigma \to 0+}} \frac{f_i(\hat{x}^i, y^i + \sigma v^i) - f_i(\hat{x}^i, y^i)}{\sigma} \tag{6}$$

(see F. H. Clarke [8],[9], or J. P. Aubin [3]). They do exist when f_i is *locally Lipschitz* with respect to x^i; if f_i is *continuously differentiable* with respect to x^i, we obtain $D_i f_i(\hat{x}^i, x^i)(v^i) = \langle \nabla_i f_i(\hat{x}^i, x^i), v^i \rangle$ where $\nabla_i f_i(\hat{x}^i, x^i)$ is the gradient with respect to x^i. If f_i is convex and continuous, we obtain

$$D_i f_i(\hat{x}^i, x^i)(v_i) = \inf_{\sigma > 0} \frac{f_i(\hat{x}^i, x^i + v^i) - f_i(\hat{x}^i, x^i)}{\sigma} . \tag{7}$$

Let us consider a time interval $[0,T]$. We assume given

$$\begin{array}{l} \text{an } \textit{initial multistrategy } x_o \in X \text{ and an} \\ \textit{initial velocity } v_o \in S(x_o) . \end{array} \tag{8}$$

We recall that a bounded function on $[0,T]$ is "regulated" if it is a uniform limit of step functions.

Problem 2. *Find* n *functions* $x^i(.) : [0,T] \to X^i$ *whose derivative* $x^i(.)$ *are regulated satisfying for all* $i = 1,\ldots,n$:

$$\begin{array}{ll} \text{i)} & \dot{x}^i(t) \in S_i(x(t)) \quad (\dot{x}^i(t) = \frac{\dot{d}x^i}{dt}(t)) \\ \text{ii)} & x^i(0) = x_0^i \text{ and } \dot{x}^i(0) = v_0^i \\ \text{iii)} & D_i f_i(\hat{x}^i(t), x^i(t))(\dot{x}^i(t)) \le 0 \end{array} \tag{9}$$

In the functions f_i *are continuously differentiable with respect to* x^i, *condition* (9)iii) *becomes*

$$\langle \nabla_i f_i(\hat{x}^i(t), x^i(t)), \dot{x}^i(t) \rangle \le 0 . \tag{10}$$

Note that in both problems 1 and 2, we require that the *trajectories remain in the strategy sets* X^i.

Definition 1.2. *We say that discrete trajectories of* (5)i) *or continuous trajectories of* (9)i) *are "noncooperative" if respectively conditions* (5)iii) *and conditions* (9)iii) *hold.*

<u>Example</u>. We illustrate these concepts in the simplest noncooperative two-person game, i.e., the Cournot duopoly game, where strategy sets $X^1 = X^2$ are equal to $[0,a]$ and where the loss functions f_1 and f_2 are defined by

$$f_1(x^1,x^2) = x^1(x^1+x^2-a) \quad \text{and} \quad f_2(x^1,x^2) = x^2(x^1+x^2-a)$$

Then sequences x_m^1 and x_m^2 form noncooperative trajectories if

$$x_m^1(x_m^1 + x_m^2 - a) \leq x_{m-1}^1(x_{m-1}^1 + x_m^2 - a)$$

$$x_m^2(x_m^1 + x_m^2 - a) \leq x_{m-1}^2(x_m^1 + x_{m-1}^2 - a)$$

Functions $x^1(t)$ and $x^2(t)$ form noncooperative trajectories if

$$\text{i)} \quad (2x^1(t) + x^2(t) - a)\dot{x}^1(t) \leq 0$$

$$\text{ii)} \quad (2x^2(t) + x^1(t) - a)\dot{x}^2(t) \leq 0$$

Let us consider the dynamical system defined by

$$\text{i)} \quad S_1(x) = \{a - 2x^1 - x^2\}$$

$$\text{ii)} \quad S_2(x) = \{a - x^1 - 2x^2\}$$

It is obvious that any solution of the differential system $\frac{dx_1}{dt} = S_1(x)$, $\frac{dx_1}{dt} = S_2(x)$ yields noncooperative trajectories.

III. STATEMENT OF EXISTENCE THEOREMS

We shall use together with continuity and/or convexity assumptions *tangential conditions* which link the dynamical system defined by S, the strategy sets X^i and the loss functions f_i.

For that purpose, we associate to the loss functions f_i the "partial tangent cones" $T^i_{f_i}(x)$ to f_i at x:

$$T^i_{f_i}(x) = \{v^i \in U^i | D_i f_i(\hat{x}^i, x^i)(v^i) \leq 0\} . \tag{1}$$

Since the distance function d_{X^i} defined by $d_{X^i}(x^i) = \inf\{\| x^i - y^i \| \mid y^i \in X^i\}$ is Lipschitz, it has a Clarke derivative $Dd_{X^i}(x^i)(v^i)$. We define the tangent cone $T_{X^i}(x^i)$ to X^i at x^i by

$$T_{X^i}(x^i) = \{v^i \in U^i | Dd_{X^i}(x^i)(v^i) \leq 0\} \tag{2}$$

(see F. H. Clarke [8],[9], or J. P. Aubin [3], §5).

So, we associate to the n-person game the cones $C(x) = \prod_{i=1}^{n} C_i(x)$ of *noncooperative velocities* defined by

$$C_i(x) = T_{X^i}(x^i) \cap T^i_{f_i}(x) . \tag{3}$$

We begin by stating an existence theorem in the discrete case.

Theorem 2.1. *Let us assume that*

i) the strategy sets X^i are *convex* and *compact*

ii) the loss functions f_i are *continuous* and *convex* with respect to x^i (4)

iii) the multifunctions S_i are *upper semicontinuous* with nonempty *closed convex* images $S_i(x)$.

We assume furthermore that

$$\forall x \in X, \ \forall i = 1,\ldots,n, \ \ S_i(x) \cap C_i(x) \neq \emptyset. \tag{5}$$

Then there exist noncooperative discrete trajectories solving problem 1.

In the continuous case, we shall prove the following result.

Theorem 2.2. *Let us assume that*

i) the strategy sets X^i are *compact*,

ii) the loss functions f_i are *locally Lipschitz* with respect to x^i and their partial Clarke derivatives $(x,v^i) \to D_i f_i(x)(v^i)$ are *continuous*, (6)

iii) the multifunctions S_i are *continuous* with nonempty *compact* images.

We assume furthermore that

$$\forall x \in X, \ \forall i = 1,\ldots,n, \quad S_i(x) \subset C_i(x) \ . \tag{7}$$

Then there exist noncooperative continuous trajectories solving problem 2.

Example (continuation). Let us consider the Cournot model defined above. Since $T_{[0,a]}(0) = \mathbb{R}_+$, $T_{[0,a]}(1) = -\mathbb{R}_+$, $T^1_{f_1}(x) = 2x^1 + x^2 - a$ and $T^2_{f_2}(x) = 2x^2 + x^1 - a$, it is clear that the map $S(x)$ defined by $S_1(x) = a - 2x^1 - x^2$ and $S_2(x) = a - 2x^2 - x^1$ satisfies the tangential conditions (5) and (7).

IV. EXISTENCE OF DISCRETE NONCOOPERATIVE TRAJECTORIES

We shall deduce the existence of discrete noncooperative trajectories from the Ky Fan inequality (see Ky Fan [10], or J. P. Aubin [2], Chapter 7, §1, Theorem 3):

Theorem (Ky Fan). *Let* X *be compact convex and* $\psi : X \times X \to \mathbb{R}$ *be a function satisfying*

i) $\forall x \in X$, $y \to \psi(x,y)$ is concave,

ii) $\forall y \in X$, $x \to \psi(x,y)$ is lower semi-continuous, (1)

iii) $\forall y \in X$, $\psi(y,y) \le 0$.

Then there exists $\bar{x} \in X$ *satisfying*

$$\sup_{y \in X} \psi(\bar{x}, y) \le 0 . \tag{2}$$

We also need to consider noncooperative games *perturbed* by continuous linear functionals $p = (p_1, \ldots, p_n) \in U^* = \prod_{i=1}^{n} U^{i*}$, defined by

$$\begin{array}{ll} \text{i) their strategy sets} & X^i \\ \text{ii) their loss functions} & f_i(\hat{x^i}, x^i) - \langle p_i, x^i \rangle . \end{array} \tag{3}$$

We recall that if X^i is closed and convex, the tangent cone $T_{X^i}(x^i)$ is the negative polar cone $N_{X^i}(x^i)^-$ of the normal cone $N_{X^i}(x^i)$ defined by $N_{X^i}(x^i) = \{p_i \in U^i$ such that $\langle p_i, x^i \rangle = \max\{\langle p_i, y^i \rangle | y^i \in X^i\}\}$. We shall use the following characterization of the noncooperative equilibria (see J. P. Aubin [2], Chapter 9, Proposition 3).

Lemma 4.1. *We suppose that assumptions* (2.4)i) *and* ii) *of Theorem* 3.1 *hold. A multistrategy* x *is a noncooperative equilibrium of the perturbed game* (3) *if and only if*

$$\forall i = 1, \ldots, n, \quad p_i \in \partial_i f_i(\hat{x^i}; x^i) + N_{X^i}(x^i) \tag{4}$$

where ∂_i denotes the subdifferential of f_i with respect to x^i.

Proof. Indeed, x is a noncooperative equilibrium if and only if, for all $j = 1, \ldots, n$,

$$\langle p_j, x^j \rangle - f_j(\hat{x^j}, x^j) = \max_{y^j \in X^j} [\langle p_j, y^j \rangle - f_j(\hat{x^j}, y^j)] . \tag{5}$$

Since the function $y^j \to f_j(\hat{x^j}, y^j)$ is convex and continuous, inequality (5) amounts to saying that

$$p_j \in \partial_j f_j(\hat{x^j}, x^j) + N_{X^j}(x^j) .$$

We denote by G the correspondence with *closed convex* values defined by

$$G(x) = \prod_{i=1}^{n} \{\lambda^i \partial_i f_i(x^{\hat{i}}, x^i)\}_{\lambda^i \geq 0} + N_{X^i}(x^i)) \; . \tag{6}$$

Now, we can characterize the tangential condition (3.5) of Theorem 3.1.

Lemma 4.2. *Let* S *be any multifunction from* X *into itself. Then the tangential condition*

$$\forall\, i = 1,\ldots,n, \quad \forall\, x \in X, \quad S_i(x) \cap T_{X^i}(x^i) \cap T^i_{f_i}(x) \neq \emptyset \tag{7}$$

implies the following property:

$\forall\, i = 1,\ldots,n,\ \forall\, \lambda^i \geq 0,\ \forall\, p_i \in U^{i*}$, for every *noncooperative equilibrium* x *associated to the loss functions* $\lambda^i f_i(x^{\hat{i}}, y^i) - \langle p_i, y^i \rangle$, *we have*

$$\inf\{\langle p_j, v^j\rangle \mid v^j \in S_j(x)\} \geq 0 \; . \tag{8}$$

The converse is true when the images $S_i(x)$ *are convex compact for all* $i = 1,\ldots,n$, $x \in X$.

Before proving Lemma 4.2, we recall that

$$T^i_{f_i}(x^{\hat{i}}, x^i) = \partial_i f_i(x^{\hat{i}}, x^i)^- \tag{9}$$

where $\partial_i f_i(x^{\hat{i}}, x^i)$ is the subdifferential of the continuous convex $y_i \to \partial_i f_i(x^{\hat{i}}, x^i)$ at x^i. [This derives from the fact that the Clarke derivative $D_i f_i(x^{\hat{i}}, x^i)(.)$ is the support function of $\partial_i f_i(x^{\hat{i}}, x^i)$.]

Proof of Lemma 4.2. Let us assume that the tangential condition (7) holds. Let us consider $p_i \in U^{i*}$, $\lambda^i \geq 0$ and a noncooperative equilibrium x associated to the loss functions (3)ii). Then $p_i \in \lambda_i \partial_i f_i(x^{\hat{i}}, x^i) + N_{X_i}(x^i)$. Therefore, if $\bar{v}^i \in S_i(x) \cap T_{X^i}(x^i) \cap T^i_{f_i}(x)$, which exists by assumption (7), we obtain $\langle p_i, \bar{v}^i\rangle \leq 0$ for $\bar{v}^i \in \partial_i f_i(x^{\hat{i}}, x^i)^-$ and $\bar{v}^i \in N_{X^i}(x^i)^-$. Since $\bar{v}^i \in S^i(x)$, $\inf\{\langle p_i, v^i\rangle \mid v^i \in S_i(x)\} \leq 0$.

Conversely, let us assume that the images $S_i(x)$ are *convex* and *compact*. We can write property (8) in the form $\sup_{p\in G(x)} \inf_{v\in S(x)} \sum_{j=1}^{n} \langle p_j, v^j\rangle \leq 0$. Since $G(x)$ is convex and $S(x)$ is *convex compact*, the "lop-sided" minimax theorem (see J. P. Aubin [1] or [2], Chapter 7, 1, Theorem 6) implies the existence of $\bar{v} \in S(\bar{x})$ such that:

$$\sup_{p\in G(x)} \inf_{v\in S(x)} \sum_{j=1}^{n} \langle p_j, v^j\rangle = \sup_{p\in G(x)} \sum_{j=1}^{n} \langle p_j, \bar{v}^j\rangle \leq 0$$

Therefore, $\bar{v}^j \in S_j(x)$ and $\bar{v}^j \in \{\lambda^i \partial_i f_i(\hat{x}^i, x^i)\}_{\lambda^i \geq 0} + N_{X^i}(x^i))^- = T^i_{f_i}(\hat{x}^i, x^i) \cap T_{X^i}(x^i)$.

Now, we shall derive Theorem 3.1 from the following Theorem.

Theorem 4.1. *We suppose that assumptions of Theorem 3.1 hold. Let* u *be given in* X.

a) *There exists* $\bar{x} \in X$ *such that*

$$\begin{aligned} &\text{i)} \quad 0 \in S(\bar{x}) \quad (\text{i.e., } \bar{x} \text{ is a critical point of } S) \\ &\text{ii)} \quad \forall i = 1,\ldots,n, \; f_i(\hat{\bar{x}}^i, \bar{x}^i) \leq f_i(\hat{\bar{x}}^i, u^i) \end{aligned} \tag{10}$$

b) *There exists* $x \in X$ *such that*

$$\begin{aligned} &\text{i)} \quad x-u \in S(x) \quad (\text{i.e., } x \in (1-S)^{-1}(u)) \\ &\text{ii)} \quad \forall i = 1,\ldots,n, \; f_i(\hat{x}^i, x^i) \leq f_i(\hat{x}^i, u^i) \end{aligned} \tag{11}$$

It is clear that Theorem 3.1 follows from the second statement of Theorem 4.1. The first statement of Theorem 4.1 amounts to saying that we can associate to any multistrategy y a strategy x that "improves" y *and* that is a critical point of the dynamical system.

Proof of Theorem. It is sufficient to prove that for any $u \in X$, we can find a solution $\bar{x} \in X$ to the problem

$$\begin{aligned} &\text{i)} \quad \varepsilon(\bar{x}-u) \in S(\bar{x}) \\ &\text{ii)} \quad \forall i = 1,\ldots,n, \; f_i(\hat{\bar{x}}^i, \bar{x}^i) - f_i(\hat{\bar{x}}^i, u^i) \leq 0 \end{aligned} \tag{12}$$

where ε is either equal to 0 (case a)) or 1 (case b)). Assume that no solution to problem (12) exists. Then, for any $x \in X$, either $\varepsilon(x-u) \notin S(x)$ or there exists i such that $f_i(x^{\hat{i}},x^i) - f_i(x^{\hat{i}},u^i) > 0$. In the first case, the Hahn-Banach theorem implies the existence of $p = (p_1,\ldots,p_n) \in U^*$ such that $\varepsilon\langle p,u-x\rangle > \sigma(-S(x),p)$, where

$$\sigma(-S(x),p) = -\inf\left\{\sum_{j=1}^{n} \langle p_j,v^j\rangle \mid v^j \in S_j(x)\right\} \tag{13}$$

We introduce the subsets

$$V_p = \{x \in X \text{ such that } \varepsilon\langle p,u-x\rangle > \sigma(-S(x),p)\}$$

and

$$V_i = \{x \in X \text{ such that } f_i(x^{\hat{i}},x^i) - f_i(x^{\hat{i}},u^i) > 0\}\ .$$

Since the multifunction S_i is upper semicontinuous, then the functions $x \to \sigma(-S(x),p)$ are upper semicontinuous and thus, *the sets* V_p *are open*. The continuity of the loss functions f_i implies that *the sets* V_i *are open*. The negation of the conclusion implies that X is covered by the open subsets V_p and V_i. Since X is compact, we can extract a finite covering $X \subset \bigcup_{i=1}^{n} V_i \cup \bigcup_{j=1}^{m} V_{p_j}$.

Let $\{\alpha_i,\beta_j\}_{\substack{1\le i\le n\\ 1\le j\le m}}$ be a continuous partition of unity with respect to this finite covering. Let ψ the function defined on $X \times X$ by

$$\psi(x,y) = \sum_{i=1}^{n} \alpha_i(x)(f_i(x^{\hat{i}},x^i) - f_i(x^{\hat{i}},x^i)) - \sum_{j=1}^{m} \beta_j(x)\langle p_j,x-y\rangle\ . \tag{14}$$

The function ψ satisfies obviously the assumptions (1) of the Ky Fan theorem. Therefore, there exists $\bar{x} \in X$ such that

$$\psi(\bar{x},y) \le 0 \quad \forall\, y \in X\ . \tag{15}$$

Let us set $\bar{\lambda}^i = \alpha_i(\bar{x})$ and $\bar{p} = \sum_{j=1}^{m} \beta_j(\bar{x})p_j$. Hence inequality (15) can be written in the form

$$\forall\, i = 1,\ldots,n,\quad \bar{\lambda}_i f_i(\hat{\bar{x}}^i,\bar{x}^i) - \langle \bar{p}_i,\bar{x}^i\rangle \le \bar{\lambda}_i f_i(\hat{\bar{x}}^i,y^i) - \langle \bar{p}_i,y^i\rangle .$$

In other words, $\bar{x}$ is a noncooperative equilibrium of the game defined by the loss functions

$$\bar{\lambda}^i f_i(\hat{x}^i,x^i) - \langle \bar{p}_i,x^i\rangle .$$

By Lemma 4.2, assumption (7) implies property (8), that can be written:

$$\sigma(-S(\bar{x}),\bar{p}) = -\inf\{\langle \bar{p},u\rangle \mid u \in S(\bar{x}) \ge 0 . \tag{16}$$

We obtain a contradiction of (15) by showing that

$$\psi(\bar{x},u) > 0 . \tag{17}$$

Indeed, there exists at least an index i or j such that $\alpha_i(\bar{x}) > 0$ or $\beta_j(\bar{x}) > 0$, since $\{\alpha_i,\beta_i\}$ is a partition of unity. If $\alpha_i(\bar{x}) > 0$, then $\bar{x} \in V_i$ and thus,

$$\bar{\lambda}^i(f_i(\hat{\bar{x}}^i,\bar{x}^i) - f_i(\hat{\bar{x}}^i,u^i) > 0 . \tag{18}$$

If $\beta_j(\bar{x}) > 0$, then $\bar{x} \in V_{p_j}$ and thus,

$$-\varepsilon\langle \bar{p}_j,\ \bar{x}-\bar{u}\rangle > \sigma(-S(x),p_j).$$

Hence (16) implies:

$$-\varepsilon\langle \bar{p},\bar{x}-\bar{u}\rangle > \sigma(-S(\bar{x}),\bar{p}) \ge 0 .$$

<u>Case where $\varepsilon = 0$</u>. We deduce first that $\beta_j(\bar{x}) = 0$ for all $j = 1,\ldots,n$. [If not, $\bar{p}$ would be different from 0 and (16) and (19) yield a contradiction.] Therefore, $\alpha_i(\bar{x}) > 0$ for at least one i and thus, (18) implies (17).

Case where $\varepsilon = 1$. At least inequality (18) or (19) holds. Therefore, they obviously imply (17).

This completes the proof of Theorem 4.1.

Remark. We can consider the case of explicit dynamical system (problem 1 bis). By techniques analogous to J. P. Aubin- - A. Cellina- J. Nohel [5], §5, we can prove existence of noncooperative trajectories of problem 1 bis under assumption (4) of Theorem 3.1 and the following assumption:

$\forall \lambda \in \mathbb{R}^n_+$, $\forall p \in U$, $\forall x$, noncooperative equilibrium for the loss functions $\lambda^i f_i(x) - \langle p_i, x^i \rangle$, $\forall y \in X$, we have
$$\sum_{i=1}^{n} \lambda^i (f_i(\hat{x}^i, x^i) - f_i(\hat{x}^i, y^i)) - \sum_{i=1}^{n} \langle p_i, x^i - y \rangle + \inf\{\sum_{i=1}^{n} \langle p_i, v^i \rangle \mid v^i \in S_i(y)\} \leq 0 .$$

V. EXISTENCE OF CONTINUOUS NONCOOPERATIVE TRAJECTORIES

The proof of Theorem 3.2 involves a technique devised by Fillipov [11] and the following lemma using the properties of the Clarke derivatives (see Aubin [3], Aubin-Clarke [6], and Clarke [8],[9]).

Lemma 5.1. *We suppose that assumptions of Theorem* 3.2 *hold. Hence there exists a function* $a(h)$ *converging to* 0 *with* h *such that, for every* $i = 1,\dots,n$, $x^i \in X^i$ *and* $v^i \in S_i(x)$, *we can find* $u^i \in U^i$ *satisfying*

$$\begin{aligned} &\text{i)} && \|u^i - v^i\| \leq a(h) \\ &\text{ii)} && D_i f_i(\hat{x}^i, x^i)(u^i) \leq a(h) \\ &\text{iii)} && y^i = x^i + hu^i \in X^i \end{aligned} \qquad (1)$$

Proof. We can associate to any $\varepsilon > 0$, $y \in X$, $w \in S(x)$ neighborhoods $N(y)$ and $N(w)$ of y and w respectively and a positive number $\sigma(y,w)$ such that, for all $x \in X \cap N(y)$, for all $v \in N(w)$, for all $i = 1,\dots,n$, for all $h \leq \sigma(y,w)$,

$$d_{X^i}(x^i+hv^i) \leq d_{X^i}(x^i) + hDd_{X^i}(y^i)(w^i) \leq h\varepsilon, \tag{2}$$

and

$$D_i f_i(\hat{x}^i,x^i)(v^i) \leq D_i f_i(\hat{y}^i,y^i)(w^i) + \varepsilon \leq \varepsilon \tag{3}$$

since $S_i(x) \subset T_{X^i}(x^i) \cap T^i_{f_i}(x)$ for all $i = 1,\ldots,n$, and since $D_i f_i(x)(v^i)$ is upper semicontinuous with respect to (x,v^i).

The graph $G(S)$ is compact, since X is compact and G is upper semicontinuous. So, it can be covered by a finite number of neighborhoods $N(y_k) \times N(w_k)$. Let $\sigma = \min_k \sigma(y_k,w_k) > 0$. We thus derive that for all $h \leq \sigma$, for all $x \in X$, for all $v \in S(x)$, and for all $i = 1,\ldots,n$, we obtain

$$d_{X^i}(x^i+hv^i) \leq h\varepsilon \quad \text{and} \quad D_i f_i(\hat{x}^i,x^i)(v^i) \leq \varepsilon .$$

Furthermore, we can find $y^i \in X^i$ such that $\|y^i-x^i-hv^i\| \leq 2\varepsilon h$. We set $u^i = \frac{y^i-x^i}{h}$ so that $y^i = x^i+hu^i \in X^i$ and that $\|u^i-v^i\| \leq 2\varepsilon$. Furthermore, there exists a constant $c > 0$ such that $D_i f_i(\hat{x}^i,x^i)(u^i) \leq D_i f_i(\hat{x}^i,x^i)(v^i) + c\|u^i-v^i\| \leq \varepsilon(1+2c)$. Hence Lemma 5.1 ensues.

Proof of Theorem 3.2. We use techniques devised by Filippov in [11] for constructing sequences of piecewise linear functions x^i_m that approximate solutions of problem 2, in such a way that they stay in a precompact set of $\mathcal{C}(0,T;U^i)$ and that their derivatives remain in a precompact set of the Banach space $\mathcal{B}(0,T;U^i)$ of bounded functions from $[0,T]$ into U^i. We consider a decreasing sequence of partitions $\mathcal{P}_m$ of $[0,T]$ made of intervals $[qh_m,(q+1)h_m]$, where q is an integer and where T/h_1 and h_{m-1}/h_m are integers. If $\tau_m = qh_m$ is any node of $\mathcal{P}_m$, we define the maps φ and ψ associating to τ_m the smallest index $j = \varphi(\tau_m)$ such that τ_m/h_{j+1} is an integer and the largest node $\psi(\tau_m)$ of the partition $\mathcal{P}_{\varphi(\tau_m)}$ strictly smaller than τ_m.

Since $S(X)$ is contained in a ball of radius M and since S is (uniformly) continuous on X we can find h_m such that

$$\begin{aligned} &\text{i)} \quad \|x-y\| \leq (M+1)h_m \Rightarrow S(u) \subset S(y) + 2^{-m}B \\ &\text{ii)} \quad a(h_m) \leq 2^{-m} . \end{aligned} \tag{4}$$

We construct by recursion a sequence of piecewise linear functions $x_m^i(t)$ defined on $[\sigma_m, \sigma_m+h_m]$ by

$$x_m^i(t) = x_m^i(\sigma_m) + (t-\tau_m)v_m^i(\sigma_m)$$

satisfying $x_m^i(0) = x_0$, $v_m^i(0) = v_0$ and, for any node σ_m,

$$\begin{aligned} &\text{i)} \quad v_m^i(\sigma_m) \in S_i(x_m(\sigma_m)) + 2^{-m}B \\ &\text{ii)} \quad x_m^i(\sigma_m + h_m) \in X^i \\ &\text{iii)} \quad D_i\hat{f}_i(x_m^i(\sigma_m), x_m^i(\sigma_m))(v_m^i(\sigma_m)) \leq 2^{-m} \\ &\text{iv)} \quad \|v_m^i(\sigma_m) - v_m^i(\psi(\sigma_m))\| \leq 2.2^{-\varphi(\sigma_m)} \end{aligned} \tag{5}$$

Indeed, assume that x_m^i is constructed on $[0,\tau_m]$. Let $\sigma_m = \psi(\tau_m)$ and $j = \varphi(\tau_m) < m$. Since

$$\begin{aligned} \|x_m^i(\tau_m)-x_m^i(\sigma_m)\| &\leq \int_{\sigma_m}^{\tau_m} \|\dot{x}_m^i(t)\| \, dt \leq (M+1)|\tau_m-\sigma_m| \\ &\leq (M+1)h_j \quad (\text{for } \tau_m \in]\sigma_m, \sigma_m+h_j]) \end{aligned}$$

we deduce from (4)i) and (5)i) that

$$v_m^i(\sigma_m) \in S_i(x_m(\sigma_m)) + 2^{-m}B \subset S_i(x_m(\tau_m)) + (2^{-m}+2^{-j})B \tag{6}$$

(We take $v_m^i(\sigma_m) = v_0^i$ if $\sigma_m = 0$). Hence there exists $w_m^i \in S_i(x_m(\tau_m))$ such that $\|v_m^i(\sigma_m)-w_m^i\| \leq 2^{-m}+2^{-j}$.

By Lemma 5.1, there exists $v_m^i(\tau_m)$ such that $\|v_m^i(\tau_m)-w_m^i\| \leq a(h_m) \leq 2^{-m}$ and such that properties (5)i), ii) and iii) are satisfied with σ_m replaced by τ_m. Property (5)iv) holds since

$$\begin{aligned} \|v_m^i(\tau_m)-v_m^i(\sigma_m)\| &\leq \|v_m^i(\tau_m)-w_m^i\| + \|v_m^i(\sigma_m)-w_m^i\| \\ &\leq 2.2^{-m} + 2^{-j} \leq 2.2^{-j} . \end{aligned}$$

Thus, we notice that $\|\dot{x}_m^i(t)\| \leq M+2^{-m} \leq M+1$ and that $x_m^i(t)$ belongs to the *compact* subset $\overline{co}(X)$ for all $t \in [0,T]$. Hence Ascoli's theorem implies that $x_m^i(.)$ remains in a compact set of $\mathcal{B}(0,T;U^i)$. Furthermore, following Filippov [11], properties (5)iv) imply that $\dot{x}_m^i(.)$ remains in a compact set of $\mathcal{B}(0,T;U^i)$. Hence subsequences (again denoted by) $x_m^i(.)$ and $\dot{x}_m^i(.)$ converge uniformly to $x^i(.)$ and $\dot{x}^i(.)$. So for any $t \in [0,T]$, there exist nodes τ_m such that $x_m^i(\tau_m)$ converges to $x^i(t)$ and $v_m^i(\tau_m)$ to $\dot{x}^i(t)$. Property (5)i) implies that $\dot{x}^i(t) \in S_i(x(t))$ for all t since the multifunctions S_i are continuous, property (5)ii) that $x^i(t) \in X^i$ for all t and property (5)iii) that $D_i f_i(x^{\hat{i}}(t), x^i(t))(\dot{x}^i(t)) \leq 0$ by the assumption of continuity of $D_i f_i(x)(v^i)$. The proof is completed.

VI. STABLE NONCOOPERATIVE EQUILIBRIA

The introduction of the binary relation "x improves y" suggests the following concept of equilibrium.

Definition 6.1. *We say that a multistrategy* $\bar{y} = (\bar{y}^1, \ldots, \bar{y}^n) \in X$ *is a "stable noncooperative equilibrium" if there is no other multistrategy* $x = (x^1, \ldots, x^n) \in X$ *that strictly improves* $\bar{y}$.

In other words, $\bar{y}$ is a stable noncooperative equilibrium if $x \in \mathcal{N}(\bar{y})$ is equivalent to

$$i = 1,\ldots,n, \quad f_i(x^{\hat{i}}, x^i) = f_i(x^{\hat{i}}, y^i).$$

Remark. Let us recall that a multistrategy $\bar{x}$ is a *noncooperative equilibrium* if and only if

$$i = 1,\ldots,n, \quad f_i(\bar{x}^{\hat{i}}, \bar{x}^i) = \min \{f_i(\bar{x}^{\hat{i}}, y^i) \mid y^i \in X^i\} .$$

Nash's theorem (see [14]) implies the existence of noncooperative equilibria under the assumptions (4)i) and ii) of Theorem 3.1.

Remark. We refer to R. Aumann [7], H. Moulin [13] and M.

Moulin-J.P. Vial [14] for related concepts and further comments. In two person games, stable non cooperative equilibria are the "prudent equilibria" introduced by H. Moulin [13].

The terminology is justified by the two following propositions.

Proposition 6.1. *Any stable noncooperative equilibrium is a noncooperative equilibrium.*

Proof. Let $\bar{y} \in X$ be a stable noncooperative equilibrium. Let $x = (\bar{y}^{\hat{i}}, x^i)$ be a multistrategy associated to any strategy $x^i \neq y^i$. Since $(x^{\hat{j}}, \bar{y}^j) = (x^{\hat{j}}, x^j)$ and since $(x^{\hat{i}}, \bar{y}^i) = (y^{\hat{i}}, y^i)$, then $f_j(x^{\hat{j}}, \bar{y}^j) = f_j(x^{\hat{j}}, x^j)$ for all $j \neq i$ and consequently, $f_i(y^{\hat{i}}, y^i) = f_i(x^{\hat{i}}, \bar{y}^i) \geq f_i(x^{\hat{i}}, x^i) = f_i(\bar{y}^{\hat{i}}, x^i)$ for x does not strictly improve $\bar{y}$. Hence $\bar{y}$ is a noncooperative equilibrium.

The following result conveys the idea of stability:

Proposition 6.2. *Let $\bar{y}$ be a stable noncooperative equilibrium. Then, for any neighborhood M of $\mathcal{N}(\bar{y})$, there exists $\eta > 0$ such that $N_\eta = \{x \in X | f_j(x^{\hat{j}}, x^j) \leq f_j(x^{\hat{j}}, \bar{y}^j) + \eta$ for all $j = 1, \ldots, n\}$ is contained in M.*

Proof. Let $K = \{x \in X$ such that $x \notin M\}$, which is *compact*. If $x \in K$, we deduce from the fact that $\bar{y}$ is a stable noncooperative equilibrium that there exist an index i and $\eta(x) > 0$ such that $f_i(x^{\hat{i}}, \bar{y}^i) < f_i(x^{\hat{i}}, x^i) + \varepsilon(x)$. We define $B(x) = \{z \in X$ such that $f_i(z^{\hat{i}}, \bar{y}^i) < f_i(z^{\hat{i}}, z^i) + \varepsilon(x)\}$. Hence $x \in B(x)$ and $B(x)$ is open by the continuity of the loss functions. Therefore, K can be covered by n such open subsets $B(x_k)$. Let $\eta = \min_{k=1,\ldots,m} \eta(x_k) > 0$. It is clear that $N_\eta \subset M$.

Note that the subsets N_η are neighborhoods of $(\bar{y})$ since the loss functions are continuous.

Remark. See L. A. Gerard-Varet and H. Moulin [12].

Theorem 4.1 implies the following consequence.

Proposition 6.3. *Suppose that assumptions of Theorem 3.1 hold. If $\bar{y}$ is a stable noncooperative equilibrium, then there exists a multistrategy $x \in X$ satisfying*

$$\begin{aligned} &\text{i)} \quad 0 \in S(\bar{x}) \quad (\text{i.e., } \bar{x} \text{ is a critical point}) \\ &\text{ii)} \quad i = 1,\ldots,n, \quad f_i(\bar{x}^{\hat{i}}, \bar{x}^i) = f_i(\bar{x}^{\hat{i}}, \bar{y}^i) \,. \end{aligned} \tag{1}$$

It is useful to introduce the following function φ defined on $X \times X$ by

$$\varphi(x,y) = \sum_{j=1}^{n} (f_j(x^{\hat{j}}, x^j) - f_j(x^{\hat{j}}, y^j)) \tag{2}$$

that is *semicontinuous with respect to* x, *concave with respect to* y and satisfying

$$\sup_{y \in X} \inf_{x \in X} \varphi(x,y) \leq 0 \leq \inf_{x \in X} \sup_{y \in X} \varphi(x,y) \tag{3}$$

since $\varphi(y,y) = 0$ for all $y \in X$.

We recall that $\bar{x}$ is a noncooperative equilibrium if and only if

$$\sup_{y \in X} \varphi(\bar{x},y) = 0$$

(see J. P. Aubin [2], Chapter 9, §1, proposition 2). We note that *any multistrategy* $\bar{y} \in X$ *satisfying*

$$0 = \inf_{x \in X} \varphi(x,\bar{y}) \tag{4}$$

is a stable noncooperative equilibrium.

Indeed, $\varphi(x,\bar{y}) < 0$ when x strictly improves y. So, if $\inf_{x \in X} \varphi(x,\bar{y}) \geq 0$, there is no x that strictly improves y.

There are situations where every noncooperative equilibrium is stable:

<u>Definition 6.2</u>. *We say that a game is "monotone" if for any pair of multistrategies* x *and* y, *we have*

$$\sum_{j=1}^{n} (f_j(x^{\hat{j}}, y^j) + f_j(y^{\hat{j}}, x^j) \leq \sum_{j=1}^{n} (f_j(x^{\hat{j}}, x^j + f_j(y^{\hat{j}}, y^j)) . \tag{5}$$

Example.

a) Two person zero-sum games are obviously monotone.

b) More generally, *two-person games satisfying the relation*

$$x^1 \in X^1, \quad x^2 \in X^2, \quad f_1(x^2,x^1) + f_2(x^1,x^2) = \\ = a_1(x^1) + a_2(x^2) \tag{6}$$

where $a_i : X^i \to \mathbb{R}$ $(i = 1,2)$ *are monotone.*

If f_1 is convex with respect to x^1, concave with respect to x_2, if a_2 is convex and if the functions f_1, a_1 and a_2 are continuous, then Nash's theorem implies the existence of noncooperative equilibria $\bar{x} = (\bar{x}^1,\bar{x}^2)$, satisfying

$$\begin{aligned} &\text{i)} \quad f_1(\bar{x}^2,\bar{x}^1) = \min\{f_1(\bar{x}^2,y^1) \mid y^1 \in X^1\} \\ &\text{ii)} \quad f_1(\bar{x}^2,\bar{x}^1) - a(\bar{x}^2) = \max\{f_1(y^2,\bar{x}^1) - a(y^2) \mid y^2 \in X^2\} \end{aligned} \tag{7}$$

For instance, the game defined by $X^1 = X^2 = [0,1]$ and $f_1(x^2,x^1) = 2x^1x^2$, $f_2(x^1,x^2) = (x^1-x^2)^2$ is monotone.

Proposition 6.4. *If the game is monotone, any noncooperative equilibrium is stable.*

Proof. Let $\bar{y}$ be a noncooperative equilibrium. Then $0 \leq -\varphi(\bar{y},x)$; since the game is monotone, we can write (5) in the form $-\varphi(\bar{y},x) \leq \varphi(x,\bar{y})$. Hence $\varphi(x,\bar{y}) \geq 0$ for all $x \in X$ and *thus,* $\bar{y}$ *is stable.*

Remark. See H. Moulin and J. P. Vial [14].

Now, we prove existence of stable noncooperative equilibria.

Proposition 6.5. *We suppose that the sets* X^i *are convex compact and that the loss functions* f_i *are convex with respect to* x^i. *Let us assume that we can associate with any mixed multistrategy* $\sum_k \lambda^k \delta(x_k)$ [1] *a multistrategy* y *such that*

[1] $\delta(x_k)$ denotes the Dirac measure at x_k; here, the λ^k are positive and satisfy $\sum \lambda^k = 1$.

$$\sum_k \sum_{j=1}^n \lambda^k f_j(x_k^{\hat{j}}, y^j) \leq \sum_k \sum_{j=1}^n \lambda^k f_j(x_k^{\hat{j}}, x_k^j) \ . \tag{8}$$

Then there exists a stable noncooperative equilibrium.

Proof. Let $\mathcal{M}(X)$ be the convex set of mixed strategies μ and $\hat{\varphi}$ the function defined on $\mathcal{M}(X) \times X$ by

$$\hat{\varphi}(\mu, y) = \sum_k \lambda^k \varphi(x_k, y) \quad \text{where} \quad \mu = \sum_k \lambda^k \delta(x_k) \ .$$

Since $\hat{\varphi}$ is linear with respect to μ, concave with respect to y, we can apply the "lop-sided" minimax theorem (see J.P. Aubin [1] or [2], Chapter 7, §1.8, theorem 6): there exists $\bar{y} \in y$ such that

$$\inf_{x \in X} \varphi(x, \bar{y}) = \inf_{\mu \in \mathcal{M}(X)} \hat{\varphi}(\mu, \bar{y}) = \inf_{\mu \in \mathcal{M}(X)} \sup_{y \in X} \hat{\varphi}(\mu, y) \ .$$

Assumption (8) amounts to saying that $\inf_{\mu \in \mathcal{M}(X)} \sup_{y \in X} \hat{\varphi}(\mu, y) \geq 0$. Hence the proposition ensues.

Remark. See H. Moulin [13].

Remark. One can check that a monotone game satisfies condition (8). Indeed, we associate with $\sum_k \lambda^k \delta(x_k)$ the multistrategy $y = \sum \lambda^k x_k$. Hence

$$\sum_k \lambda^k \varphi(x_k, \sum_\ell \lambda^\ell x_\ell) \geq \sum_{k,\ell} \lambda^k \lambda^\ell \varphi(x_k, x_\ell) =$$

$$= \frac{1}{2} \sum_{k,\ell} \lambda^k \lambda^\ell (\varphi(x_k, x_\ell) + \varphi(x_\ell, x_k)) \geq 0.$$

REFERENCES

1. Aubin, J. P., Un théorème de minimax pour une classe de fonctions. *C. R. Acad. Sci. 244*, 455-458 (1972).
2. Aubin, J. P., Mathematical Methods of Game and Economic Theory. Amsterdam, North-Holland, (1978).
3. Aubin, J. P., Microcours. Gradient généralisés de Clarke. CRM n. 703. Université de Montréal, (1977).

4. Aubin, J. P. and Nohel, J., Existence de trajectories monotones de systèmes dynamiques multivoques discrets. *C. R. Acad. Sci. 282*, 267-270 (1976).

5. Aubin, J. P., Cellina, A., and Nohel, J., Monotone trajectories of multivalued dynamical systems.

6. Aubin, J. P., and Clarke, F. H., Monotone invariant solutions to differential inclusions. *J. London Math. Soc.*

7. Aumann, R., Subjectivity and correlation in randomized strategies. *J. Math. Econ. 1*, (1974).

8. Clarke, F. H., Generalized gradient and applications. *Trans. Amer. Math. Soc. 205*, 247-262 (1975).

9. Clarke, F. H., Generalized gradients of Lipschitz functionals. *Advances in Math.*

10. Ky Fan, A minimax inequality and applications, *in* "Inequalities III" (Schishia, ed.), pp. 103-113. New York, Academic Press, (1972).

11. Filippov, A. F., On the existence of solutions of multivalued differential equations. *Math. Zametki 19*, 307-313 (1971).

12. Gerard-Varet, L. A., and Moulin, H., Improving upon Cournot equilibrium by correlation. *J. Math. Econ.*

13. Moulin, H., Correlated and prudent equilibria.

14. Moulin, H., and Vial, J. P., Strategically zero-sum games. *Int. J. Game Theory.*

15. Nash, J., Non cooperative games. *Annals of Math. 54*, (1951).

PROCESSUS DE CONTRÔLE AVEC CONTRÔLE INITIAL

Andrea Bacciotti[1]

Istituto Matematico
Università di Siena
Siena, Italy

I. INTRODUCTION

Du point de vue mathématique un processus de contrôle est la donnée d'une équation différentielle de la forme

$$\dot{x} = f(x,u), \quad x \in \mathbb{R}^n, \quad u \in \mathbb{R}^m \tag{1}$$

dépendant du paramètre u, et d'un ensemble $\mathcal{U}$ de fonctions de contrôle $t \mapsto u(t)$, qui s'appelle ensemble des contrôles admissibles. Le choix d'un contrôle $u(\cdot) \in \mathcal{U}$ transforme (1) en

$$\dot{x} = f(x,u(t)), \quad x \in \mathbb{R}^n, \quad t \geq 0 \tag{1'}$$

et permet donc de guider l'évolution du point représentatif x du système. Supposons que pour tout $u(\cdot) \in \mathcal{U}$ et pour tout $x_o \in \mathbb{R}^n$ l'équation (1') admette une solution unique

[1]Present address: Istituto Matematico, Università di Siena, Via del Capitano 15, Siena, Italy.

ISBN 0-12-186280-1

$t \mapsto x(t;x_o,u(\cdot))$, et dénotons

$$R(T,x_o) = \{x \in \mathbb{R}^n : x = x(T;x_o,u(\cdot)),\ \ u(\cdot) \in \mathcal{U}\}$$

l'ensemble des points de $\mathbb{R}^n$ qui sont atteignables de x_o à l'instant $T \geq 0$ le long des trajectoires du système. L'application multivoque $t \mapsto R(t,x_o)$ n'est pas en général croissante (dans le sens de l'inclusion): pour éliminer ce défaut, qui pourrait soulever des difficultés dans la suite, on considère au lieu de $R(T,x_o)$, l'ensemble

$$R([0,T],x_o) = \bigcup_{0\leq t\leq T} R(t,x_o)$$

des points atteignables en temps non supérieur à T. Plusieurs auteurs on étudié la propriété

$$x_o \in \text{int } R([0,T],x_o) \quad \text{quel que soit} \quad T > 0 \tag{2}$$

avec des appellatifs différents: *contrôlabilité locale* ([1]), *N-locale contrôlabilité* ([2],[3]), *stabilité contrôlée* ([4], [5] , *autoaccessibilité* ([6],[7]), ou bien en disant que un système qui la possède est *propre* ([8],[9]).

Sur l'importance de (2) on peut faire les suivantes observations:

(a) à côté de (1), considérons le système "symétrique" défini par la fonction $(x,u) \mapsto -f(x,u)$, et supposons que $f(x_o,u) = 0$ pour un certain $u(\cdot) \in \mathcal{U}$. Si (1) a la propriété (2), alors le système "symétrique" a la propriété suivante: il existe un voisinage de la position d'équilibre x_o les points duquel peuvent être transférés en x_o dans tout instant préfixé. Cela justifie la dénomination "stabilité contrôlée".

(b) le problème du temps minimum peut être appelé "bien posé" lorsque il y a existence, unicité et continuité des solutions par rapport aux données ([10],[11]). La proprieté (2) est remarquable du point de vue de la continuité de la fonction du temps minimum ([12],[13]).

(c) du point de vue géométrique, le principe du maximum de Pontryagin est une condition nécessaire pour que une trajectoire du système se trouve sur la frontière de l'ensemble des points atteignables; si l'intérieur de cet ensemble est vide, le principe du maximum ne donne aucune information utile. La propriété int $R([0,T], x_o) \neq \emptyset$ est appelé accessibilité ([14]). La (2) entraîne évidemment la propriété de accessibilité; cela justifie la dénomination "auto-accessibilité".

(d) soit le système (1) linéaire; sous la condition (2) le principe du maximum est une condition suffisante, outre que nécessaire, d'optimalité pour le problème du temps minimum ([9]).

A côté de (2), on considère souvent aussi la propriété

$$x_o \in \text{int } R(x_o), \quad \text{où} \quad R(x_o) = \bigcup_{t \geq 0} R(t, x_o) \ . \tag{3}$$

Dénotons par $C(x_o)$ l'ensemble des points dont on peut atteindre x_o; il est clair que si le système (1) a la propriété (3), alors le système "symétrique" défini en (a) a la propriété

$$x_o \in \text{int } C(x_o) \ . \tag{3'}$$

La propriété (3) (ou, plus communément, la (3') référée au système (1)) est presque unanimement appelée *contrôlabilité locale* ([3],[15],[16],[17],[18],[19],[34],[35]). On doit remarquer que un système qui a la propriété (3), n'a pas en général la propriété (3'): pour un exemple, voir [36].

Si $x_o = 0$, la (3') s'appelle aussi *nulle-contrôlabilité* ([37]). La propriété suivante est un peu plus intéréssante que (3): quel que soit le voisinage U de x_o, il existe un voisinage $V \subset U$ de x_o tel que tout point $x_1 \in V$ est atteignable de x_o le long d'une trajectoire que ne sort pas de U. Cette propriété est connue comme *contrôlabilité en petit* ([20],[21]) ou même *locale-locale contrôlabilité* ([3], [22]).

Pour compléter le panorama, notons que souvent le terme "localement controlable" est utilisé aussi pour indiquer: 1) qu'un point situé sur une trajectoire de référence appartient à l'intérieur de l'ensemble des points atteignables ([23],[24],[38]); 2) que le système est complètement contrôlable sur un domaine donné ([39]); 3) qu'il existe $T_1 > 0$ tel que $x_o \in \text{int } R(T_1, x_o)$ ([40]).

A' propos de (2) et (3) on connaît un certain nombre de conditions suffisantes, que nous partagerons en deux classes: conditions du type algébrique et conditions du type géométrique. (Nous introduisons cette subdivision pour notre convenance: elle ne se rapporte pas aux méthodes employées, mais seulement à la forme dans laquelle ces conditions sont exprimées).

Le prototype des conditions algébriques est le célèbre critère de Kalman, qui est valable pour un système linéaire

$$\dot{x} = Ax + Bu, \qquad x \in \mathbb{R}^n, \qquad u \in \Omega \subset \mathbb{R}^m \tag{4}$$

lorsque $\mathcal{U}$ est la classe des fonctions intégrables à valeurs dans le compact Ω, $x_o = 0$ et $0 \in \text{int } \Omega$: la (2) vaut si et seulement si

$$\text{rang } [B, AB, A^2B, \ldots, A^{n-1}B] = n . \tag{5}$$

Si en (4) on a $\Omega = \mathbb{R}^m$, la (5) constitue un critère de contrôlabilité complète ([8],[9],[25],[26],[27]); la connaissance d'un critère de contrôlabilité complète pour (4) est importante aussi dans l'étude des systèmes non linéaires ([17], [16],[27],[18],[23],[19]): il est bien connu par exemple que le système (1) a la propriété (3) si sa "linéarisation" (c'est-à-dire le système du type (4) avec $A = \left.\frac{\partial f}{\partial x}\right|_{(x,u) = (0,0)}$ et $B = \left.\frac{\partial f}{\partial u}\right|_{(x,u) = (0,0)}$) est complètement contrôlable ([34]).

Le récent développement de la théorie géométrique du contrôle a permis d'établir d'autres conditions du type algébrique qui peuvent être considerées des généralisations de (5), et qui sont valables pour le système (1), lorsque les contrôles $t \mapsto u(t)$ sont des fonctions différentiables ou analytiques par morceaux. La méthode consiste à représenter le système par le moyen d'une famille de champs de vecteurs D; dénotée par $J(D)$ l'algèbre de Lie engendrée par D, on définit d'une façon naturelle rang $J(D)(x)$, le rang de $J(D)$ en chaque point x. Nous rappelerons les résultats suivants:

(Chow-Lobry) ([28]): soit D une famille symétrique (c'est-à-dire, telle que si le champs de vecteurs $X(\cdot) \in D$, alors même $-X(\cdot) \in D$) de champs de vecteurs de classe C^{∞}. Si rang $J(D)(x_o) = n$, alors (3) vaut. (6)

(Brunovsky) ([29]): soit D une famille impaire (c'est-à-dire, telle que si $X(\cdot) \in D$, il existe $Y(\cdot) \in D$ tel que $X(x) = -Y(-x)$ quel que soit $x \in \mathbb{R}^n$) de champs de vecteurs analytiques. Si rang $J(D)(0) = n$, alors $0 \in \text{int } R(0)$. (7)

Parmi les conditions suffisantes du type géométrique, nous rappelerons:

(Lee-Markus) ([15]): supposons que la fonction $(x,u) \mapsto f(x,u)$ qui paraît en (1) soit de classe C^1, que $f(0,0) = 0$ et que $\mathcal{U}$ soit l'ensemble des fonctions mesurables à valeurs dans le compact Ω, $0 \in \text{int}\,\Omega$. S'il existe un vecteur $v \in \mathbb{R}^n$ tel que le vecteur $w = \frac{\partial f}{\partial u}\Big|_{(x,u)=(0,0)} \cdot v$ n'appartiene pas à un sous-espace invariante de dimension $k \leq (n-1)$ de la matrice $\frac{f}{x}\Big|_{(x,u)=(0,0)}$, alors $0 \in \text{int } R(0)$. (8)

En réalité, le critère énoncé en (8) est le même que (5), appliqué à la "linéarisation" de (1). Sur la base de ce critère, en [30] on obtient un résultat sur la contrôlabilité complète des systèmes bilinéaires.

(Petrov) ([16],[31]): soit $\mathcal{U}$ un ensemble de fonctions à valeurs en Ω comprenant les fonctions constantes par morceaux; supposons qu'ils existent k points $(k > n)$ $u_1,\ldots,u_k \in \Omega$ tels que les vecteurs $f(x_o,u_1), \ldots, f(x_o,u_k)$ constituent une base positive de $\mathbb{R}^n$. Alors (2) vaut. Le fait que les vecteurs $f(x_o,u_1),\ldots,f(x_o,u_k)$ (9)

constituent une base positive de $\mathbb{R}^n$ peut être énoncé dans les deux formes équivalentes suivantes ([16]):

quel que soit $v \in \mathbb{R}^n$, il existe $u \in \Omega$ tel que $\langle v, f(x_o, u)\rangle > 0$; (9')

l'origine de $\mathbb{R}^n$ appartient à $\mathrm{int\ co}\,\{f(x_o,\Omega)\}$. (9")

La dernière forme permet de préciser le résultat de Petrov: si $0 \notin \mathrm{co}\,\{f(x_o,\Omega)\}$, alors (2) ne vaut pas tandis que si $0 \in \partial\mathrm{co}\,\{f(x_o,\Omega)\}$, le cas est incertain, et l'on doit recourir à un test d'ordre supérieur ([1],[2]).

On connaît bien peu de conditions qui, outre que suffisantes, soient même nécessaires pour la (2) ou la (3), par rapport aux systèmes non linéaires: on sait par exemple que le critère (6) est même nécessaire si les champs de D sont analytiques. D'autres conditions nécessaires et suffisantes valables en certains cas particuliers se trouvent en [32] et [33].

Dans ce travail on considère le cas d'un processus de contrôle où l'évolution du point représentatif dépend non seulement du choix d'un contrôle $u(\cdot) \in \mathcal{U}$, mais aussi du choix d'un état initial v dans un ensemble $V \subset \mathbb{R}^n$ donné. Suivant la dénomination introduite en [26], on dira que V est l'ensemble des contrôles initiaux et que $\mathcal{U}$ est l'ensemble des contrôles permanents.

C'est facile à imaginer une situation pratique où les contrôles initiaux se présentent: supposons, par exemple, d'avoir des batteries contre avions réparties le long d'une ligne, et de poursuivre le but d'abattre un avion ennemi le plus tôt possible.

Dans un problème avec des contrôles initiaux, la (2) prend la forme

$$V \subset \mathrm{int}\, R([0,T],V), \text{ quel que soit } T > 0 \tag{10}$$

où $R([0,T],V)$ représente l'ensemble des points atteignables en temps non superieur à T le long d'une trajectoire $t \mapsto x(t;v,u(\cdot))$ telle que $u(\cdot) \in \mathcal{U}$ et $v \in V$.

L'auteur de cette note ne connaît aucun travail où l'on obtienne conditions suffisantes pour (10), avec V non ponctuel. Les théorèmes énoncés dans la prochaine section constituent un premier résultat à propos de (10) et peuvent être considérés une généralisation du théorème de Petrov énoncé dans la forme (9'). On remarque que systèmes avec la propriété (10) sont étudiés en [41]: en [41] la (10) est appelés *stabilité forte*.

II. RESULTATS PRINCIPAUX

Soit Ω un compact de $\mathbb{R}^m$ et soit V un sous-ensemble propre de $\mathbb{R}^n$, fermé, d'intérieur non vide et tel que ∂V est une variété différentiable de dimension $(n-1)$ et de classe C^1. Sous ces hypothèses, pour chaque $x \in \partial V$ il existe une unique normale unitaire sortante $\nu(x)$ à V en x; la fonction $x \mapsto \nu(x)$ est continue sur ∂V.

Supposons que la fonction $(x,u) \mapsto f(x,u)$ qui paraît en (1) et sa dérivée partielle $(x,u) \mapsto \frac{\partial f}{\partial x}(x,u)$ soient continues en $\mathbb{R}^n \times \Omega$. Enfin, soit $\mathcal{U}$ un ensemble quelconque de fonctions mesurables à valeurs en Ω, contenant toutes le fonctions constantes par morceaux.

Les exemples 1 et 2 de la dernière section montrent que les conditions ci-dessus classées du type algébrique ne semblent pas utilisables dans le cas où V contient plus qu'un point. Au contraire, les critères classés du type géométrique semblent plus aptes à la nouvelle situation; les théorèmes suivantes généralisent le critère de Petrov, dans la forme (9').

Théorème 1. Supposons que les précédent hypothèses sur V, Ω, et $(x,u) \mapsto f(x,u)$ soient vérifiées. Si quel que soit $x_o \in \partial V$ on a

$$\max_{u \in \Omega} \langle \nu(x_o), f(x_o,u) \rangle > 0 \tag{11}$$

alors la (10) vaut.

Théorème 2. Supposons que les précédentes hypothèses sur

V, Ω, $\mathcal{U}$ et $(x,u) \mapsto f(x,u)$ soient vérifiées. S'il existe $x_o \in \partial V$ tel que

$$\max_{u \in \Omega} \langle \nu(x_o), f(x_o,u) \rangle < 0 \tag{12}$$

alors la (10) ne vaut pas.

Le cas où il existe $x_o \in \partial V$ tel que

$$\max_{u \in \Omega} \langle \nu(x_o), f(x_o,u) \rangle = 0$$

est incertain: cela est montré par les exemples 3 et 4 de la dernière section.

III. DEMONSTRATIONS

Démontrons le théorème 1. Puisque la fonction $t \mapsto R([0,t],V)$ est croissante (dans le sens de l'inclusion), on a $\operatorname{int} V \subset V \subset R([0,T],V)$ quel que soit $T \geq 0$. Donc il suffit de montrer que $\partial V \subset \operatorname{int} R([0,T],V)$ quel que soit $T > 0$. Par le théorème des fonctions inverses, ∂V peut être représentée près de x_o, comme l'image d'une application différentiable $\phi : y = (y_1,\ldots,y_{n-1}) \mapsto \phi(y)$ définie dans un voisinage U de l'origine de $\mathbb{R}^{n-1}$ et telle que $\phi(0) = x_o$ et $\det \left.\frac{\partial \phi}{\partial y}\right|_{y=0} \neq 0$ ($\frac{\partial \phi}{\partial y}$ dénote la matrice jacobienne).
On note que chaque voisinage suffisamment petit de x_o en $\mathbb{R}^n$ est divisé par ∂V en deux parties,desquelles une est contenue en V, et que $\nu(x_o)$ est orthogonal au sous-espace de $\mathbb{R}^n$ engendré par les colonnes de $\left.\frac{\partial \phi}{\partial y}\right|_{y=0}$

Soit $u_o \in \Omega$ tel que le produit scalaire en (11) soit positif. Posons $t \mapsto u_o(t) \equiv u_o$ et dénotons par $t \mapsto x(t,y)$ la solution $t \mapsto x(t;(y),u_o(\cdot))$ de (1), c'est-à-dire la solution correspondant au contrôle $t \mapsto u_o(t) \equiv u_o$ et telle que $x(0,y) = \phi(y)$. Démontrons que l'application différentiable

$$(t,y) \mapsto x(t,y) \; : \; \mathbb{R}^n \mapsto \mathbb{R}^n \tag{14}$$

est de rang maximum en $0 \in \mathbb{R}^n$. On a

$$\left.\frac{\partial x(t,y)}{\partial t}\right|_{(t,y)=(0,0)} = \left.\frac{\partial x(t,0)}{\partial t}\right|_{t=0} = f(x_o,u_o)$$

$$\left.\frac{\partial x(t,y)}{\partial y}\right|_{(t,y)=(0,0)} = \left.\frac{\partial x(0,y)}{\partial y}\right|_{y=0} = \left.\frac{\partial \phi(y)}{\partial y}\right|_{y=0} .$$

Puisque $y \mapsto \phi(y)$ est de rang maximum en $y = 0$, les $(n-1)$ colonnes de $\frac{\partial x}{\partial y}$ sont linéairement indépendentes. Si $f(x_o,u_o)$ était une combination linéaire de ces colonnes, il serait orthogonal à $\nu(x_o)$, en contradiction avec (11). Donc (14) est de rang maximum en $0 \in \mathbb{R}^n$. Soit α un nombre positif suffisamment petit; la (14) transforme l'ouvert $(-\alpha,\alpha) \times U$ de $\mathbb{R}^n$ dans un voisinage W de x_o: la partie de W décrite lorsque $t \in (-\alpha,0]$ est contenue en V, donc en $R([0,\alpha],V)$ et la partie de W décrite lorsque $t \in (0,\alpha)$ est contenue en $R([0,\alpha],V)$ parce que ses points sont atteignables en temps inférieur à α. En étant α arbitraire et $t \to R([0,t],V)$ croissante, on conclut que $x_o \in W \subset R([0,T],V)$ quel que soit $T > 0$.

Avant de démontrer le théorème 2, établissons le

Lemme. La fonction $x \mapsto M(x) = \max_{u \in \Omega} \langle \nu(x), f(x,u) \rangle$ est continue en chaque point $x \in \partial V$.

Démonstration. Puisque Ω est compact et $u \mapsto f(x,u)$ est continue, quel que soit $x \in \partial V$ il existe $\omega(x) \in \Omega$ tel que $M(x) = \langle \nu(x), f(x,\omega(x)) \rangle$.

Démontrons d'abord la semi-continuité inférieure. Soit $\tilde{x} \in \partial V$ et soit $\varepsilon > 0$. En étant $x \mapsto f(x,u)$ continue, il existe $\delta_1 > 0$ tel que

$$\|y - \tilde{x}\| \leq \delta_1, \quad y \in \partial V \quad \Rightarrow \quad \|f(y,\omega(\tilde{x})) - f(\tilde{x},\omega(\tilde{x}))\| \leq \frac{\varepsilon}{2} ;$$

des plus, il existe $L > 0$ tel que

$$\|y - \tilde{x}\| \leq \delta_1, \quad y \in \partial V \quad \Rightarrow \quad \|f(y,\omega(\tilde{x}))\| < L .$$

Puisque même $x \mapsto \nu(x)$ est continue, il existe $\delta_2 > 0$ tel que

$$\|y-\tilde{x}\| \leq \delta_2, \quad y \in \partial V \Rightarrow \|\nu(y)-\nu(\tilde{x})\| \leq \frac{\varepsilon}{2L} .$$

Donc, si $\delta = \min(\delta_1, \delta_2)$, $\|y-\tilde{x}\| \leq \delta$, $y \in \partial V$ entraîne

$$|\langle \nu(y)-\nu(\tilde{x}), f(y,\omega(\tilde{x}))\rangle| \leq \|\nu(y)-\nu(\tilde{x})\| \cdot \|f(u,\omega(\tilde{x}))\| \leq \frac{\varepsilon}{2} .$$

et

$$|\langle \nu(\tilde{x}), f(y,\omega(\tilde{x})) - f(\tilde{x},\omega(\tilde{x}))\rangle|$$

$$\leq \|\nu(x)\| \cdot \|f(y,\omega(\tilde{x})) - f(\tilde{x},\omega(\tilde{x}))\| \leq \frac{\varepsilon}{2} .$$

D'autre part on a

$$M(y) = \langle \nu(y), f(y,\omega(y))\rangle = \max_{u \in \Omega} \langle \nu(y), f(y,u)\rangle$$

$$\geq \langle \nu(y), f(y,\omega(\tilde{x}))\rangle + \langle \nu(\tilde{x}), f(y,\omega(\tilde{x}))\rangle - \langle \nu(\tilde{x}), f(y,\omega(\tilde{x}))\rangle$$

$$+ \langle \nu(\tilde{x}), f(\tilde{x},\omega(\tilde{x}))\rangle - \langle \nu(\tilde{x}), f(\tilde{x},\omega(\tilde{x}))\rangle$$

$$= M(\tilde{x}) + \langle \nu(y) - \nu(\tilde{x}), f(y,\omega(\tilde{x}))\rangle$$

$$+ \langle \nu(\tilde{x}), f(y,\omega(\tilde{x})) - f(\tilde{x},\omega(\tilde{x}))\rangle \geq M(\tilde{x}) - \varepsilon .$$

Supposons maintenant que $x \mapsto M(x)$ ne soit pas semi-continue supérieurement dans un point $\tilde{x} \in \partial V$. Soit $\{\delta_k\}$ une suite de nombres positifs, $\delta_k \to 0$. Quel que soit k, il existe un point $y_k \in \partial V$, $\|y_k - \tilde{x}\| < \delta_k$ tel que

$$M(y_k) > M(\tilde{x}) + \varepsilon \quad \text{pour un certain} \quad \varepsilon > 0. \tag{15}$$

La suite $\{y_k\}$ converge a $\tilde{x}$. Puisque Ω est compact, la suite $\{\Omega(y_k)\}$ admet une sous-suite $\{\Omega(y_{k_i})\}$ qui converge à un certain $\omega \in \Omega$. En étant $(x,u) \mapsto f(x,u)$ continue, $f(y_{k_i}, \omega(y_{k_i})) \to f(\tilde{x},\omega)$, donc la limite

$$\lim_i \nu(y_{k_i}), f(y_{k_i}, \omega(y_{k_i}))\rangle$$

existe et vaut $\langle \nu(\tilde{x}), f(\tilde{x},\omega)\rangle \geq M(\tilde{x}) + \varepsilon$ par la (15). Cela est absurde, par la définition de $M(\tilde{x})$.

Démontrons le théorème 2. Par le lemme précédent, la (12) entraîne l'existence d'un boule $B(x_o,\rho)$ de centre x_o et rayon ρ telle que

$$x \in B(x_o,\rho) \cap \partial V \quad = \quad \max_{u \in \Omega} \langle \nu(x), f(x,u)\rangle < 0 \ .$$

Les points du type $x(s) = x_o + s \cdot \nu(x_o)$, $0 < s < \rho/2$ peuvent être atteints seulement en sortant de $B(x_o,\rho)$, donc seulement en temps supérieur à $\frac{\rho}{2L}$, où $L = \max_{(x,u) \in \overline{B(x_o,\rho)} \times \Omega} f(x,u)$.

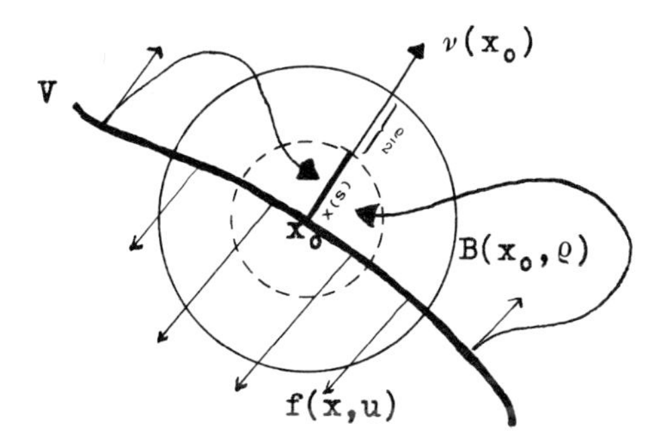

Exemples.

Le premier exemple montre qu'un système linéaire, avec $0 \in \text{int } V$ et $0 \in \text{int } \Omega$, peut avoir la propriété (10) mais non la (5). Le deuxième exemple montre le vice versa.

Exemple 1. Soit donné le système

$$\dot{x} = x$$
$$\dot{y} = u(t)$$

avec $V = \{(x,y) : x^2 + y^2 \leq 1\}$ et $\Omega = \{u : -1 \leq u \leq 1\}$. La matrice de Kalman est $\begin{pmatrix} 0 & 0 \\ 1 & 0 \end{pmatrix}$. L'ensemble $R([0,t],V)$ est montré dans la figure: il contient V dans son intérieur, quel que soit $T > 0$.

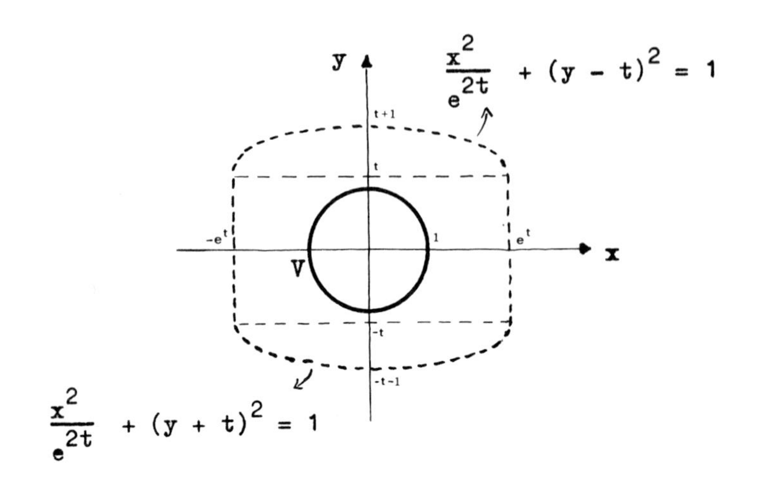

Exemple 2. Soit donné le système

$$\dot{x} = -x + u_1$$
$$\dot{y} = -y + u_2$$

avec $V = \{(x,y) : x^2 + y^2 \leq 1\}$ et $\Omega = \{(u_1,u_2) : u_1^2 + u_2^2 \leq 1\}$.
La matrice de Kalman est $\begin{pmatrix} 1 & 0 & -1 & 0 \\ 0 & 1 & 0 & -1 \end{pmatrix}$, mais quel que soit $T > 0$ on a $R(T,V) \subset V$ et donc $R([0,T],V) = V$.

Les deux exemples suivants montrent que les cas (13) est incertain.

Exemple 3. Soit donné le système

$$\dot{x} = u(t)$$
$$\dot{y} = 0$$

avec $V = \{(x,y) : y \leq x^2\}$ et $\Omega = \{u : -1 \leq u \leq 1\}$.
En $(x,y) = 0$ on a $\max \langle \nu(0), f(0,u) \rangle = 0$.
Les trajectoires du système avec $t \mapsto u(t) \equiv 1$ transforment la parabole $y = x^2$ dans la parabole $y = (x - t)^2$.
Avec $t \mapsto u(t) \equiv -1$ on obtient la parabole $y = (x + t)^2$.
Donc $V \subset \text{int } R([0,T],V)$, quel que soit $T > 0$.

Exemple 4. Considérons le même système de l'exemple 3, mais avec $V = \{(x,y) : y \geq x^2\}$. On voit aisément que $0 \notin \text{int } R([0,T],V)$ quel que soit $T > 0$.

BIBLIOGRAPHIE

1. Sussmann, A sufficient condition for local controllability. A' paraître dans *SIAM J. on Control and Opt.*
2. Petrov, *Diff. Urav. 4*, 1218-1232 (1968).
3. Petrov, *Diff. Urav. 12*, 2214-2222 (1976).
4. Hermes, *Annali di Matematica pura e applicata CXIV*, 103-119 (1977).
5. Hermes, High order controlled stability and controllability, *in* "Dynamical Systems, Proceedings of Univ. of Florida Int. Symposium" (Bednarek-Cesari, eds.). Academic Press, New York, (1977).
6. Bacciotti, *Ricerche di Automatica 7*, 189-197 (1976).
7. Costal, J. B., Note sur autoaccessibilité et structure métrique de familles symétriques de champs de vecteurs. A' paraître.
8. La Salle, Contributions to Theory of Nonlinear Oscillations, Studies n. 5, 1-24, Princeton (1960).
9. Hermes and La Salle, Functional Analysis and Time Optimal Control. Academic Press, New York (1969).
10. Kirillova, *Izv. Matematika 4*, 113-126 (1958).
11. Kirillova, *Uspehi Mat. Nauk XVII*, 141-146 (1962).
12. Petrov, *Diff. Urav. 6*, 373-374 (1970).
13. Bacciotti, Sulla continuità della funzione di tempo minimo. A' paraître dans *Bollettino U.M.I.*
14. Sussman and Jurdjevic, *Journal Diff. Eq. 12*, 95-116 (1972).
15. Lee and Markus, *Arch. Rational Mech. Anal. 8*, 36-58 (1961).
16. Petrov, *Diff. Urav. 4*, 606-617 (1968).
17. Markus, *SIAM J. on Control 3*, 78-90 (1965).
18. Yorke, *SIAM J. on Control 10*, 336-340 (1972).
19. Chukwu, *SIAM J. on Control 13*, 807-816 (1975).
20. Karasev, *Diff. Urav. 3*, 2122-2126 (1967).
21. Mitrokhin, *Diff. Urav. 10*, 1406-1411 (1974).

22. Hermes and Haynes, *SIAM J. on Control 8*, 450-460 (1970).
23. Hermes, *SIAM J. on Control 12*, 252-261 (1974).
24. Knobloch, Local controllability in nonlinear systems, *in* "Dynamical Systems, Proceedings of Univ. of Florida Int. Symposium" (Bednarek-Cesari, eds.). Academic Press, New York, (1977).
25. Kalman, Ho and Narendra, *Contrib. to Diff. Eq. 1*, 189-213 (1963).
26. Conti, Problemi di controllo e di controllo ottimale. UTET, Torino (1974).
27. Lee and Markus, Foundations of optimal control theory. J. Wiley, New York (1967).
28. Lobry, *SIAM J. on Control 8*, 573-605 (1970).
29. Brunovsky, Local controllability of odd systems, *in* "Mathematical Control Theory. Proceedings of a Conference, Zakopane, 1974" (Dolecki, Olech and Zabczyk, eds.). Banach Center Publ. 1. Polish Scientific Publishers, Warszawa, (1976).
30. Rink and Mohler, *SIAM J. on Control 6*, 477-486 (1968).
31. Petrov, *Prikl. Mat. Meh. 34*, 820-826 (1970).
32. Hermes, High order algebraic conditions for controllability, *in* "Mathematical System Theory. Proceeding Algebraic Methods in System Theory, Udine, 1975" (Marchesini and Mitter, eds.). Lecture Notes in Economics and Mathematical Systems n. 131. Springer Verlag, New York, (1976).
33. Petrov, *Diff. Urav. 5*, 962-963 (1969).
34. Kalman, *Trans. ASME J. Basic Eng., Ser. D 86*, 21-22 (1962).
35. Mirza and Womack, *IEEE Trans. on Automatic Control*, 531-535 August (1972).
36. Gronski, *Math. System Theory 10*, 285-287 (1977).
37. Brammer, *SIAM J. on Control 10*, 339-353 (1972).
38. Sec, *C. R. Acad. Sc. Paris 286*, 413-415 (1978).
39. Davison and Kunze, *SIAM J. on Control 8*, 489-497 (1970).
40. Saperstone and Yorke, *SIAM J. on Control 9*, 253-262 (1971).
41. Blagodatskikh, *Izv. Akad. Nauk SSSR 38*, 615-624 (1974).

DETERMINATION AND APPLICATION OF VEKUA RESOLVENTS

Karl Wilhelm Bauer[1]

Technische Universität Graz
Institut für Mathematik I

I. N. Vekua [6] proved the formula

$$V(z,\bar{z}) = \quad (z) + \int_{z_1}^{z} \Gamma_1(z,\bar{z};t,\overline{z_0})\,\varphi(t)\,dt + \int_{\bar{z}_1}^{\bar{z}} \Gamma_2(z,\bar{z};z_0,\tau)\,\varphi^*(\tau)\,d\tau , \tag{1}$$

$z_0, z_1 \in D$, $\varphi(z)$ hol. in D,

giving all the solutions of the differential equation

$$V_{\bar{z}} = C(z,\bar{z})\bar{V} , \quad z = x + iy , \tag{2}$$

analytic in a fundamental domain D, if C is an analytic function of $x,y \in \mathbb{R}$. The resolvents Γ_1 and Γ_2 are given by

$$\begin{aligned} \Gamma_1 &= W_1 + iW_2 , \\ \Gamma_2 &= W_1 - iW_2 . \end{aligned} \tag{3}$$

---

[1] Present address: Technische Universität Graz, Institut für Mathematik, Graz, Austria.

ISBN 0-12-186280-1

$W_k(z,\zeta;t,\tau)$, $k = 1,2$, are solutions of the differential equation

$$V_\zeta = C(z,\zeta)V^* ,$$

$$z = x+iy, \quad \zeta = x-iy, \quad x,y \in \mathbb{C}, \quad V^*(\zeta,z) = \overline{V(\bar{\zeta},z)},$$

where

$$W_1\Big|_{\zeta=\tau} = \frac{1}{2}\,C(z,\tau) , \qquad W_2\Big|_{\zeta=\tau} = \frac{i}{2}\,C(z,\tau) . \tag{4}$$

In general the determination of the functions W_1 and W_2 is difficult .

Here the question arises, if it is possible to determine the Vekua resolvents Γ_1 and Γ_2 directly. Using (3) it follows, that Γ_1 and Γ_2 are solutions of the differential equation

$$w_{z\zeta} - \frac{C_z}{C}\,w_\zeta - CC^*w = 0 . \tag{5}$$

If we know one solution $w(z,\zeta;t,\tau)$ of (5), where

$$w\Big|_{\zeta=\tau} = C(z,\tau) , \qquad w\Big|_{z=t} = C(t,\tau) , \tag{6}$$

it follows

$$\Gamma_2 = w , \qquad C(z,\zeta)\,\Gamma_1^* = w_\zeta . \tag{7}$$

We apply this method [1] in case of a differential equation of type (1), where C satisfies the differential equation

$$m^2(\log C)_{z\bar{z}} + \varepsilon C\bar{C} = 0 , \qquad m > 0, \qquad \varepsilon = \pm 1 . \tag{8}$$

Using

$$C = e^{u+iv} , \qquad u,v \text{ realvalued},$$

it follows, that u satisfies the Liouville equation

$$m^2u_{z\bar{z}} + \varepsilon e^{2u} = 0$$

and

$$V_{z\bar{z}} = 0 .$$

So we get the representation

$$C = \frac{m|f'|}{1+\varepsilon f\bar{f}} \frac{g}{\bar{g}} , \tag{9}$$

$f(z)$, $g(z)$ hol., $(1+\varepsilon f\bar{f})f'g \neq 0$.

Substituing

$$U(\xi,\bar{\xi}) = \frac{V(z,\bar{z})}{g(z)\sqrt{f'(z)}} , \qquad \xi = f(z)$$

we get

$$U_{\bar{\xi}} = \frac{m}{1+\varepsilon\xi\bar{\xi}} \bar{U} .$$

Therefore instead of (1) we consider the normed differential equation

$$V_{\bar{z}} = \frac{m}{1+\varepsilon z\bar{z}} \bar{V} . \tag{1*}$$

In this case equation (5) takes the form

$$\omega^2 w_{z\zeta} + \varepsilon\zeta\omega w_{\zeta} - m^2 w = 0 , \qquad \omega = 1+\varepsilon z\zeta . \tag{5*}$$

To get a solution w with the properties (6) we set

$$w = \frac{m}{1+\varepsilon\zeta z} H(\lambda) , \qquad \lambda = \lambda(z,\zeta;t,\tau) ,$$

$$H\Big|_{\zeta=\tau} = H\Big|_{z=t} = 1 .$$

It follows

$$\omega^2\lambda_z\lambda_\zeta H'' + \left[\omega^2\lambda_{z\zeta} + \frac{\varepsilon\omega(\zeta-\tau)}{1+\varepsilon\tau z}\lambda_\zeta\right]H' - m^2 H = 0 .$$

To get a Gauss equation we suppose

$$\omega^2\lambda_z\lambda_\zeta = a_o + a_1\lambda + a_2\lambda^2 ,$$

$$\omega^2\lambda_{z\zeta} + \frac{\varepsilon\omega(\zeta - \tau)}{1 + \varepsilon\tau z}\lambda_\zeta = b_o + b_1\lambda ,$$

$$a_o, a_1, a_2, b_o, b_1 \in \mathbb{C} .$$

So we are led to the hypergeometric equation

$$\lambda(\lambda-1)H'' + [(\alpha+\beta+1)\lambda-\gamma]H' + \alpha\beta H = 0 ,$$

$$\alpha = m\sqrt{-\varepsilon} , \quad \beta = -\alpha , \quad \gamma = 1 .$$

Using (7) we obtain the Vekua resolvents of (1^*) by

$$\Gamma_2 = \frac{m}{1 + \varepsilon\tau z} F(m\sqrt{-\varepsilon}, -m\sqrt{-\varepsilon}, 1 ; \frac{\varepsilon(z-t)(\zeta-\tau)}{(1 + \varepsilon z\zeta)(1 + \varepsilon t\tau)}) ,$$

$$\Gamma_1 = \frac{\varepsilon(\zeta-\tau)F'(\lambda)}{(1 + \varepsilon z\zeta)(1 + \varepsilon t\tau)}$$

where $F(\alpha,\beta,\gamma;\lambda)$ denotes the hypergeometric function.

In case $\varepsilon = -1$, $m \in \mathbb{N}$ the hypergeometric function F reduces to a polynomial in λ . Here it is possible to represent the solutions of (1^*) by a differential operator. We get

$$V = \frac{1}{z-1}\sum_{k=0}^{m} \frac{(-1)^{m-k}(2m-1-k)!}{k!(m-k)!}\left[\frac{(z-1)(\bar{z}-1)}{1-z\bar{z}}\right]^{m-k}\left[mr^k\Psi - (m-k)\overline{r^k\Psi}\right],$$

$$\Psi(z) \text{ hol.,} \qquad r = (z-1)^2\frac{d}{dz} .$$

APPLICATIONS

A. Axially symmetric gravitational field problem. In connection with the field equations governing the gravitational field of a uniformly rotationg axially symmetric source F. J. Ernst [3] got the differential equation

$$W_{z\bar{z}} + \frac{1}{2(1-z\bar{z})}\left[\frac{1-z}{1-\bar{z}}W_z + \frac{1-\bar{z}}{1-z}W_{\bar{z}}\right] - \frac{2W_zW_{\bar{z}}}{W+\bar{W}} = 0 . \qquad (10)$$

A.V. Bitsadze and V.I. Paškovskii [2] proved, that every solution W satisfying the equation

$$W_{\bar{z}} = \frac{1}{2(1-z\bar{z})} \frac{1-z}{1-\bar{z}} (W + \bar{W})$$

is a solution of (10). Substituing

$$W = V \sqrt{\frac{2(1-z\bar{z})}{(1-z)(1-\bar{z})}}$$

we get

$$V_{\bar{z}} = C\bar{V} , \qquad C = \frac{1}{2(1-z\bar{z})} \frac{1-z}{1-\bar{z}} .$$

This coefficient C is a solution of (8) and we obtain C in the representation (9) by

$$\varepsilon = -1 , \quad m = \frac{1}{2} , \quad f = z , \quad g = 1-z .$$

B. Pseudo-holomorphic functions. In connection with the representation of pseudo-holomorphic functions of several complex variables A. Koohara [5] got the differential equation

$$G_{\bar{z}} = \frac{\bar{K}K_{\bar{z}}}{1-K\bar{K}} G - \frac{\overline{K_z}}{1-K\bar{K}} \bar{G} . \tag{11}$$

Here it is of interest to find functions $K(z,\bar{z})$ such that the solutions of (11) can be determined. We set

$$K = \bar{f} \left(\frac{\bar{f}}{f}\right)^{\eta m} , \quad f(z) \text{ hol.,} \quad (1-f\bar{f})ff' \neq 0$$

$$\eta = \pm 1 , \qquad m > 0 .$$

Substituting

$$G = \frac{V}{(1-f\bar{f})^{1+\eta m}}$$

we get

$$V_{\bar{z}} = C\bar{V} \ , \quad C = \frac{\eta m \overline{f'}}{1-f\bar{f}} \left(\frac{f}{\bar{f}}\right)^{1+\eta m} \ .$$

We obtain this coefficient C in the representation (9) by

$$\varepsilon = -1 \ , \quad g = \frac{f^{1+\eta m}}{\sqrt{\eta f'}} \ .$$

C. Pseudo-analytic functions. G. Jank and K.-J. Wirths [4] proved, that a sharp maximum principle is valid for all pseudo-analytic functions V satisfying

$$V_{\bar{z}} = \frac{\gamma_{\bar{z}}}{\gamma} \frac{h}{\bar{h}} \bar{V} \ ,$$

$h(z)$ hol., $\neq 0$,

γ realvalued, $\neq 0$, $\gamma \in C^2$, γ^{-2} subharmonic.

We get a coefficient

$$C = \frac{\gamma_{\bar{z}}}{\gamma} \frac{h}{\bar{h}}$$

with these properties by (9), where

$$\varepsilon = -1 \ , \quad g = \frac{i(f+1)h}{\sqrt{f'}} \ .$$

Then γ takes the form

$$\gamma = \left[\frac{1 - f\bar{f}}{(f+1)(\bar{f}+1)}\right]^m$$

and γ^{-2} is a subharmonic function.

REFERENCES

1. Bauer, K. W., Bestimmung und Anwendung von Vekua-Resolventen, *Monatsh. f. Math. 85*, 89-97 (1977).
2. Bitsadze, A. V. and Paškovskii V. I., On the Theory of the Maxwell-Einstein Equations, *Dokl. Akad. Nauk. SSSR, 216*, 762-764 (1974).
3. Ernst, F. J., New Formulation of the Axially Symmetric Gravitational Field Problem, I: *Phys. Rev.167*, 1175-1178 (1968), II: *Phys. Rev. 168*, 1415-1417 (1968).
4. Jank, G. and Wirths, J., Generalized Maximum Principles in Certain Classes of Pseudo-Analytic Functions, *in* "Function Theoretic Methods in Differential Equations" (R.P. Gilbert and R. Weinacht, eds.), pp. 63-67. Research Notes in Mathematics, vol. 8, Pitman Publ., London, (1976).
5. Koohara, A., Representation of Pseudo-Holomorphic Functions of Several Complex Variables, *J. Math. Soc. Japan 27*, 257-277 (1976).
6. Vekua, I. N., New Methods for Solving Elliptic Equations, J. Wiley and Sons, New York,(1968).

GENERALIZED HOPF BIFURCATION

S. R. Bernfeld[1,2]

Department of Mathematics
University of Texas at Arlington
Arlington, Texas

L. Salvadori[3]

Istituto di Matematica
Università di Trento
Trento, Italy

I. INTRODUCTION

Let us consider the differential equation

$$\dot{x} = f_0(x), \tag{1.1}$$

where $f_0 \in C^\infty[B^n(r_0), R^n]$, $f_0(0) = 0$ and $B^n(r_0) = \{x \in R^n: \|x\| < r_0\}$. Assume the Jacobian matrix $f_0'(0)$ has a complex conjugate pair of eigenvalues $\pm i$ and that any other eigenva-

---

[1] Research partially supported by Italian Council of Research (C.N.R.) and by U.S. Army Research Grant DAAG 29-77-G0062.

[2] Present address: Department of Mathematics, University of Texas at Arlington, Arlington, Texas.

[3] Present address: Istituto di Matematica, Università di Trento, Trento, Italy.

ISBN 0-12-186280-1

lue λ satisfies $\lambda \neq mi$, $m = 0,\pm 1,\pm 2,\ldots$.

For those $f \in C^{\infty}[B^n(r_0),R^n]$ close to f_0 (with respect to an appropriate topology) consider the perturbed differential equation

$$\dot{x} = f(x). \tag{1.2}$$

In [2] N. Chafee has considered the problem of determining the number of nonzero periodic orbits of (1.2) lying near the origin and having period T close to 2π for each f close to f_0. Using the alternative method as described by Hale [3], Chafee constructed a determining equation $\Psi(\xi,f) = 0$ where $\xi = \rho^2$ is a measure of the amplitude of the periodic orbits of (1.2). Letting k (finite) be the multiplicity of the root $\xi = 0$ of the equation $\Psi(\cdot,f) = 0$, Chafee proved among others the following two properties: (a) there exists a neighborhood $\mathcal{N}$ of f_0 and a number $r_1 > 0$ such that for any $f \in \mathcal{N}$ equation (1.2) has no more than k nontrivial periodic orbits in $B^n(r_1)$ with period close to 2π; (b) for any integer j, $0 \leq j \leq k$, for any neighborhood $\mathcal{N}^*$ of f_0, $\mathcal{N}^* \subseteq \mathcal{N}$, and for any $r_2 \in (0,r_1]$ there exists an $f \in \mathcal{N}^*$ such that (1.2) has exactly j nontrivial periodic orbits in $B^n(r_2)$ with period close to 2π.

Since the construction of the determining equation requires the use of fixed point theorems, its form is only known implicitely. Thus the determination of the number k is a new problem that needs to be solved.

In this paper we consider the case $n = 2$ and show that properties (a) and (b) occur if and only if the origin of (1.1) is either $(2k+1)$-asymptotically stable or $(2k+1)$-completely unstable; that is the origin is asymptotically stable in the future or in the past and this property is recognizable in a suitable sense by the terms of f_0 of degree $\leq 2k+1$. This result gives a complete answer to the problem since there is a constructive procedure to determine k. This is exactly the classical method of Poincaré which is an algebraic procedure to construct either a power series satisfying formally the condition of being a first integral of (1.1) or a positive definite function whose derivative along the solutions of (1.1) is definite in sign.

Now denote by S_j, $0 \le j \le k$, the set of those $f \in \mathcal{N}^k$ for which (1.2) has exactly j nontrivial periodic orbits lying in $B^2(r_1)$. Our techniques allow us to construct a broad class of functions lying in the interior of S_j. Indeed these functions can be found by determining polynomials of order k that have j simple real roots. In this way one provides a method for determining a class of functions that enjoy the structural property of preserving the number of periodic orbits near the origin under small perturbations.

Using different techniques Andronov and Leontovich ([1], Ch. IX) previously obtained the sufficient part of the previous result, that is properties (a) and (b) occur if the origin of (1.1) is either (2k+1)-asymptotically stable or (2k+1)-completely unstable. Their techniques do not seem to provide a constructive procedure for the determination of those f which have the above structural property.

For the case $k = 1$ we have given an explicit condition, based on the Poincaré procedure, to determine whether there exists one periodic or no periodic orbits of (1.2).

We also consider the case $k = \infty$ and obtain an infinite dimensional analogue of our previous results.

Finally we remark that generalizations to higher order systems may be obtained by use of the center manifold theorem (see Kelley [4]) which we do not address ourselves to here.

II. RESULTS

Under the hypotheses given on f_0, by applying a coordinate transformation, equation (1.1) in R^2 can be written as

$$\begin{aligned} \dot{x} &= -y + X_0(x,y), \\ \dot{y} &= x + Y_0(x,y) , \end{aligned} \tag{2.1}$$

where $X_0(0,0) = Y_0(0,0) = 0$ and $X_0'(0,0) = Y_0'(0,0) = 0$. We shall again refer to the right handside of (2.1) as f_0. Similarly equation (1.2) can be written as

$$\dot{x} = \alpha x - \beta y + X(x,y) ,$$
$$\dot{y} = \alpha y + \beta x + Y(x,y) , \tag{2.2}$$

where $X(0,0) = Y(0,0) = 0$ and $X'(0,0) = Y'(0,0) = 0$ and again we refer to the right hand side of (2.2) as f.

The following definitions and comments were introduced in the paper of Negrini and Salvadori [7].

Definition. Let h be an integer ≥ 2. The solution $x \equiv y \equiv 0$ of (2.1) is said to be h-asymptotically stable (resp. h-completely unstable) if

(i) for every $\tau_1, \tau_2 \in C^\infty[B^2(r_0), R]$ of order greater than h the solution $x \equiv y \equiv 0$ of

$$\dot{x} = -y + X_{02} + X_{03} + \ldots + X_{0h} + \tau_1$$

$$\dot{y} = x + Y_{02} + Y_{0_3} + \ldots + Y_{0h} + \tau_2$$

is asymptotically stable [resp. completely unstable]. Here X_{0i}, Y_{0i}, $2 \leq i \leq h$, represents the ith term of the MacLaurin expansion of X_0 and Y_0. Recall that $X_{01} \equiv Y_{01} \equiv 0$;

(ii) property (i) is not satisfied when h is replaced by any integer $m \in \{2,3,\ldots,h-1\}$.

The properties in the above definition can occur only when h is odd.

Moreover it has been shown that the h-asymptotic stability or h-complete instability can be recognized by means of a classical procedure due to Poincaré [8]. In particular consider a polynomial of the form

$$F(x,y) = x^2 + y^2 + F_3 + \ldots + F_m , \tag{2.3}$$

where m is an even integer and F_j is a homogeneous polynomial of degree j. Let $\dot{F}$ be the derivative of F along the solutions of (2.1). The Poincaré procedure is an algebraic method for the determination of m, F and a constant G_m such that

$$\dot{F} = G_m(x^2+y^2)^{m/2} + \chi(x,y) \tag{2.4}$$

where χ is of order $> m$. Obviously (2.4) is satisfied for $m = 2$ if and only if $G_2 = 0$. In addition if (2.4) can be satisfied for $m = \bar{m}$ with $G_{\bar{m}} = 0$, then (2.4) can be satisfied for $m = \bar{m} + 2$ and G_m is uniquely determined.

Proposition. [7]. Let $h \geq 3$ be odd. The solution $x \equiv y \equiv 0$ of (2.1) is h-asymptotically stable [resp. h-completely unstable] if and only if $G_r = 0$ for $r \in \{2,\ldots,h-1\}$ and $G_{h+1} < 0$ [resp. $G_{h+1} \quad 0$].

We are now able to present our results.

Theorem A. Let k be any integer such that $k \geq 1$. Then the property that the origin is for (2.1) either h-asymptotically stable or h-completely unstable, with $h = 2k+1$, is equivalent to the following conditions:

(i) there exist a neighborhood $\mathcal{N}$ of f_0 and a number $r_1 > 0$ such that for any $f \in \mathcal{N}$ equation (2.2) has no more than k nonzero periodic orbits in $B^2(r_1)$;

(ii) for any integer j, $0 \leq j \leq k$, for any neighborhood $\mathcal{N}^*$ of f_0, $\mathcal{N}^* \subseteq \mathcal{N}$, and for any $r_2 \in (0,r_1]$ there exists a function $f \in \mathcal{N}^*$ such that equation (2.2) has exactly j nonzero periodic orbits in $B^2(r_2)$.

In addition if the solution $x \equiv y \equiv 0$ of (2.1) is either h-asymptotically stable or h-completely unstable with $h = 2k+1$ then

(iii) for any $r \in (0,r_1]$ there exists a neighborhood $\mathcal{N}_r \subseteq \mathcal{N}$ of f_0 such that if $f \in \mathcal{N}_r$, and if Γ is a periodic orbit of (2.2) lying in $B^2(r_1)$, then Γ lies in $B^2(r)$;

(iv) letting S_j, $0 \leq j \leq k$, be the set of functions $f \in \mathcal{N}$ for which there exist exactly j nontrivial periodic orbits of (2.2) in $B^2(r_1)$, we have that S_j has a nonempty interior with f_0 lying on its boundary.

Theorem B. Assume that the solution $x \equiv y \equiv 0$ of (2.1) is either 3-asymptotically stable or 3-completely unstable. Then there exist a neighborhood N of f_0, $N \subseteq \mathcal{N}$, and an $\varepsilon \in (0,r_1]$ such that we have $f \in S_o(N,\varepsilon)$ if $\alpha G_4 \geq 0$ and $f \in S_1(N,\varepsilon)$ if $\alpha G_4 > 0$.

Remarks. We thus see the conclusions of Theorem A depend on the order of the first nonvanishing Poincaré constant associated with f_0. In Theorem B one observes that we answer a question posed by Chafee in Section 7 of [2] by providing a method to determine the sign of $\psi'(\cdot, f_0)$ which is the same as that of G_4.

Finally we obtain a result which extends Theorem A to the case when $k = \infty$.

Theorem C. If the origin is neither h-asymptotically stable nor h-completely unstable for every $h \geq 3$, then in any neighborhood N of f_0, for any integer $j \geq 0$, and for any number $r > 0$ there exists $f \in N$ such that equation (2.2) has j nontrivial periodic solutions lying in $B^2(r)$.

III. OUTLINE OF PROOFS

In this section we outline the proofs of our results. Complete details and extensions will appear in a forthcoming paper.

We first make precise the topology we shall use on the space $C^\infty[B^2(r_0), R^2]$. Define the metric as follows

$$\|f\| = \sum_{\ell=0}^{\infty} \frac{1}{2^\ell} \frac{\|f\|^{(\ell)}}{1 + \|f\|^{(\ell)}} ,$$

where $\|f\|^{(\ell)}$ denotes the usual C^ℓ-supremum norm of f on $B^2(r_0)$. Then $C^\infty[B^2(r_0), R^2]$ is a metric linear space under $\| \ \|$.

Converting (2.2) into polar form by letting $x = r\cos\vartheta$, $y = r\sin\vartheta$, we have

$$\begin{aligned} \dot{r} &= \alpha r + X^*(r,\vartheta)\cos\vartheta + Y^*(r,\vartheta)\sin\vartheta \\ r\dot{\vartheta} &= \beta r + Y^*(r,\vartheta)\cos\vartheta - X^*(r,\vartheta)\sin\vartheta \end{aligned} \tag{3.1}$$

where $X^*(r,\vartheta) = X(r\cos\vartheta, r\sin\vartheta)$ and $Y^*(r,\vartheta) = Y(r\cos\vartheta$

$r \sin\vartheta)$. Since β is close to one and $X^*(r,\vartheta)$, $Y^*(r,\vartheta)$ are $o(r)$ we have the existence of $\bar{r} > 0$ and $b > 0$ such that $\dot{\vartheta} > b$ for all $r \in [0,\bar{r}]$. For every $r_0 \in [0,\bar{r})$ and $\vartheta_0 \in R$ the orbit of (2.2) passing through (r_0,ϑ_0) will be represented by the solution $r(f,\vartheta,r_0,\vartheta_0)$ of

$$\frac{dr}{d\vartheta} = R(f,r,\vartheta), \qquad r(\vartheta_0) = r_0 ,$$

where $R(f,r,\vartheta) = \dfrac{\alpha r + X^*(r,\vartheta)\cos\vartheta + Y^*(r,\vartheta)\sin\vartheta}{\theta(f,r,\vartheta)}$ with

$$\theta(f,r,\vartheta) = \beta + \frac{Y^*(r,\vartheta)\cos\vartheta - X^*(r,\vartheta)\sin\vartheta}{r} \quad \text{for } r > 0,$$

$$\theta(f,0,\vartheta) = \beta .$$

By continuous dependence arguments it follows that for a sufficiently small neighborhood M of f_0 and for sufficiently small $\bar{c} > 0$ the solution $r(f,\vartheta,c,0)$ of (3.2) will exist on $[0,2\pi]$ for all $f \in M$ and all $c \in [0,\bar{c})$.

We now can define the displacement function for (2.2), $V(f,c)$, given by

$$V(f,c) = r(f,2\pi,c) - c$$

where $r(f,2\pi,c) \equiv r(f,2\pi,c,0)$. Since $R(\cdot,\cdot,\cdot) \in C^\infty$ whe have

$$\begin{aligned} r(f,\vartheta,c) &= u_1(f,\vartheta)c + u_2(f,\vartheta)c^2 + \dots \\ &\quad + u_k(f,\vartheta)c^k + \eta(f,\vartheta,c) , \end{aligned} \tag{3.3}$$

where k is any positive integer and $\eta(f,\vartheta,\cdot)$ is of order $> k$. Now $u_i(f,\cdot) \in C^\infty$ with $u_1(f,0) = 1$, $u_i(f,0) = 0$, $i = 2,3,\dots,k$, and $\eta(f,0,c) = 0$. If we insert (3.3) into (3.2) then we get for any fixed f a system of differential equations for $u_i(f,\vartheta)$ given by

$$\begin{aligned} \frac{du_1}{d\vartheta} &= \frac{\alpha}{\beta} u_1 , \\ \frac{du_i}{d\vartheta} &= \frac{\alpha}{\beta} u_i + U_i(u_1,u_2,\dots,u_{i-1},\vartheta) \qquad i = 2,3,\dots \end{aligned} \tag{3.4}$$

We define $\hat{V}(f,c) = V(f,c)/c$ for $c \neq 0$, and $\hat{V}(f,0) = u_1(f,2\pi)-1$.

We see that the orbit through $(c,0)$ is closed if and only if $V(f,c) = 0$, that is either $c = 0$ or $\tilde{V}(f,c) = 0$.

The following result of Negrini and Salvadori [7] establishes a relationship between h-asymptotic stability of (2.1) and $V(f_0,0)$; namely,

Theorem 3. Let $h \geq 3$ be odd. Then the solution $x \equiv y \equiv 0$ of (2.1) is h-asymptotically stable [resp. h-completely unstable] if and only if $\dfrac{\partial^i V(f_0,0)}{\partial c^i} = 0$ for $i \in \{1,2,\dots,h-1\}$ and $\dfrac{\partial^h V(f_0,0)}{\partial c^h} < 0$ [resp. $\dfrac{\partial^h V(f_0,0)}{\partial c^h} > 0$].

Thus the displacement function for (2.1) when the origin is h-asymptotically stable or h-completely unstable has the form

$$V(f_0,c) = gc^h + \eta(f_0,c),$$

where g is a constant different than zero.

We suppose now that the origin is either h-asymptotically stable or h-completely unstable for (2.1). Then for every f in a neighborhood of f_0, by means of a transformation T due to Takens [9, Lemma 2.4] equation (2.2) can be written

$$\begin{aligned} \dot{x} &= a_0x - y + a_1x(x^2+y^2) + \dots + a_kx(x^2+y^2)^k + \Phi(f,x,y) \\ \dot{y} &= a_0y + x + a_1y(x^2+y^2) + \dots + a_ky(x^2+y^2)^k + \Psi(f,x,y) \end{aligned} \tag{P}$$

where $k = (h-1)/2$, $a_0,\dots,a_k$ are constants depending on f such that

$$a_j(f_0) = 0, \quad j = 0,1,\dots,k-1, \quad a_k(f_0) = a, \tag{3.5}$$

with $a < 0$ [resp. $a > 0$] in the case of h-asymptotic stability [resp. h-complete instability]. Moreover Φ and Ψ are of order $> k$ in (x,y) and T is a C^∞ diffeomorphism in (x,y,w) where w is the set of coefficients of the terms of degree $\leq 2k+1$ in the MacLaurin expansion of f. The relationships (3.5) are a consequence of the equivalence proved in [7] between the $(s,-)$ [resp. $(s,+)$] Taken's singularity [9] and the assumed h-asymptotic stability [resp. h-complete instability] of the origin for $f = f_0$. We write for future reference the system (P) when $f = f_0$, as

$$\begin{aligned} \dot{x} &= -y + ax(x^2+y^2)^{\frac{h-1}{2}} + \Phi(f_0,x,y) \\ \dot{y} &= x + ay(x^2+y^2)^{\frac{h-1}{2}} + \Psi(f_0,x,y) \end{aligned} \tag{U}$$

Our analysis before shows that the displacement function for (U), again denoted by $V(f_0,c)$, is given by

$$V(f_0,c) = u_{2k+1}(f_0,2\pi)c^{2k+1} + \rho(f_0,c)$$

where $\rho(f_0,c) = o(c^{2k+1})$; the displacement function for (P) denoted by $V(f,c)$ is given by

$$\begin{aligned} V(f,c) = {} & (u_1(f,2\pi)-1)c + u_3(f,2\pi)c^3 + \dots + \\ & + u_{2k+1}(f,2\pi)c^{2k+1} + \rho(f,c), \end{aligned}$$

where $\rho(f,c) = o(c^{2k+1})$. Moreover $V(f,\cdot)$ is C^∞. Analogously we shall denote by $\tilde{V}(f,c)$ the function $\tilde{V}$ for (P).

<u>Lemma 1</u>. Consider the system (P), without the higher order terms, given by

$$\begin{aligned} \dot{x} &= a_0x - y + a_1x(x^2+y^2) + a_2x(x^2+y^2)^2 + \dots + a_kx(x^2+y^2)^k, \\ \dot{y} &= a_0y + x + a_1y(x^2+y^2) + a_2y(x^2+y^2)^2 + \dots + a_ky(x^2+y^2)^k, \end{aligned} \tag{3.6}$$

where we denote the right hand side of (3.6) by $\bar{f}$. The displacement function is given by

$$\begin{aligned} V(\bar{f},c) = {} & (u_1(\bar{f},2\pi)-1)c + u_3(\bar{f},2\pi)c^3 + \dots + \\ & + u_{2k+1}(\bar{f},2\pi)c^{2k+1} + \xi(\bar{f},c), \end{aligned}$$

where $\xi(\bar{f},c) = o(c^{2k+1})$ and the constants $u_{2i+1}(\bar{f},2\pi)$, $i = 0,1,\dots,k-1$ are homeomorphic (in fact C^∞ diffeomorphic) functions of $(a_0,a_1,\dots,a_i)$ in a neighborhood of $(0,0,\dots,0)$ with $u_{2i+1}(\bar{f},2\pi)(0,0,\dots,0) = 0$, $i \neq 0$, and $u_1(\bar{f},2,\pi)(0) = 1$. Moreover $u_{2k+1}(\bar{f},2\pi)$ is a homeomorphic (C^∞ diffeomorphic) function of $(a_0, a_1, \ldots, a_k)$ in a neigborhood of $(0,0,\ldots,a)$ such that $u_{2k+1}(\bar{f},2\pi)(0,0,\ldots,a_k) = 2\pi a_k$.

Lemma 2. Consider the scalar function

$$f(\mu,z) = (z-\mu)(z-2\mu)(z-3\mu)\ldots(z-j\mu)(z^2+\mu^2)^{\frac{k-j}{2}} + \Phi(\mu,z)$$

where $\Phi(\mu,\cdot)$ is monotone in a right neighborhood of $z = 0$, $\Phi(\cdot,\cdot) \in C^{\infty}$ and $\Phi(\mu,\cdot) = o(z^k)$. Then for μ sufficiently small $f(\mu,z)$ has exactly j real simple roots in a sufficiently small neighborhood of $z = 0$.

We now present a few proposition with a rough idea of the method of proof. These propositions will lead to the conclusions of Theorems A, B and C.

Proposition 1. Assume the origin of (U) is h-asymptotically stable [resp. h-completely unstable]. Then there exist a neighborhood $\mathcal{N}$ of f_0 and a neighborhood of the origin, $B^2(r_1)$, such that for $f \in \mathcal{N}$ there exist a most k nontrivial periodic solutions of (P) in $B^2(r_1)$, where $h = 2k+1$.

Outline of proof. The function $\tilde{V}(f,c)$ for (P) is given by

$$\tilde{V}(f,c) = (u_1(f,2\pi)-1)+u_3(f,2\pi)c^2 + \ldots + u_{2k+1}c^{2k} + \tilde{\rho}(f,c).$$

By applying Rolle's Theorem and using the symmetry of the polynomial part of $\tilde{V}(f,c)$ we arrive at the conclusion that the equation $\tilde{V}(f,\cdot) = 0$ has at most k positive roots in a sufficiently small neighborhood of $c = 0$.

This gives us part (i) of Theorem A. Except when stated otherwise we shall assume the origin of (U) is h-asymptotically stable or h-completely unstable.

Proposition 2. For any integer j such that $0 \leq j \leq k$, for any neighborhood N of f_0, and for any neighborhood $B^2(r_2)$ of the origin there exists an $f \in N$ such that there exist j periodic orbits lying in $B^2(r_2)$.

Outline of proof. Using Lemma 1 we assume for the a_i's in (P) continuous functions $a_i(\mu)$ in a right interval of $\mu = 0$ and moreover we perturb the terms in f_0 of degree $> h$ (by using a fixed parameter τ), so that

$$\tilde{V}(f,c) = 2\pi a(c^2-\mu)(c^2-2\mu)\ldots(c^2-j\mu)(c^4+\mu^2)^{\frac{k-j}{2}} + \tilde{\rho}(\mu,\tau,c),$$

where $\tilde{\rho}(\mu,\tau,c)$ is monotone in c for sufficiently small positive c and for all μ sufficiently small. Then an application of Lemma 2 leads to the result.

Using the above idea we see that S_j has a nonempty interior for we may modify the a_k's of (P) slightly and still obtain exactly j positive roots of $\tilde{V}(f,c)$. This leads us to parts (ii) and (iv) of Theorem A. To prove part (iii) it is sufficient to consider the following proposition:

Proposition 3. Assume that r_1 from Proposition 1 is given. Then there exists an $\bar{r} < r_1$ such that for any $r \in (0,\bar{r})$ there exists a neighborhood $\mathcal{N}_r$ of f_0 so that for any $n \in \mathcal{N}_r$, $\tilde{V}(f,c)$ has no zeros in $[r,\bar{r}]$.

Outline of Proof. We show that there exists an $\bar{r}$ such that for any $r < \bar{r}$ we have

$$|V(f_0,c)| \geq \gamma r^{2k},$$

whenever $r \leq |c| \leq \bar{r}$; here γ is a positive constant. Using the continuity of $V(\cdot,\cdot)$ we have the existence of a neighborhood $\mathcal{N}_r$ of f_0 such that for $f \in \mathcal{N}_r$ and for $r\leq|c|\leq\bar{r}$

$$|V(f,c)| \geq \frac{1}{2}\gamma r^{2k}.$$

An outline of the proof of the necessity part in Theorem A will be given later.

Outline of Proof of Theorem B. In the case $k = 1$ the function

$$\tilde{V}(f,c) = a_0 + a_2 c^2 + \tilde{\rho}(f,c)$$

where $\tilde{\rho}(f,c) = o(c^2)$, $a_0 = e^{2\pi\frac{\alpha}{\beta}} - 1$ and a_2 has the same sign as the Poincaré constant G_4. It then follows that if $a_2 a_0 \geq 0$ there is no positive root of $\tilde{V}(f,\cdot) = 0$ and if $a_2 a_0 < 0$ there exists one positive root. Since the sign of a_0 is the same as that of α the conclusion of Theorem B follows.

Remark. It is easy to prove that if $k = 1$ the non zero periodic orbits occurring when f is varying in a sufficiently small neighborhood of f_0 are all attracting [resp. repelling] if the origin for the unperturbed system is 3-asymptotically stable [resp. 3-completely unstable].

In order to obtain Theorem C we need the following

Proposition 4. Suppose the origin of (2.1) is not h-asymptotically stable or h-completely unstable for any integer $h \geq 3$. Then for any neighborhood $\bar{N}$ of f_0, for any neighborhood $B^2(r_1)$ of the origin, and for any integer j there exists $f \in \bar{N}$ having j nontrivial periodic orbits in $B^2(r_1)$.

Outline of proof. Consider the following perturbation of (2.1)

$$\dot{x} = -y - X_0(x,y) - bx(x^2+y^2)^{\frac{j-1}{2}},$$

$$\dot{y} = x + Y_0(x,y) - by(x^2+y^2)^{\frac{j-1}{2}},$$

where b is any positive constant. This system is j-asymptotically stable since an application of the Poincaré procedure yields

$$\dot{F} = -b(x^2+y^2)^{\frac{j+1}{2}} + \zeta(x,y),$$

where $\zeta(x,y)$ has order $> j+1$. Then using Proposition 2, we have the conclusion of Proposition 4.

In order to prove the necessity of f_0 being h-asymptotically stable or h-completely unstable in Theorem A it is sufficient to combine the proof of Theorem C with the sufficiency part of Theorem A.

REFERENCES

1. Andronov, A. A., Leontovich, E. A., Gordon, I. I., and Maier, A. G., Theory of Bifurcation of Dynamic Systems on a Plane, Halsted Press, New York, (1973).
2. Chafee, N., Generalized Hopf Bifurcation and Perturbation in a Full Neighborhood of a Given Vector Field. *Indiana Univ. Math. Journal 27*, 173-194 (1978).
3. Hale, J., Ordinary Differential Equations, Interscience, New York, (1969).
4. Kelley, A., Appendix C of "Transversal mapping and flows" by Abraham, R., and Robbin, J., Benjamin, (1967).
5. Marchetti, F., Negrini, P., Salvadori, L., and Scalia, M., Liapunov Direct Method in Approaching Bifurcation Problems. *Ann. Mat. Pura Appl., (iv), CVIII*, 211-225 (1976).
6. Marsden, J. E., and McCracken, M., The Hopf Bifurcation and its Applications, Springer Verlag, New York, (1976).
7. Negrini, P., and Salvadori, L., Attractivity and Hopf bifurcation. *Nonl. Anal., Theory, Math. and Appl. 3*, 87-89 (1979).
8. Sansone, G., and Conti, R., Nonlinear Differential Equations, MacMillan, New York, (1964).
9. Takens, F., Unfolding of Certain Singularities of Vector Fields: Generalized Hopf Bifurcations. *J. Diff. Equations 14*, 476-493 (1973).

PERTURBATION OF LINEAR DIFFERENTIAL EQUATIONS BY A HALF-LINEAR TERM DEPENDING ON A SMALL PARAMETER

I. Bihari[1]

Mathematical Institute
Hungarian Academy of Sciences
Budapest, Hungary

The purpose of the present talk is to show how easily approximate solution can be obtained provided the perturbation is a half-linear term. Two examples will be considered.

1. First let us take the autonomous system

$$\begin{aligned} u' &= v + \varepsilon f(u,v) \\ v' &= -u + \varepsilon g(u,v) \end{aligned} \qquad (' = \frac{d}{dt}) \tag{1}$$

where ε is a small parameter, f and g are half-linear and analytic on the circle $u^2 + v^2 = 1$, i.e.

$$f(\lambda u, \lambda v) = \lambda f(u,v), \quad g(\lambda u, \lambda v) = \lambda g(u,v), \quad \forall\ \lambda, u, v.$$

We raise the problem of finding the first, second, ... approximations of the general solution of (1) which are accurate up to order $\varepsilon, \varepsilon^2, \ldots$ having εt an arbitrary value. This is not

[1]Present address: Mathematical Institute, Hungarian Academy of Sciences, Realtanoda n. 13-15, 1053 Budapest V, Hungary.

ISBN 0-12-186280-1

possible for expansions of the form $u = \sum_{n=0}^{\infty} \varepsilon^n u_n(t)$, $v = \sum_{n=0}^{\infty} \varepsilon^n v_n(t)$ on account of appearing of secular terms.
The origin is a critical point and concerning the corresponding homogeneous system

$$x' = y, \quad y' = -x \tag{2}$$

we have to do with the so called critical case being the eigenvalues of the matrix on the right side pure imaginary. The solution is $x = a \cos(t+t_o)$, $y = -a \sin(t+t_o)$, [a, t_o are constant] and the orbits are circles covered in negative direction.-- For a general nonlinear perturbating term the method, running back to Krylov-Bogoliubov-Mitropolski, consists in assuming the solution of (1) in the form

$$\begin{aligned} u &= a\cos\psi + \varepsilon u_1(a,\psi) + \varepsilon^2 u_2(a,\psi) + \ldots \\ v &= -a\sin\psi + \varepsilon v_1(a,\psi) + \varepsilon^2 v_2(a,\psi) + \ldots \end{aligned} \tag{3}$$

where u_i, v_i are 2π periodic in ψ, $a(t)$ and $\psi(t)$ are not constant anymore but satisfy the equations

$$\begin{aligned} a' &= \varepsilon A_1(a) + \varepsilon^2 A_2(a) + \ldots \\ \psi' &= 1 + \varepsilon B_1(a) + \varepsilon^2 B_2(a) + \ldots \end{aligned} \tag{4}$$

and A_i, B_i must be determined in such a way that (3) be solution of (1). Applying this approach no secular term appear.-- Now making a consequent and exhaustive use of the half-linearity of f and g the method can be strongly simplified by the following facts:

(i) It is sufficient to assume f and g to be analytic on the unit-circle [instead of the whole plane].

(ii) It is enough to suppose u_i, v_i, A_i to be linear in a and B_i to be constant, i.e. instead of (3)-(4) the solution will be assumed in the form

$$\begin{aligned} u &= a[\cos\psi + \varepsilon u_1(\psi) + \varepsilon^2 u_2(\psi) + \ldots], \\ v &= a[-\sin\psi + \varepsilon v_1(\psi) + \varepsilon^2 v_2(\psi) + \ldots], \end{aligned} \tag{5}$$

$$a' = a(\varepsilon A_1 + \varepsilon^2 A_2 + \ldots)\ ,$$

$$(A_i\ B_i \text{ are constant}) \qquad (6)$$

$$\psi' = 1 + \varepsilon B_1 + \varepsilon^2 B_2 + \ldots\ ,$$

(iii) The Fourier expansions of such functions as $f(\cos\psi, -\sin\psi)$ etc. will involve terms only of the form $\cos(2n+1)\psi$, $\sin(2n+1)\psi$. Now $f(u,v) = a\ f(\cos\psi + \varepsilon u_1 + \varepsilon^2 u_2 + + \ldots, -\sin\psi + \varepsilon v_1 + \varepsilon^2 v_2 + \ldots)$, $g(u,v) = \ldots$. Let us expand $f(\cos\psi + \varepsilon u_1 + \ldots, -\sin\psi + \varepsilon v_1 + \ldots)$ in Taylor series about the point $\Omega = (\cos\psi, -\sin\psi)$ of the unit circle. By putting (5)-(6) and these expansions in (1) and comparing the coefficients of $\varepsilon, \varepsilon^2, \ldots$ on both sides of the equations the following equations will be obtained to determine the first approximation

$$A_1\cos\psi - B_1\sin\psi + \dot{u}_1 - v_1 = f(\Omega)\ ,$$

$$(7)$$

$$-A_1\sin\psi - B_1\cos\psi + \dot{v}_1 + u_1 = g(\Omega)\ .$$

An easy consideration shows that the Fourier expansion of $f(\Omega)$ is of the form

$$f(\Omega) = \sum_{n=0}^{\infty} [f_n \cos(2n+1)\psi + F_n \sin(2n+1)\psi]$$

and that of $g(\Omega)$ is a similar one. Let us expand also u_1 and v_1 in Fourier series

$$u_1 = \sum_{n=0}^{\infty} (\alpha_n \cos n\psi + \beta_n \sin n\psi)\ ,$$

$$v_1 = \sum_{n=0}^{\infty} (\gamma_n \cos n\psi + \delta_n \sin n\psi)$$

and put it in (7) and compare the coefficients of $\cos n\psi$, $\sin n\psi$ on both sides, obtaining

$$A_1 = \frac{1}{2}(f_o - G_o) = \lambda, \qquad \beta_1 - \gamma_1 = \frac{1}{2}(f_o + G_o)\ ,$$

$$B_1 = -\frac{1}{2}(F_o + g_o) = \mu, \qquad \alpha_1 + \delta_1 = \frac{1}{2}(g_o - F_o)\ ,$$

where g_n, G_n are the Fourier coefficients of $g(\Omega)$. Thus A_1 and B_1 are determined and among $\alpha_1,\beta_1,\gamma_1,\delta_1$ we have two relations, i.e. two of them are undetermined for a while and can be determined by further conditions. Furthermore

$$\alpha_{2k} = \beta_{2k} = \gamma_{2k} = \delta_{2k} = 0$$

and e.g.

$$\alpha_{2k+1} = \frac{g_k - (2k+1)F_k}{4k(k+1)}\ , \quad \ldots$$

which shows that the series of u_1 and v_1 are convergent. From the equations

$$a' = \varepsilon a\lambda\ , \qquad \psi' = 1 + \varepsilon\mu$$

we have as first approximation of a and ψ

$$a = a_o\, e^{\varepsilon\lambda t}\ , \quad \psi = (\psi_o + t) + \varepsilon\mu t \qquad (a_o, \psi_o \text{ are constant})$$

and u_1, v_1 can be calculated easily.

E.g. if $f = \dfrac{u^3}{u^2+v^2}$, $g = \dfrac{v^3}{u^2+v^2}$ then

$$a = a_o\, e^{\frac{3}{4}\varepsilon t}\ , \quad \psi = t + \psi_o$$

$$u_1 = \alpha_1 \cos\psi + \beta_1 \sin\psi + \frac{1}{16}\sin 3\psi\ ,$$

$$v_1 = \beta_1 \cos\psi - \alpha_1 \sin\psi - \frac{1}{16}\cos 3\psi$$

where α_1 and β_1 are undetermined. The $a = a_o\, e^{\frac{3}{4}\varepsilon t}$ will be the complete amplitude of the basic harmonics $\cos\psi$ and $-\sin\psi$ if and only if $\alpha_1 = 0$ and these basic harmonics bear no phase-shift too if $\beta_1 = 0$. Then the first approximation is a follows

$$u = a\,(\cos\psi + \frac{\varepsilon}{16}\sin 3\psi)\ , \qquad a = a_o\, e^{\frac{3}{4}\varepsilon t}$$

$$v = a\,(-\sin\psi - \frac{\varepsilon}{16}\cos 3\psi)\ , \qquad \psi = t + t_o$$

and

$$r^2 = u^2 + v^2 = a^2 \; (1 + \frac{\varepsilon}{8} \sin 4\psi + \frac{\varepsilon^2}{256})$$

which is in good agreement with the exact solution, namely the last gives

$$r^2 = a_o^2 \, e^{\frac{3}{2}\varepsilon t} (1 + \frac{\varepsilon}{8} \sin 4\psi + \mathcal{O}(\varepsilon^2))$$

provided $\alpha \sim -\psi$ where $\mathrm{tg}\alpha = \frac{v}{u}$.
The calculation of the second approximation is not more difficult. On the other hand the convergence of the series (5)-(6) is an open question.

If we interchange the above values of f and g, then it can be proved that the solutions are 2π periodic, the orbits are closed and the first approximation reads as

$$u = a_o \cos\psi ,$$

$$v_o = a_o\left[(-1 + \frac{3}{4}\varepsilon) \sin\psi + \frac{5}{16}\varepsilon \sin 3\psi\right] ,$$

$$a_o = \text{const}$$
$$\psi = t+t_o$$

whence

$$u^2 + \frac{v^2}{(1 - \frac{3}{4}\varepsilon)^2} = a_o^2\left[1 + \mathcal{O}(\varepsilon)\right],$$

i.e. the orbits are nearly ellipses.

2. Without going in details let it be remarked that the discussion of the equation

$$\ddot{x} + \omega^2 x = \varepsilon f(\gamma t, x, \dot{x}) \qquad (\cdot = \frac{d}{dt})$$

for an $f(\theta,u,v)$ wich is half-linear in u and v, 2π periodic in θ and analytic on the cylinder $u^2 + \frac{v^2}{\omega^2} = 1$, is much easier than in the general case, moreover both for rational or irrational $\frac{\gamma}{\omega}$.

ON SOME CAUCHY PROBLEMS ARISING IN COMPUTATIONAL METHODS

Arrigo Cellina[1]

Istituto di Matematica Applicata
Università di Padova

Purpose of the present talk is to discuss some open problems concerning the methods for finding fixed points of a self mapping of the unit disk and to report some work done jointly with C. Sartori on them.

§ 1. Let S^N be the standard N simplex, a subset of E^{N+1}, let $\mathcal{G}$ be any triangulation of it and V the collection of all the vertexes of $\mathcal{G}$. Call S_i^N the face opposite to the i-th vertex of S^N and call a labeling $E : V \longrightarrow \{0,1,\ldots,N\}$ admissible if $x \in V \cap S_i^N$ implies $E(x) \neq i$. Sperner's Lemma guarantees the existence of some element of $\mathcal{G}$ whose set of vertexes receives all the labels $\{0,1,\ldots,N\}$. When we have a map $f : S^N \longrightarrow S^N$ we can use (continuity and) Sperner's Lemma to prove the existence of fixed points by assigning to the vertexes admissible labels such that the existence of a subsimplex carrying the full set of labels is equivalent to the existence of some point x missing its image $f(x)$ by some preassigned number ε.

---

[1] Present address: Istituto di Matematica Applicata, Via Belzoni, 3, 35100 Padova, Italy.

RECENT ADVANCES IN DIFFERENTIAL EQUATIONS

ISBN 0-12-186280-1

Two such labels have been traditionally considered [3],[4]. We can consider for each $x \in V$ the set $\{i \mid f^i(x) - x^i \leq 0$ and $x^i \geq 0\}$, where we denote by x^i the i-th component of $x \in E^{N+1}$. Any label $E(x) \in \mathcal{U}(x)$ is admissible and the existence of a simplex with a full set of labels, for a triangulation with a sufficiently small mesh, is equivalent to the existence of some x missing its image by ε. Assume we start on a face of S^N with a simplex carrying N labels, say $\{0,1,\ldots,N-1\}$. It is well known that we can follow a path of simplexes, always carrying this set of labels, that either ends on the boundary of S^N or leads to a fully indexed simplex. Hence whenever we can guarantee the existence of a path that does not lead back to the boundary of S^N we have not only an existence argument for fixed points but also an algorithm to find them.

Assume that the mesh of the triangulation is so small that the "direction of the motion" $(f(p)-p)/\|f(p)-p\|$ can be considered to be essentially constant on each subsimplex of $\mathcal{G}$, coinciding with the same vector computed at b, the barycenter of this simplex. Then carrying all the labels from 0 to $N-1$ means that $f(b)-b$ has the components from 0 to $N-1$ negative, hence that the direction of the motion is essentially constant along the simplexes connected by the path we are following or, in other words, that we are generating such a path by connecting points whose direction of the motion equals a given unit vector (in our example, for instance, $(-1/\sqrt{N}, -1/\sqrt{N}, \ldots, 0)$. This method could be called a "constant direction" method.

Induction on the dimension of the space is used to prove the existence of at least one path that does not lead back to the boundary of S^N. This proves in a non constructive way the existence of a fully indexed subsimplex. Two ways could be used to make it constructive. One is by extending S^N to a larger simplex and by extending the labeling to this new simplex in such a way that no path can reach the boundary of this larger simplex, the other by changing the rules at the boundary of S^N so that a path leading back to the boundary can be continued on the boundary until it reaches a new almost fully indexed simplex, so that the process can start again.

The second method considers, instead of $f(p)-p$, the half line issuing from $f(x)$ through x and defines $H(x)$ as the point of intersection of this half line with the boundary of S^N. Set v_i to be the i-th element of the orthonormal basis of E^{N+1} and consider $\mathcal{U}(x) = \{i \mid \langle v_i, H(x)\rangle > 0\}$. Then it is easy to see that any label $E(x) \in \mathcal{U}(x)$ is at once admissible and satisfies the requirement of implying the existence of fixed points through the existence of fully labeled subsimplexes. Since at the boundary $H(x) = x$, the only i not in $\mathcal{U}(x)$ is the i corresponding to the vertex opposed to the face containing x. By choosing a rule on the boundary of S^N like $E(x) = \inf\{\mathcal{U}(x)\}$ only one simplex carries the set of indexes from 0 to $N-1$, i.e. there are no paths leading back to the boundary. In this way this method avoids the main difficulty of the preceding one, at the expenses of a by far less straightforward construction of the labels. For a sufficiently small triangulation, H of the barycenter b of a simplex carrying the labels $0,\dots,N-1$ is on the face opposite to the N-th vertex. Hence a path connects points x such that $H(x)$ is essentially constant (in one of the $N+1$ faces, in our example). We could call this a method of "constant target". A further discussion of these methods can be found in [1] and [7].

§ 2. The use of Sard's Lemma enables one to give continuous versions of these methods. Let S be the unit disk. Let $D(f)-I$ be the Jacobian of $f(x)-x$, and consider the equation

$$(D(f) - I)\frac{dx}{dt} = f(x) - x \,. \tag{1}$$

Then an easy computation shows that $(f(x)-x)/\|f(x)-x\|$ is a (vector) first integral i.e. that along solution the "direction of the motion" is constant. By setting the (implicit) initial value

$$(f(x^o) - x^o)/\|f(x^o) - x^o\| = v \tag{2}$$

one has a solution such that this vector constantly equals v. We can prove that a solution exists until it leaves S and, in the case it remains in S, its ω-limit set is in the fixed point set of f. We call the above (1) and (2) an implicit value problem since even if v, a unit vector, is given, x^o satisfying (2) is not and in general (2) does not uniquely determine x^0. Had we only one solution of (2) there would not be solutions to (1), (2) returning to ∂S. We remark that for a self mapping f os S, the vector $f(x)-x$ is inward, i.e. satisfies the usual subtangentiality condition

$$\underline{\lim}\ d_S(x + tv) / t = 0 \tag{3}$$

but this is not so for the vector x' as defined by (1). As a consequence, unless very special conditions are assumed on the behaviour of $f(x)$ on ∂S, solutions to (1), (2) can very well leave S. Proving the weak invariance of S, i.e. the existence of at least one solution to (1), (2) remaining in S is equivalent to proving the existence of fixed points of f.

A possibility, suggested by the discrete case, is to extend f to a larger disk S_ε so that the field x' is subtangential to the new boundary and then pass to the limit. Another approach could be to modify the field on ∂S so that solutions do not leave S. However one ought to be able, as done in the discrete case by induction, to prove directly the weak invariance of S with respect to (1), (2). To this author's knowledge this has not been done.

The fixed target method has the advantage, at the expenses of some other difficulties that we shall discuss later, of avoiding the invariance problem. In fact the differential version of this method is

$$D(H)\ \frac{dx}{dt} = 0 \qquad x(0) = \xi^0. \tag{4}$$

This equation has solutions such that the total derivative of H with respect to time is zero, i.e. along solutions H is constant, equal to $\xi^0 \in \partial S$. An easy argument shows that at the boundary of S a vector x' satisfying (4) has the direction of $f(x)-x$ i.e. that (3) holds. Hence the existence of solutions remaining in S, and so of fixed points of f, can be proved. In this respect, i.e. to give a clean existence proof, this second method has no comparision with the fixed direction method.

§ 3. Sard's Lemma assures that the set of regular values of a map Φ, i.e. those x's such that $\Phi^{-1}(x)$ consists only of points on which the Jacobian of Φ has maximal rank, is of full measure. Equivalently, the set of critical values has measure zero. Applying this result to problem (1), (2) and to problem (4) we see that in (2) the allowed set of directions v is a subset of full measure of the unit sphere and that in (4) the set of target points ξ^o is a subset of full measure of ∂S. However, sets of full measure can be fairly weird and one wonders what kind of continuous dependence on the data and on the numerical integration errors one should expect.

Let us consider the following hypothesis (Hypothesis GH). On a fixed point x^* of f in S, the Jacobian of $f(x)-x$ has maximal rank.

Smale calls the above "genericity hypothesis" since the class of differentiable mappings satisfying it is of second cathegory. In [6] he proves that

Theorem 1. Under assumption GH the set of critical values of $f(x)-x$ is a compact zero dimensional subset.

The above result means that the regular values v of problem (1), (2) beside being of full measure, are open in the unit sphere.

Let us consider problem (4). Solutions to it are paths connecting points such that the half line from $f(x)$ through x meets ∂S at ξ^o, the target. We ask about the properties of the set of admissible targets i.e. the set of regular values of D(H), and the stability properties of the solutions of (4) through them. However there is no reason to restrict the perturbed target to lay in ∂S instead of on a neighborhood in E^N. In this case we face the problem that although the path is well defined and meaningful, H is not, since H depends on ∂S, and that the Cauchy problem (4) is meaningless for $\xi^o \notin \partial S$. In [2] the following results are proven:

Theorem 2. Under assumption (GH) the set of critical values is a compact zero dimensional subset of E^N.

Theorem 3. Let ξ^o be a regular value of H. Let the solution to (4) exist on $(0,\omega)$. Then, under assumption (GH), for every $T < \omega$, for every $\varepsilon > 0$, there exists δ such that: whenever $\| \xi - \xi^o \| < \delta$ there exist function u_ξ such that: the solution to

$$x'_\xi = u_\xi(x_\xi), \qquad x_\xi(0) = \xi \tag{5}$$

a) exists on $|0,T)$ and $x_\xi(0) = \xi$

b) the ω-limit set of (5) is in the fixed point set of f .

REFERENCES

1. Cellina, A., Metodi costruttivi nella teoria del punto fisso, *in* "Applicazioni del teorema del punto fisso all'analisi economica". Accademia Nazionale dei Lincei, Roma, (1978), 147-158.
2. Cellina, A., and Sartori, C., The Search for Fixed Points under Perturbations. *Rend. Sem. Mat. Padova 59*, 199-208 (1978).
3. Cohen, D. I. A., On the Sperner Lemma, *J. Comb. Theory 2*, 585-587 (1967).
4. Hirsch, M. W., A proof of the nonretractability of a cell onto its boundary, *Proc. A.M.S. 14*, 364-365 (1963).
5. Kellog, R. B., Li, T. Y. and Yorke, J. A., A constructive proof of the Brouwer fixed point theorem and computational results, *SIAM J. Num. An. 13*, 473-483 (1976).
6. Smale, S., Price adjustment and global Newton method, *J. Math. Econ. 3*, 1-14 (1976).
7. Todd, M. J., The computation of fixed points and applications. Lecture Notes in Economics and Math. Systems, 124, Springer Verlag, Berlin, (1976).

COMPARISON RESULTS AND CRITICALITY IN SOME COMBUSTION PROBLEMS

Jagdish Chandra[1]

U. S. Army Research Office
Research Triangle Park
North Carolina

Paul Wm. Davis[2,3]

Department of Mathematics
Worcester Polytechnic Institute
Worcester, Massachusetts

I. INTRODUCTION

The purpose of this lecture is to demonstrate how comparison theorems for systems of parabolic differential inequalities lead to useful information in the study of phenomena modelled by combined reaction-diffusion equations. In particular, we

---

[1]Present address: U. S. Army Research Office, Research Triangle Park, North Carolina, 27709.

[2]Present address: Department of Mathematics, Worcester Polytechnic Institute, Worcester, Massachusetts, 01609.

[3]Research supported by the U. S. Army Research Office under grant number DAAG29-76-G0237.

ISBN 0-12-186280-1

will be concerned with a model problem in combustion theory. Comparison theorems developed for parabolic systems, generally, make some sort of monotonicity assumptions on the nonlinearities in the problem (see, for instance, [6], Chapter 10). However, the very nature of the physics precludes this possibility in the combustion problem. Specifically, the nonlinear terms in such a model cannot possibly be quasimonotone because temperature and fuel concentration must feed back upon one another in contrary ways.

The study of reaction-diffusion equations has aroused considerable interest in recent years. This is because a variety of interesting phenomena such as chemical kinetics, biochemical processes, population dynamics and ecological systems, can be modelled by systems of combined reaction-diffusion equations. For instance, in a series of interesting papers [2, 3, 7], sufficient conditions have been derived for the existence of positively invariant sets for nonlinear diffusion equations. These results, in turn, lend themselves to formulation of comparison theorems for systems of reaction-diffusion equations. Comparison techniques developed yield estimates which are independent of space variable. In a recent paper [1], we have demonstrated an approach which easily circumvents many of the technical restrictions inherent in the above results. In this paper, we concentrate, however, on how these comparison results can be utilized in specific applications. In Section 2, we describe the model problem. Section 3 reviews the comparison result that will be repeatedly used in this paper. Section 4 outlines calculations of criticality phenomena in thermal explosion theory.

II. A MODEL PROBLEM

A simple model governing the combustion of a single specie is described by the following system of equations |4|:

$$\begin{aligned} \frac{\partial T}{\partial t} &= k_1 \Delta T + \lambda y e^{-\frac{E}{RT}} \\ \frac{\partial y}{\partial t} &= k_2 \Delta y - y e^{-\frac{E}{RT}} \end{aligned} \tag{1}$$

Here T denotes the temperature and y the concentration of the combustible substance. The positive constant λ is the heat of reaction, and the constants k_1 and k_2 are respectively the thermal and material diffusion coefficients. For the sake of simplicity, in the above model we assume a first order reaction, however, our analysis could be extended to higher order reactions for multiple species. The Arrenhius reaction rate factor is $\exp(-\frac{E}{RT})$, where E denotes the activation energy (assumed to be large) and R is the universal gas constant. We will consider an initial-boundary value problem for the system (1) on a bounded domain Ω for $t > 0$. The initial conditions are always given by

$$T(x,0) = T_0(x) \;, \quad y(x,0) = y_0(x) \;, \quad x \in \Omega \;. \tag{2}$$

We shall assume throughout this paper that $T_0(x) \geq \beta$, $x \in \Omega$.

It is convenient to consider the following non-dimensional variables. Define

$$\alpha = \frac{E}{R\beta} \;, \quad \varepsilon = e^{-\alpha} \;, \quad z = \alpha\left(\frac{T-\beta}{\beta}\right) \;,$$

$$f(z) = \exp\left\{\frac{z}{1+\frac{z}{\alpha}}\right\} .$$

With these transformations, the system (1) becomes

$$\begin{aligned} \frac{\partial z}{\partial t} &= k_1 \Delta z + P(z,y) \\ \frac{\partial y}{\partial t} &= k_2 \Delta y + S(z,y) \;, \end{aligned} \tag{3}$$

where $P(z,y) \equiv H\, y\, f(z)$, $S(z,y) = -\varepsilon\, y\, f(z)$ and $H = \frac{\alpha \lambda \varepsilon}{\beta} > 0$.

III. A COMPARISON RESULT

Let $u(x,t)$, $v(x,t)$ denote the vector-valued functions $(u^1,\ldots,u^n)$, $(v^1,\ldots,v^n)$ for x in some m-dimensional domain Ω and $t \geq 0$. Let $f = (f^1,\ldots,f^n)$ with each $f^i(x,t,u)$ continuously dependent on its arguments. Further, let L^i be a strongly elliptic linear differential operator,

$$L^i = \sum_{k,\ell=1}^{m} a^i_{k,\ell}(x) \frac{\partial^2}{\partial x_k \partial x_\ell} + \sum_{k=1}^{m} b^i_k(x) \frac{\partial}{\partial x_k},$$

whose coefficients are bounded in Ω, and let

$$Lu = (L^1 u^1, \ldots, L^n u^n) .$$

Let B_r denote one of the following boundary operators:

$$B_1 u \equiv c(x)\, u(x,t), \quad c > 0$$

$$B_2 u \equiv d(x) \frac{\partial u(x,t)}{\partial \nu}, \quad d > 0, \quad \text{or}$$

$$B_3 u \equiv c(x)\, u(x,t) + d(x) \frac{\partial u(x,t)}{\partial \nu},$$

$$c \geq 0, \quad d \geq 0, \quad c^2 + d^2 > 0,$$

for $x \in \Omega$. Here, $\frac{\partial}{\partial \nu}$ denotes any outward directional derivative on $\partial \Omega$.

We will need the following result established in [1].

Proposition. Let $f(x,t,u) \geq g(x,t,u)$ for $x \in \Omega$, $t \geq 0$ where g is quasi-monotone non-decreasing in u. Suppose

$$\frac{\partial u}{\partial t} = Lu + f(x,t,u)$$

$$\frac{\partial v}{\partial t} \leq Lv + g(x,t,v),$$

with $u(x,0) \geq v(x,0)$, $B_r u \geq B_r v$, $x \in \partial \Omega$ for $r = 1,2$ or 3, and v depends continuously upon its data and its differential equation. Then, $u(x,t) \geq v(x,t)$ for $x \in \Omega$, $t > 0$.

Now, define

$$\underline{P}(z,y) = \inf \{P(z,\theta) : y \leq \theta\}$$

$$\underline{S}(z,n) = \inf \{S(\theta,y) : z \leq \theta\} .$$

Similarly, we could define $\bar{P}$, $\bar{S}$, replacing infimum by supremum and reversing the inequalities.

From the definitions of P and S, since

$P(0,y) \geq 0$ and $S(z,0) \geq 0$,

it is an easy consequence of the above proposition that if (z,y) is any solution of the system (3) satisfying non-negative initial and boundary conditions,

$$(0,0) \leq (\underline{z},\underline{y}) \leq (z,y), \tag{4}$$

where $(\underline{z},\underline{y})$ is a solution of (3) with $\underline{P}$, $\underline{S}$ replacing P, S, and $(\underline{z},\underline{y})$ satisfy the same initial and boundary data as (z,y).

IV. SOME CRITICALITY CALCULATIONS

Of fundamental interest in thermal explosion theory is the description of temperature and fuel-concentration history. Numerous approximate and numerical techniques employed to study this problem usually attempt to describe the combustion processes only during the ignition phase of the explosion when the temperature is only slightly above the initial value and when only a small fraction of the fuel has been consumed.

In [8], for instance, the sub-critical case, where the temperature and concentration remain below certain critical value, is analyzed using perturbation techniques. In this case the combustion proceeds very slowly on a time scale of order ε, with temperature remaining relatively close to β, and the temperature and concentration decaying as $\exp(-\varepsilon t)$. As can be seen easily, these facts are a direct consequence of the comparison result applied to (1) or (3).

On the other hand, if temperature and concentration are initially above certain critical values, the reaction proceeds very quickly, with a rapid build up to high temperature and a concomitant consumption of the combustible material, thus producing an explosion. In [5], this supercritical problem has been analyzed in the case of spatial homogeneity. There the author considers the complete solution of the resulting system of ordinary differential equations using a matched asymptotic expansion procedure with several time-scales. Specifically, the solution is described in six successive time zones. Each of these is characterized by a distinct physical process. Of course, the most interesting of all are the extremely rapid processes which characterize a truly explosive event.

In this section, we describe a simple approach based on our comparison result that elucidates the dependence of the "final" temperature on the initial state. This approach reveals the range of initial phenomena by demonstrating explicitly the functional relation between the initial and final temperatures under various circumstances.

Recall that ν denotes an outward pointing normal. We will always assume the following boundary condition on y.

$$\frac{\partial v}{\partial \nu} = 0 , \qquad x \in \partial \Omega .$$

For z either of two following boundary conditions will be invoked:

$$z(x,t) = 0 , \qquad x \in \partial \Omega , \tag{5}$$

$$\frac{\partial z(x,t)}{\partial \nu} = 0 , \qquad x \in \partial \Omega . \tag{6}$$

Now recall that

$$\underline{P}(z,y) = \bar{P}(z,y) = Hyf(z) ,$$

$$\underline{S}(z,y) = -\varepsilon y f(\infty) = -y$$

and

$$\bar{S}(z,y) = -\varepsilon y f(0) = -\varepsilon y .$$

Then for $x \in \Omega$ and $t > 0$, the proposition gives,

$$\underline{y}(x,t) \leq y(x,t) \leq \bar{y}(x,t) \tag{7}$$

where $\underline{y}(x,t)$ is the solution of the initial-boundary problem

$$\frac{\partial \underline{y}}{\partial t} = \Delta y - y$$

$$y(x,0) \le y_0(x) \ , \qquad x \in \Omega$$

$$\frac{\partial \underline{y}}{\partial \nu} = 0 \ , \qquad x \in \partial \Omega$$

and $\bar{y}(x,t)$ is the solution of the initial-boundary problem

$$\frac{\partial \underline{y}}{\partial t} = \Delta y - \varepsilon y$$

$$y(x,0) \ge y_0(x) \ , \qquad x \in \Omega \ ,$$

$$\frac{\partial \underline{y}}{\partial \nu} = 0 \ , \qquad x \in \partial \Omega \ .$$

If $y_0(x)$ is independent of x (say $\equiv y_0$), then the explicit bounds depending only on t are given by

$$y_0 \ e^{-t} \le y(x,t) \le y_0 \ e^{-\varepsilon t} \ . \tag{8}$$

The corresponding bounding problems for $z(x,t)$ are

$$\frac{\partial z}{\partial t} = \Delta \underline{z} + Hy_0 f(\underline{z}) e^{-t} \ .$$

$$\underline{z}\ (x,0) \le z_0(x) \ , \qquad x \in \Omega$$

$$\underline{z}\ (x,t) \le 0 \ , \qquad x \in \partial \Omega$$

or $\dfrac{\partial z(x,t)}{\partial \nu} \le 0 \ , \qquad x \in \partial \Omega$

and

$$\frac{\partial \bar{z}}{\partial t} = \Delta\ \bar{z} + Hy_0 f(\bar{z})\ e^{-\varepsilon t}$$

$$\bar{z}(x,0) \ge z_0(x) \ , \qquad x \in \Omega$$

$$\bar{z}(x,t) \ge 0 \ , \qquad x \in \partial \Omega$$

or $\dfrac{\partial \bar{z}(x,t)}{\partial \nu} \ge 0 \ , \qquad x \in \partial \Omega \ .$

Note that the lower bounding problem $\underline{z}$ in the case of the Dirichlet boundary condition is not of interest because satisfaction of both the initial and the boundary conditions forces $\underline{z}(x,t) \equiv 0$. On the other hand, the upper bounding problems provides nontrivial information to study the functional dependence of the "final" temperature on the initial state.

For the sake of definiteness, consider the original problem (3) with z subject both to the spatially uniform initial condition $z(\mathbf{x},0) \equiv z_0$ and to the Neumann boundary condition (6). The comparison result quickly reveals that the solution $\underline{z}(x,t)$ of the lower bounding problem is itself bounded by the solution $\Phi(t)$ of

$$\Phi'(t) = Hy_0 f(\Phi(t))\, e^{-t}, \qquad \Phi(0) = z_0. \tag{9}$$

Letting $z_\infty = \lim_{t \to \infty} \Phi(t)$, we find from (9) that

$$\int_{z_0}^{z_\infty} d\Phi / f(\Phi) = Hy_0, \tag{10}$$

a relation defining a lower bound z_∞ on the final temperature in terms of the initial temperature z_0.

Some numerical solutions of (10) for various parameter values of physical interest are plotted in the figure. At each vertical asymptote, the final bounding temperature has become highly sensitive to z_0, indicating a super-critical threshold for the lower bounding temperature. These threshold values, which are listed in the table, are themselves upper bounds on the true super-critical initial temperature. These ideas and their physical implications will be explored in more detail elsewhere.

CAPTION FOR FIGURE

A plot of values of (z_0, z_∞) which satisfy (10) for $\alpha = 80$ and various values of Hn_0. The sharp asymptote defines the critical temperature for this bounding problem.

TABLE

Hy_0	Upper bound on super-critical initial temperature for $\alpha = 80$
2×10^{-2}	19.85
2×10^{-3}	16.27
2×10^{-4}	12.95
2×10^{-5}	9.86
2×10^{-6}	6.97
2×10^{-7}	4.26

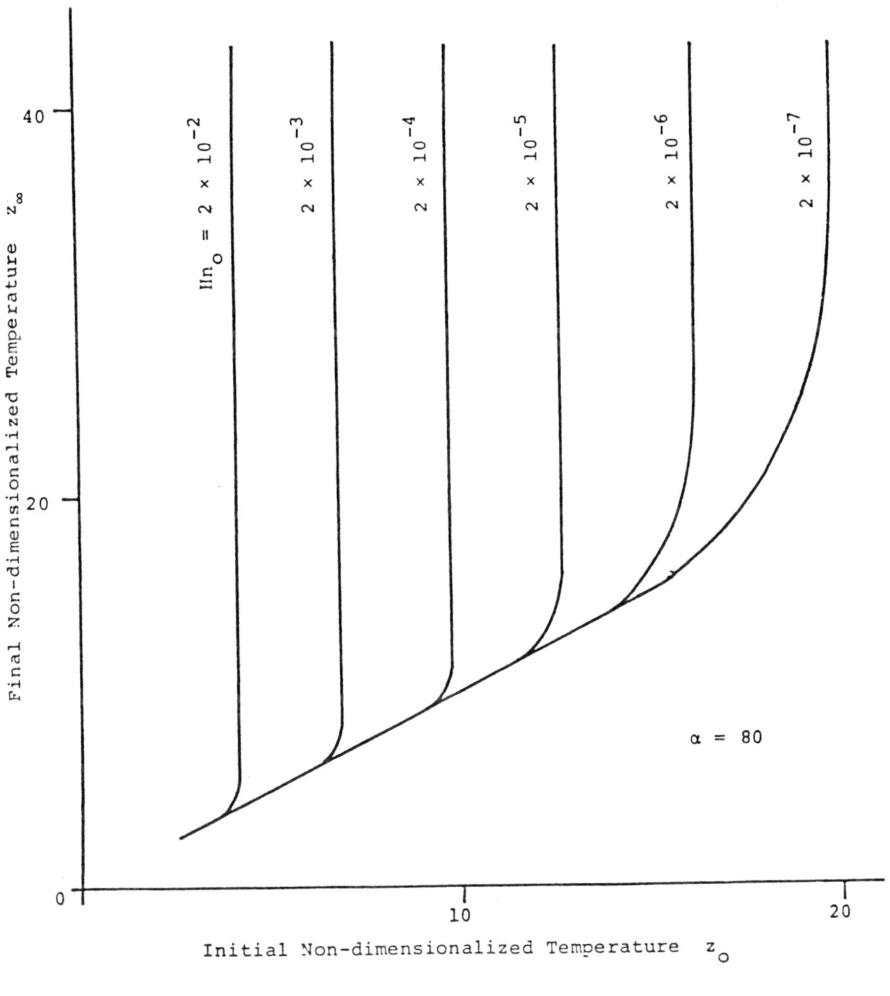

Final Non-dimensionalized Temperature z_∞

40

20

0

$Hn_o = 2 \times 10^{-2}$

2×10^{-3}

2×10^{-4}

2×10^{-5}

2×10^{-6}

2×10^{-7}

$\alpha = 80$

10 20

Initial Non-dimensionalized Temperature z_o

REFERENCES

1. Chandra,J. and Davis, P. W., Comparison theorems for systems of reaction-diffusion equations, Proc. Internat. Conference on Nonlinear Analysis, Academic Press (To appear).
2. Chueh, K. N., Conley, C. C. and Smoller, J. A., Positively invariant regions for systems of nonlinear diffusion equations, *Indiana Univ. Math. J. 26*, 373-392 (1977).
3. Conway, E. D. and Smoller,J. A., A comparison technique for systems of reaction-diffusion equations, *Comm. in Partial Diff. Eqs. 2*, 679-697 (1977).
4. Frank-Kamenetskii, D. A., Diffusion and heat exchange in chemical kinetics, Plenum,(1969).
5. Kassoy, D. R., The supercritical spatially homogeneous thermal explosion: initiation to completion, *Quart. Journ. Mech. Appl. Math. 30*, 71-89 (1977).
6. Lakshmikantham, V. and Leela, S., Differential and integral inequalities, V. II, Academic Press, (1969).
7. Redheffer, R. and Walter, W., Invariant sets for systems of partial differential equations, *Arch. Rat. Mech. Anal.* (To appear).
8. Sattinger, D. H., A nonlinear parabolic system in the theory of combustion, *Quart. Appl. Math. 33*, 47-61 (1975).

BOUNDARY VALUE PROBLEMS FOR PAIRS OF ORDINARY DIFFERENTIAL OPERATORS

Earl A. Coddington[1]

Department of Mathematics
University of California
Los Angeles, California

I. INTRODUCTION

This is a survey of some recent joint work with H. S. V. de Snoo. Some of the details have been reported on at the Equadiff 4 Conference in Prague in August 1977, and others at the Equadiff 78 Conference in Florence in May 1978.

Recall that when we simultaneously diagonalize two hermitian matrices K, H, where H is positive definite, $H > 0$, we consider the finite-dimensional Hilbert space $\mathfrak{H}_H = \mathbb{C}^n$ with the inner product given by $(f,g) = Hf \cdot g = g^* Hf$. Here we view $\mathbb{C}^n$ as the set of one-column matrices f with components $f_1, \ldots, f_n$, $f_j \in \mathbb{C}$. Then $A = H^{-1}K$ is selfadjoint in $\mathfrak{H}_H$ and the spectral theorem applied to A yields the desired simultaneous diagonalization, with H diagonalized to the unit matrix. It is also clear that we could consider an arbitrary K and the corresponding A in $\mathfrak{H}_H$.

---

[1]Present address: Department of Mathematics, University of California, Los Angeles, California.

ISBN 0-12-186280-1

Here we consider the situation when we replace K and H by two ordinary differential operators L, M on an interval ι, of orders n, ν, respectively, where $M > 0$ in some sense. Two problems immediately arise: (i) there are many choices for a selfadjoint H generated by M, and thus many choices for $\mathfrak{H}_H$, and (ii) once a choice for H has been made there are many choices for the analog of A. A naive approach is the following. Let us identify operators with their graphs, and put

$$L_1 = \{\{f,Lf\} \mid f \in C^n(\iota)\} , \quad M_1 = \{\{g,Mg\} \mid g \in C^\nu(\iota)\},$$

and then consider $A = M_1^{-1}L_1$. This is given by the usual composition

$$\begin{aligned} A &= \{\{f,g\} \mid \{f,h\} \in L_1, \ \{h,g\} \in M_1^{-1}, \text{ some } h\} \\ &= \{\{f,g\} \mid \{f,h\} \in L_1, \ \{g,h\} \in M_1, \ \text{some } h\} \\ &= \{\{f,g\} \mid f \in C^n(\iota), \ g \in C^\nu(\iota), \ h = Lf = Mg\}. \end{aligned}$$

We note first of all that

$$A(0) = \{g \mid \{0,g\} \in A\} = \{g \mid Mg = 0\} ,$$

so that A is possibly not the graph of a (single-valued) operator. The point of this survey is to indicate that this naive approach can be made to work. It leads in a natural way to multi-valued operators and Hilbert spaces $\mathfrak{H}_H$ whose inner products contain boundary terms as well as integral terms.

II. THE DIFFERENTIAL OPERATOR M

Let $\iota = (a,b)$ be an arbitrary open real interval, which may be unbounded (i.e., $a = -\infty$, $b = +\infty$, or both, are allowed). We shall consider systems of equations, so that $L^2(\iota)$ will denote the set of all one-column matrix-valued functions $f : \iota \to \mathbb{C}^m$ such that

$$\int_{\imath} |f|^2 < \infty, \qquad |f|^2 = f^* f,$$

with the inner product and norm

$$(f,g)_2 = \int_{\imath} g^* f, \qquad \|f\|_2 = ((f,f)_2)^{1/2} .$$

If J is any subinterval of $\imath$ we write

$$(f,g)_{2,J} = \int_J g^* f$$

provided the integral makes sense, and we denote by $L^2_{loc}(\imath)$ the set of all $f : \imath \to C^m$ such that $(f,f)_{2,J} < \infty$ for every compact subinterval $J \subset \imath$.

We consider a formal differential operator M of even order $\nu = 2\mu$ on $\imath$,

$$M = \sum_{k=0}^{\nu} Q_k D^k , \qquad D = d/dx ,$$

which is formally selfadjoint, $M = M^+$, where

$$M^+ = \sum_{k=0}^{\nu} (-D)^k Q_k^* .$$

Here the Q_k are $m \times m$ complex matrix-valued functions on $\imath$ such that $Q_k \in C^k(\imath)$, and $Q_\nu(x)$ is invertible for all $x \in \imath$. Moreover we *assume* that

$$(Mf,f)_2 \geq 0, \qquad f \in C_0^\infty(\imath), \tag{2.1}$$

where $C_0^\infty(\imath)$ is the set of all $f \in C^\infty(\imath)$ with compact support. This positivity assumption implies that ν must be even. If we put

$$M_0 = \{\{f, Mf\} \mid f \in C_0^\infty(\imath)\} ,$$

then the condition (2.1) can be written simply as $M_0 > 0$. It is always possible to write such an M in the form

$$M = \sum_{j=0}^{\mu} \sum_{k=j-1}^{j+1} (-1)^j D^j Q_{jk} D^k ,$$

where Q_{jk} are matrices satisfying $Q_{jk}^* = Q_{kj}$.

III. THE HILBERT SPACE $\mathfrak{H}_H$

Let $H = H^*$ be a selfadjoint extension of M_0 and *assume*

(A_1) (*Local Inequality*). For each compact subinterval $J \subset \iota$ there is a constant $c(J) > 0$ such that

$$(Hf,f)_2 = (Mf,f)_2 \geq (c(J))^2 (f,f)_{2,J}, \qquad f \in \mathfrak{D}(H),$$

where $\mathfrak{D}(H)$ is the domain of H.

We let $\mathfrak{H}_H$ be the completion of $\mathfrak{D}(H)$ with the inner product $(f,g) = (Hf,g)_2$, $f,g \in \mathfrak{D}(H)$. It is a Hilbert space, and we further *assume*

(A_2) the identity map on $\mathfrak{D}(H)$ extends to an injection of $\mathfrak{H}_H$ into $L^2_{loc}(\iota)$.

Then we can identify $\mathfrak{H}_H$ as a subset of $L^2_{loc}(\iota)$ and the local inequality extends to $\mathfrak{H}_H$,

$$\|f\| \geq c(J) \|f\|_{2,J}, \qquad f \in \mathfrak{H}_H .$$

We have $C_0^\infty(\iota) \subset \mathfrak{D}(H) \subset \mathfrak{H}_H$, and we put $\mathfrak{H}_M(C_0^\infty(\iota))^c$, the closure of $C_0^\infty(\iota)$ in $\mathfrak{H}_H$. It can be shown that $\mathfrak{H}_H = \mathfrak{H}_H \oplus \mathfrak{N}_H$, an orthogonal sum, where

$$\mathfrak{N}_H = \{f \in C^\nu(\iota) \cap \mathfrak{H}_H \mid Mf = 0\} .$$

It is clear that $\dim \mathfrak{N}_H \leq \nu m$. On $C_0^\infty(\iota)$ the inner product is a Dirichlet integral,

$$(f,g) = \int_\iota \sum_{j=0}^{\mu} \sum_{k=j-1}^{j+1} (D^j g^*) Q_{jk} (D^k f), \qquad f,g \in C_0^\infty(\iota) ,$$

and we have $\mathfrak{H}_M = \mathfrak{H}_{M_F}$, where M_F is the Friedrichs extension of M_0.

Let $L_0^2(\iota)$ denote the set of all $f \in L^2(\iota)$ which have compact support. The injection $\mathfrak{H}_H \to L^2_{loc}(\iota)$ implies an injection $G : L_0^2(\iota) \to \mathfrak{H}_H$ with the following properties:

$$\begin{aligned}
&(f, G\alpha) = (f,\alpha)_2, \quad f \in \mathfrak{H}_H, \quad \alpha \in L_0^2(\iota),\\
&GM\varphi = \varphi, \quad \varphi \in C_0^\infty(\iota),\\
&MG\alpha = \alpha, \quad \alpha \in L_0^2(\iota),\\
&(\mathcal{R}(G))^c = \mathfrak{H}_H.
\end{aligned} \tag{3.1}$$

Here $\mathcal{R}(G)$ denotes the range of G.

Some selfadjoint extensions $H \supset M_0$ may satisfy

(A_1') (*Global Inequality*) there is a constant $c > 0$ such that

$$(Hf,f)_2 \geq c^2(f,f)_2, \quad f \in \mathfrak{D}(H).$$

Then the identity map on $\mathfrak{D}(H)$ extends to an injection of $\mathfrak{H}_H$ into $L^2(\iota)$, and we have

$$\|f\| \geq c\|f\|_2, \quad f \in \mathfrak{H}_H.$$

The injection G has an extension, call it G also, to an injection $G : L^2(\iota) \to \mathfrak{H}_H$ satisfying (3.1) with $L_0^2(\iota)$ replaced everywhere by $L^2(\iota)$. In fact, we have $G = H^{-1}$, which is a bounded operator on all of $L^2(\iota)$, and $\mathfrak{H}_H = \mathfrak{D}(H^{1/2})$, where $H^{1/2}$ is the positive square root of H.

If $\iota = (a,b)$ is a finite interval $\bar{\iota} = [a,b]$, and $Q_k \in C^k(\bar{\iota})$, with $Q_\nu(x)$ invertible for all $x \in \bar{\iota}$, then we say M is in the regular case. Then the local inequality (A_1) implies the global inequality (A_1'). In this case $\mathfrak{H}_H$ is the set of all $f : \iota \to \mathbb{C}^m$ such that $f^{(\mu-1)}$ is absolutely continuous on $\bar{\iota}$, $f^{(\mu)} \in L^2(\iota)$, and where f satisfies certain homogeneous boundary conditions in $f, f', \ldots, f^{(\mu-1)}$, the so-called essential boundary conditions for H. The inner product on $\mathfrak{H}_H$ involves the Dirichlet integral as well as boundary terms in $f, f', \ldots, f^{(\mu-1)}$.

IV. SUBSPACES A DETERMINED BY L,M

Let M be as above, and H a fixed selfadjoint extension of M_0 satisfying (A_1), (A_2). If $\mathfrak{H} = \mathfrak{H}_H$ we let $\mathfrak{H}^2 = \mathfrak{H} \oplus \mathfrak{H}$, viewed as a Hilbert space. We now consider another formal differential operator L of order n on ι

$$L = \sum_{k=0}^{n} P_k D^k ,$$

where $P_k \in C^k(\iota)$, and $P_n(x)$ is invertible for $x \in \iota$ if $n > \nu$. All cases, $n < \nu$, $n \geq \nu$, are allowed. The maximal linear manifolds for L,M and L^+,M are defined by

$$T = \{\{f,g\} \in \mathfrak{H}^2 \mid f \in C^r(\iota), \quad g \in C^\nu(\iota), \quad Lf = Mg\} ,$$

$$T^+ = \{\{f,g\} \in \mathfrak{H}^2 \mid f \in C^r(\iota), \quad g \in C^\nu(\iota), \quad L^+f = Mg\} ,$$

where $r = \max(n,\nu)$. The minimal linear manifolds are given by

$$S = \{\{f,GLf\} \mid f \in C_0^\infty(\iota)\}, \qquad S^+ = \{\{f,GL^+f\} \mid f \in C_0^\infty(\iota)\} .$$

Clearly, S,S^+ are (the graphs of) operators, but T,T^+ need not be. Due to the third equality in (3.1) we have $S \subset T$, $S^+ \subset T^+$. The maximal subspaces (closed linear manifolds) for L, M and L^+, M are defined by $T_1 = T^c$, $T_1^+ = (T^+)^c$, and the minimal subspaces are defined as $T_0 = S^c$, $T_0^+ = (S^+)^c$. These subspaces have the following properties:

(i) $T_0 \subset T_1 = (T_0^+)^*$, $\quad T_0^+ \subset T_1^+ = T_0^*$,

(ii) $T_1(0) = T_1^+(0) = T(0) = T^+(0) = \mathfrak{N}_H$,

(iii) $(T_1 \ominus T_0) \cup (T_1^+ \ominus T_0^+) \subset C^r(\iota) \times C^r(\iota)$,

(iv) $\nu(T_1 - \ell I) = \nu(T - \ell I) = \{f \in C^r(\iota) \cap \mathfrak{H} \mid (L - \ell M)f = 0\}$,

$\nu(T_1^+ - \bar{\ell}I) = \nu(T^+ - \bar{\ell}I) = \{f \in C^r(\quad) \quad \mid (L^+ - \bar{\ell}M)f = 0\}$,
for all $\ell \in \mathbb{C} \setminus \mathbb{C}_e$, where $\mathbb{C}_e$ is the set of all $\ell \in \mathbb{C}$ such that the leading coefficient of $L - \ell M$ is not invertible for some $x \in \iota$.

The adjoint of a subspace is defined joust as for an operator, and ν denotes "null space". We note that $\mathbb{C}_e = \emptyset$ if $n > \nu$ and $\mathbb{C}_e = \{0\}$ if $n < \nu$, whereas

$$C_e = \bigcup_{x \in \iota}' \{\ell \mid \det(P_\nu(x) - \ell Q_\nu(x)) = 0\}, \qquad n = \nu.$$

The property (iii) shows that the nonsmooth elements in T_1, T_1^+ arise from those which are in T_0, T_0^+, respectively.

The principal objects of study are intermediate subspace A satisfying

$$T_0 \subset A \subset T_1, \quad T_0^+ \subset A^* \subset T_1^+ .$$

Not $t = \dim(T_1/T_0) = \dim(T_1 \ominus T_0) \leq 2rm$, and if $\dim(A/T_0) = d$, then A, A^* may be characterized as follows:

$$A = T_0 \oplus M_1 = T_1 \cap (M_1^+)^*, \qquad A^* = T_0^+ \oplus M_1^+ = T_1^+ \cap M_1^*,$$

where M_1, M_1^+ are subspaces satisfying

$$M_1 \subset T_1 \ominus T_0, \qquad M_1^+ \subset T_1^+ \ominus T_0^+,$$

$$\dim M_1 = d, \qquad \dim M_1^+ = t - d,$$

$$M_1^+ \subset M_1^* .$$

The relation $A = T_1 \cap (M_1^+)^*$ shows that A is obtained from T_1 by the imposition of $t - d$ generalized boundary conditions

$$A = \{\{f,g\} \in T_1 \mid (g, \alpha_j^+) - (f, \beta_j^+) = 0, \quad j = 1,\ldots,t-d\},$$

where $\{\alpha_j^+, \beta_j^+\}$, $j = 1,\ldots,t-d$, is a basis for M_1^+.

Now A, A^* in general contain nonsmooth elements, so that these generalized boundary conditions can not be written as boundary conditions of the usual sort. However, smooth versions A_T, A_T^+ of A, A^* exist, where

$$A_T = A \cap T, \qquad A_T^+ = A^* \cap T^+,$$

and $A_T, A_T^+ \subset C^r(\iota) \times C^r(\iota)$, and moreover $(A_T)^c = A$,

$(A_T^+)^c = A^*$. These smooth versions can, in many cases, be characterized by boundary conditions of the usual type. In particular, if L,M are in the regular case, then A_T, A_T^+ can be explicitly characterized in terms of such boundary conditions, and the conditions for the eigenvalues of A involve the parameter ℓ in a linear fashion.

V. RESOLVENTS

Let A be a subspace as above. Its resolvent set $\rho(A)$ is the set of all $\ell \in C$ such that $(A - \ell I)^{-1}$ exists as a bounded operator on all of $\mathfrak{H}$, and its resolvent R_A is the operator-valued function defined via

$$R_A(\ell) = (A - \ell I)^{-1}, \quad \ell \in \rho(A).$$

For $\ell \in \rho(A) \cap (\mathbb{C} \setminus \mathbb{C}_e)$ it can be proved that $R_A(\ell)G: L_0^2(\iota) \to \mathfrak{H}$ is an integral operator

$$R_A(\ell)G\alpha(x) = \int_\iota k(x,y,\ell)\alpha(y)dy, \quad \alpha \in L_0^2(\iota),$$

with a smooth kernel k, and, since $(R_A(\ell))^* = R_{A^*}(\bar{\ell})$, we have $R_{A^*}(\bar{\ell})G$ is also such an integral operator. The same line of reasoning shows that G is an integral operator with a smooth kernel. If $M = I$ $(\nu = 0)$, then it can be shown that $R_A(\ell)$ is an integral operator of Carleman type.

In the regular case more can be said. In this case, $R_A(\ell)$ is compact if $n > \nu$, $R_{A^{-1}}(\ell^{-1})$ is compact if $n < \nu$, whereas $R_A(\ell) - (Q_\nu^{-1}P_\nu - \ell I)^{-1}$ is compact if $n = \nu$; all for $\ell \in \rho(A) \cap (\mathbb{C} \setminus \mathbb{C}_e)$. This implies that the Riesz-Schauder results are applicable. In the particular case $L = L^+$ we have $A = A^*$, $\mathbb{C}_e \subset \mathbb{R}$, and we obtain an eigenfunction expansion result which gives the direct analog of the simultaneous diagonalization of two hermitian matrices K, H, with $H > 0$.

SEMILINEAR ELLIPTIC EQUATIONS AT RESONANCE: HIGHER EIGENVALUES AND UNBOUNDED NONLINEARITIES

Djairo G. de Figueiredo[1,2]

Dept.o de Matematica
Universidade de Brasilia
Brasilia, Brasil

1. Let L be a uniformly strongly elliptic operator of order $2m$ with smooth coefficients acting on real-valued functions defined in a bounded domain Ω in R^N. Since we intend to study the Dirichlet problem we may then consider L as an operator in $L^2(\Omega)$ with domain $D(L) = H^{2m} \cap H_0^m(\Omega)$. Such an operator L is an unbounded closed linear operator in $L^2(\Omega)$ with finite dimensional nullspace $N(L)$. Assuming also that the coefficients of L are symmetric, as we do, it follows that L is self-adjoint and its spectrum is made up of real eigenvalues $\lambda_1 < \lambda_2 < \dots \lambda_n \to +\infty$, with corresponding finite dimensional eigenspaces $N(L-\lambda_i)$, $i = 1,2,\dots;$ moreover $(L-\lambda)^{-1}$ which is a well defined operator from $R(L-\lambda)$ into $D(L) \cap R(L-\lambda)$ is compact. From now on we suppose that 0 is

[1]The author acknowledges a grant of the "Conselho Nacional do Desenvolvimento Cientifico e Tecnologico" (CNP_q) of Brasil for a stay at the Courant Institute of Mathematical Sciences where this research was conducted.

[2]Present address: Dept.o de Matematica, Universidade de Brasilia, Campus Universitario, Asa Norte, Brasilia, DF CEP 70910.

ISBN 0-12-186280-1

a higher (i.e. $\lambda_1 < 0$) eigenvalue of L, and let α, be the smallest positive eigenvalue of $-L$.

We consider a nonlinearity given by a function $g : \Omega \times R \to R$ satisfying Caratheodory's conditions and the following basic assumption regarding its growth:

(g) there are a positive number $\gamma < \alpha$ and L^2-functions $b(x)$, $c(x)$ and $d(x)$ such that

$$|g(x,u)| \leq \gamma|u| + b(x), \quad u \in R, \quad x \in \Omega \tag{1}$$

$$ug(x,u) \geq -c(x)|u| - d(x), \quad u \in R, \quad x \in \Omega. \tag{2}$$

Theorem 1. *Assume* L *and* g *as above. Then there exists a* $u \in H^{2m}(\Omega) \cap H_0^m(\Omega)$ *satisfying*

$$Lu + g(x,u) = h, \tag{3}$$

for a given $h \in L^2(\Omega)$, *provided*

$$(h,v) < \int_{v>0} g_+(x)v(x)dx + \int_{v<0} g_-(x)v(x)dx, \quad \forall\, v \in N(L),\ v \neq 0$$

where

$$g_-(x) = \limsup_{u\to-\infty} g(x,u), \qquad g_+(x) = \liminf_{u\to+\infty} g(x,u).$$

Observe that g in the above theorem could be an unbounded function with any linear growth $< \alpha$. This means that eventually we could have a situation like this

$$0 = \lim_{u\to+\infty} \frac{g(x,u)}{u} < \beta < \lim_{u\to-\infty} \frac{g(x,u)}{u} < \alpha$$

where β is as close to α as we wish. So we have here resonance at a higher eigenvalue (or the "nonlinearity jumping off an eigenvalue" in the terminology of Fučik); in this connection see the papers of Fučik [8] and Dancer [7]. We remark that in this paper the nonlinearity does not "cross" an eigenvalue; this follows from the fact that $\gamma < \alpha$ and (2) above.

The case of bounded nonlinearities has been extensively

studied in the literature starting with the pioneering work of Landesman and Lazer [11]; some interesting recent research appears in Fučik and Krbec [9] and Hess [10]. The case of nonlinearities with a sublinear growth and satisfying (2) above was investigated in Brezis and Nirenberg [6]. Unbounded nonlinearities g without condition (2) above lead to other difficulties, and to treat them some additional hypotheses have been imposed on g by several authors in recent papers, e.g. Ambrosetti and Prodi [4], Ambrosetti and Mancini [2],[3], Amann, Ambrosetti and Mancini [1], Berger and Podolak [5].

Instead of restricting ourselves to the case of differential operators we preferred to follow the work of Brezis and Nirenberg [6] and in Section 2 we consider more general equations of the type (3) given by

$$Au + Bu = h \tag{4}$$

where A is some (unbounded) linear operator in a Hilbert space H and B is some nonlinear operator. The problem taken up in [6] was the characterization of the $R(A+B)$. In this paper we pursue a more modest goal, but we are able to consider more general nonlinearities as far as growth at infinity is concerned.

2. We shall study the solvability of the equation (4) in $H = L^2(\Omega)$. As in [6] we assume the following condition on A:

Property I. $A : D(A) \subset H \to H$ *is a closed densely defined linear (not necessarily bounded) operator such that* $R(A) = N(A)^{\perp}$. *Consequently we have that* $N(A) = N(A^*)$, $H = R(A) \oplus N(A)$ *and* A *is a bijective mapping from* $D(A) \cap R(A)$ *onto* $R(A)$, *whose inverse we denote by* A^{-1}. *It is also assumed that* A^{-1} *is compact.*

As usual $N(A)$ denotes the nullspace of A, $R(A)$ its range and A^* the adjoint of A.

The following assertion is an immediate consequence of Property I: there exists α_o such that

$$\alpha_o \|Qu\| \leq \|Au\| , \qquad u \in D(A) \tag{5}$$

where Q is the orthogonal projection on $R(A)$ and $\|\cdot\|$ is the norm of H. And from (5) it follows readily that

$$(Au,u) \geq -(1/\alpha_o)\,\|Au\|^2, \qquad u \in D(A) \tag{6}$$

where $(\ ,\ )$ is the inner-product in H. Let us denote by α the largest positive constant α_o for which (6) holds:

$$(Au,u) \geq -(1/\alpha)\,\|Au\|^2, \qquad u \in D(A)\ . \tag{7}$$

The interest in this class of linear operators A lies mostly in the fact that $A = \pm L$ with $D(A) = H^{2m}(\Omega) \cap H_0^m(\Omega)$, where L is the elliptic operator of Section 1, satisfies Property I. In this case α is the smallest strictly positive eigenvalue of $-L$ (resp. L) accordingly to $A = L$ (resp. $A = -L$).

The nonlinear operator B is the Niemytskii mapping given by $Bu = g(x,u)$, where $g : \Omega \times R \to R$ is a function satisfying Carathéodory's conditions and the condition (g) above regarding its growth.

Inequality (1) implies that B is bounded and continuous from $L^2(\Omega)$ into $L^2(\Omega)$. As a consequence of (2) we have that

$$\begin{aligned} g_-(x) &\equiv \limsup_{u\to-\infty} g(x,u) \leq c(x) \\ g_+(x) &\equiv \liminf_{u\to+\infty} g(x,u) \geq -c(x) \end{aligned} \tag{8}$$

and so $-\infty \leq g_-(x) < +\infty$ and $-\infty < g_+(x) \leq +\infty$ for $x \in \Omega$. We note that when we write $x \in \Omega$ we actually mean $x \in \Omega$ (a.e.).

Now we state our main result.

Theorem 2. *Suppose that* A *satisfies Property* I *and that* $N(A)$ *is finite dimensional. Assume also that* B *is a mapping defined by a function* $g(x,u)$ *satisfying Carathéodory's conditions and condition* (g). *Conclusion:*

(a) *If*

$$(h,v) < \int_{v>0} g_+(x)v(x)\,dx + \int_{v<0} g_-(x)v(x)\,dx, \tag{9}$$

for all $0 \neq v \in N(A)$, *then* $h \in \operatorname{Int} R(A+B)$.

(b) *If*

$$(h,v) \leq \int_{v>0} g_+(x)v(x)dx + \int_{v<0} g_-(x)v(x)dx \tag{10}$$

for all $0 \neq v \in N(A)$, *then* $h \in \overline{R(A+B)}$.

Remarks: 1) The above theorem with some stronger assumption on the behavior of g at $\pm\infty$ was proved in [6; Theorem IV.5]. Namely it is assumed there, in addition to the hypothesis above, that one of the following conditions hold:

either (a) $g_-(x) \leq g_+(x)$, $x \in \Omega$,

or (b) $|g(x,u)| \leq \gamma_1|u| + b(x)$, $\gamma_1 > 0$, $x \in \Omega$, $u \in R$.

Our proof contains a new idea (a decomposition of g as stated in Lemma 2 below), and provides a simpler proof of the Brezis-Nirenberg theorem. The above theorem was originally proved by the author under an additional restriction on g enabling us a certain decomposition of g needed in Step 3' of the proof; we thank Ed Joer for suggesting us a modification of that decomposition so as to eliminate the need of our extra assumption.

2) It follows from the remarks after the statement of Property I that Theorem 1 is an immediate corollary of Theorem 2. Also the following result is a corollary of Theorem 2.

Theorem 3. *Let* L *be as in Section* 1 *and let* α *be the smallest positive eigenvalue of* L. *Suppose that* $g : \Omega \times R \to R$ *is a function satisfying Carathéodory's conditions and the following assumptions:*

$$|g(x,u)| \leq \gamma|u| + b(x), \quad u \in R, \quad x \in \Omega$$

$$ug(x,u) \leq c(x)|u| + d(x), \quad u \in R, \quad x \in \Omega$$

where $0 < \gamma < \alpha$, *is a constant and* b, c *and* d *are* L^2-*functions in* Ω. *Then, for given* $h \in L^2(\Omega)$, *the Dirichlet problem for*

$$Lu + g(x,u) = h$$

has a solution provided the inequality

$$(h,v) > \int_{v>0} \left[\limsup_{u\to+\infty} g(x,u)\right] v(x)\,dx + \int_{v<0} \left[\liminf_{u\to-\infty} g(x,u)\right] v(x)\,dx$$

holds for all $v \in N(L)$, $v \neq 0$.

Lemma 1. *Let* $g(x,u)$ *be a function satisfying Carathéodory's condition and condition* (g). *Then for each* $\gamma' > \gamma$ *there are positive constants* c_1 *and* c_2 *such that*

$$(Bu,u) \geq \frac{1}{\gamma'} \|Bu\|^2 - c_1\|u\| - c_2. \tag{11}$$

If, in addition, $g(x,u)u \geq 0$ *for all* $x \in \Omega$ *and all* $u \in R$, *then* (11) *holds with* $c_1 = 0$.

Proof. Using (1) and (2) we obtain

$$ug \geq \frac{1}{\gamma} |g|^2 - \frac{b}{\gamma} |g| - 2c|u| - 2d.$$

Now we use the inequality $pq \leq \varepsilon p^2 + (1/4\varepsilon)q^2$ with $\varepsilon = 1 - (\gamma/\gamma')$, $p = |g|$, $q = b$, and it follows from the above inequality that

$$ug \geq \frac{1}{\gamma'} |g|^2 - 2c|u| - \frac{1}{4\varepsilon\gamma} b^2 - 2d.$$

Integrating we then obtain (11). The second part of the lemma is a consequence of the fact that $c_1 = 2\|c\|$. Q.E.D.

Lemma 2. *Let* $g(x,u)$ *be a function satisfying Carathéodory's conditions and condition* (g). *Then for each* $R > 0$ *there is a decomposition*

$$g(x,u) = g_R(x,u) + q_R(x,u)$$

of g *by functions* g_R *and* q_R *satisfying Carathéodory's conditions and the following additional conditions:*

$$uq_R(x,u) \geq 0, \quad x \in \Omega, \quad u \in R,$$

q_R *satisfies inequality* (1) *with same; there is an* L^2*-function* $a_R(x)$ *such that*

$$|g_R(x,u)| \le a_R(x), \quad x \in \Omega, \quad u \in R$$

$$g_{R,-}(x) \equiv \limsup_{u\to-\infty} g_R(x,u) = \begin{cases} -R & \text{if } g_-(x) \le -R \\ g_-(x) & \text{if } g_-(x) > -R \end{cases}$$

and

$$g_{R,+}(x) \equiv \liminf_{u\to+\infty} g_R(x,u) = \begin{cases} R & \text{if } g_+(x) \le R \\ g_+(x) & \text{if } g_+(x) > R \end{cases}$$

Proof. Given $R > 0$ define

$$\hat{g}_R(x,u) = \begin{cases} \sup\{g(x,u),-R\} & \text{if } u \le -1 \\ \inf\{g(x,u),R\} & \text{if } u \ge 1 \end{cases}$$

and

$$\hat{q}_R(x,u) = g(x,u) - \hat{g}_R(x,u) \qquad x \in \Omega, \quad |u| \ge 1.$$

Next we define $q_R(x,u)$ as $\hat{q}_R(x,u)$ if $|u| \ge 1$ and as $q_R(x,u) = |u|\hat{q}_R(x, u/|u|)$ if $|u| \le 1$. And finally we define $g_R = g - q_R$. It is a straightforward calculation to check that all conditions of the lemma 2 are satisfied. Q.E.D.

Proof of Theorem 2. Step 1. For each positive integer n there exists $u_n \in D(A)$ such that

$$Au_n + \frac{1}{n} Pu_n + Bu_n = h, \tag{12}$$

where P is the orthogonal projection over $N(A)$. This is proved using the Leray-Schauder fixed point theorem, see step 1 in the proof of Theorem III.2 of [6].

Step 2. There exists a constant $c > 0$ such that

$$\|\frac{1}{n} Pu_n\| \le c, \quad \text{for all } n. \tag{13}$$

From now on all positive constants independent of n are designated by the same letter c; of course, they depend on A, B and h. To prove (13) we proceed as follows. Take the inner product of (12) with u_n and estimate using (7) and (11)

$$-\frac{1}{\alpha}\|Au_n\|^2 + \frac{1}{n}\|Pu_n\|^2 + \frac{1}{\gamma'}\|Bu_n\|^2 \leq c + c\|u_n\|,$$

where $\gamma < \gamma' < \alpha$. To estimate Bu_n use the equation (12) to get

$$\left(\frac{1}{\gamma''} - \frac{1}{\alpha}\right)\|Au_n\|^2 + \frac{1}{n}\|Pu_n\|^2 - c\left\|\frac{1}{n}Pu_n\right\|^2 \leq c + c\|u_n\|, \tag{14}$$

where $\gamma' < \gamma'' < \alpha$. The last term in the left side of (14) is absorbed by the second term for large n. Besides that we estimate $\|u_n\| \leq \|Pu_n\| + \|Qu_n\|$ and use (5). All this in (14) gives

$$c\|Au_n\|^2 + \frac{1}{n}\|Pu_n\|^2 \leq c + c\|Pu_n\| + c\|Au_n\|.$$

Finally we use the fact that given $as^2 - bs$ with $a > 0$ there are numbers $0 < a' < a$ and $c > 0$ such that $as^2 - bs > a's^2 - c$ for all $s \in R$, and obtain

$$c\|Au_n\|^2 + \frac{1}{n}\|Pu_n\|^2 \leq c + c\|Pu_n\|. \tag{15}$$

Now omitting the first term in (15) we readily prove (13).

Step 3. Suppose that (Pu_n) contains a bounded subsequence. In view of (15) it follows that (Au_n) also contains a bounded sequence, and using the fact that A^{-1} is compact we obtain a convergent subsequence of (Qu_n). All this together with the finite dimensionality of $N(A)$ implies that (u_n) contains a convergent subsequence; we use the same notation for the subsequence and write that $u_n \to u \in H$. So $Bu_n \to Bu$ and $Au_n \to h-Bu$. Since A is closed, it follows that $u \in D(A)$ and $Au = h-Bu$. Thus we have proved that $h \in R(A+B)$ on the assumption that (Pu_n) contains a bounded subsequence.

Step 3'. On the other hand suppose that $\|Pu_n\| \to \infty$, and let us prove that such an assumption leads to a contradiction. First we observe that (15) and (5) imply that

$$\lim_{n\to\infty} \frac{\|Qu_n\|}{\|Pu_n\|} = 0 \; . \tag{16}$$

Let $v_n = u_n / \|Pu_n\|$. It follows from (16) that (v_n) is a bounded sequence. As in step 3 we may assume, passing to a subsequence, that $v_n \to v$. Again (16) implies that $\|v\| = 1$. It also follows from (15), using (12) and (13), that

$$\|Bu_n\|^2 \leq c + c\,\|Pu_n\|$$

and so $\|Bu_n\| / \|Pu_n\| \to 0$. Dividing the equation (12) by $\|Pu_n\|$ and going to the limit we obtain that $Av_n \to 0$. Thus $v \in N(A)$.

Now let us denote by G_R and Q_R the Niemytskii operators defined respectively by the functions g_R and q_R of the decomposition of g in Lemma 2. So $Bu = G_R u + Q_R u$. Taking the inner product of (18) with u_n and using Lemma 1 to the operator Q_R we get

$$-\frac{1}{\alpha}\|Au_n\|^2 + \frac{1}{n}\|Pu_n\|^2 + \frac{1}{\gamma'}\|Q_R u_n\|^2 + (G_R u_n, u_n) \leq (h,u_n) + c$$

where $\gamma < \gamma' < \alpha$. Using the equation (12) to estimate the third term in the left side of the above inequality we get

$$\frac{1}{n}\|Pu_n\|^2 + (G_R u_n, u_n) \leq (h,u_n) + c. \tag{17}$$

Dividing by $\|Pu_n\|$ and taking limits we obtain by Fatou's lemma that

$$\int_{v>0} g_{R,+}(x)v(x)\,dx + \int_{v<0} g_{R,-}(x)v(x)\,dx \leq (h,v). \tag{18}$$

Now letting $R \to +\infty$ and using (8) we obtain

$$\int_{v>0} g_{+}(x)v(x)\,dx + \int_{v<0} g_{-}(x)v(x)\,dx \leq (h,v)$$

which contradicts (9). So on the assumption of (9) we proved that $h \in R(A+B)$. In fact we have to prove that $h \in \operatorname{Int} R(A+B)$, and this is proved next.

Step 4. Assume (9). We claim that there is a $\beta > 0$ such that

$$\ell(v) \equiv \int_{v>0} g_+ v + \int_{v<0} g_- v - (h,v) \geq \beta > 0 \tag{19}$$

for all $v \in N(A)$, $\|v\| = 1$. Suppose this were not so. Then there would exist $v_n \in N(A)$, $\|v_n\| = 1$, $v_n \to v$ and $\ell(v_n) \to 0$. Using (8) we write

$$\ell(v_n) = \int_{v_n>0} (g_+ + c)v_n + \int_{v_n<0} (g_- - c)v_n - (h,v_n) - (c,v_n).$$

Taking limits and using Fatou's lemma

$$0 = \lim \ell(v_n) \geq \int_{v>0} (g_+ + c)v + \int_{v<0} (g_- - c) - (h,v) - (c,v)$$

which contradicts (9). Thus in view of (19) every $h + f$, for any $f \in H$ with $\|f\| < \beta$, would satisfy (9) and by steps 3 and 3', $h + f \in R(A+B)$. So part (i) of the teorem is proved.

Step 5. Assume (10). Passing to the limit in (17) as in step 3 but maintaining the term $\frac{1}{n}\|Pu_n\|$ we obtain

$$\limsup \frac{1}{n}\|Pu_n\| + \int_{v>0} g_{2,+}v + \int_{v<0} g_{2,-}v \leq (h,v).$$

Finally letting $R \to +\infty$ and using (10) we obtain that $\frac{1}{n}\|Pu_n\| \to 0$. Equation (12) then implies that $h \in \overline{R(A+B)}$.

Q.E.D.

REFERENCES

1. Amann, H., Ambrosetti, A. and Mancini, G., Elliptic equations with noninvertible Fredholm linear part and bounded nonlinearities. To appear.
2. Ambrosetti, A. and Mancini, G., Existence and multiplicity results for nonlinear elliptic problems with linear part at resonance. *J. Diff. Eq.*, to appear.

3. Ambrosetti, A. and Mancini, G., Theorems of existence and multiplicity for nonlinear elliptic problems with noninvertible linear part. *Ann. Scuola Norm. Sup. Pisa*, to appear.
4. Ambrosetti, A. and Prodi, G., On the inversion of some differentiable mappings with singularities between Banach spaces. *Ann. Mat. Pura App., IV, 93*, 231-247 (1972).
5. Berger, M. S. and Podolak, E., On nonlinear Fredholm operator equations. *Bull. A. M. S. 80*, 861-864 (1974).
6. Brezis, H. and Nirenberg, L., Characterizations of the ranges of some nonlinear operators and applications to boundary value problems. *Ann. Scuola Norm. Sup. Pisa, Ser. IV, V*, 115-175 (1978).
7. Dancer, E. N., On the Dirichlet problem for weakly nonlinear elliptic partial differential equations. *Proc. Royal Soc. Edinburgh 76A*, 283-300 (1977).
8. Fučik, S., Boundary value problems with jumping nonlinearities. *Časopis Pest. Mat. 101*, 69-87 (1976).
9. Fučik, S. and Krbec M., Boundary value problems with bounded nonlinearity and general null-space of the linear part. *Math. Z.*, to appear.
10. Hess, P., Nonlinear perturbations of linear elliptic and parabolic problems at resonance: existence of multiple solutions. To appear.
11. Landesman, E. A. and Lazer A. C., Nonlinear perturbations of linear elliptic boundary value problems at resonance. *J. Math. Mech. 19*, 609-623 (1970).

COUNTABLE SYSTEMS OF ORDINARY DIFFERENTIAL EQUATIONS

Klaus Deimling[1]

Gesamthochschule Paderborn

The title refers to systems of the type

$$x_i' = f_i(t, x_1, x_2, \ldots) \quad \text{for every} \quad i \geq 1 , \tag{1}$$

or with ordinary differential operators L_i of finite order instead of the first derivative.

1. The history of such systems dates back to the origins of functional analysis around 1900 at least, and the purpose of the first papers known to us has been to show that the techniques available during this period, namely Picard's successive approximation method and Hilbert's quadratic forms on l^2, may be applied to obtain existence and uniqueness theorems for the initial value problem (1), $x_i(0) = c_i$ for $i \geq 1$, where t is real and the x_i and f_i are real valued.

With a series of papers written around 1927, A. Wintner was one of the first who also looked for solutions of (1) in l^2 which are analytic in a neighborhood of the initial value. To this end he had to define analyticity of functions of infinitely many complex variables which he did by means of the trunca-

---

[1]Present address: Fachbereich 17 der GH, Warburger Str. 100, D-4790 Paderborn

ISBN 0-12-186280-1

tions $f_i(t,x_1,\ldots,x_N,0,0,\ldots)$ for $i \leq N$. He also proved implicite function theorems and he had applications to some problems in celestial mechanics; see e.g. Wintner [22]. In recent years there has been much activity in the field of analytic maps between abstract, mainly locally convex spaces, see e.g. Nachbin [14] and Noverraz [15], but up to now we have not seen a general discussion and applications of analyticity in connection with (1). Some counterexamples which indicate necessary limitations are given e.g. in the papers of N. Arley and N. Borchsenius [1] and E. Hille [10].

The literature dealing with (1) enlarged considerably in the thirties. One source has been the application of Fourier's separation method to initial/boundary value problems for parabolic and hyperbolic equations. Another one has been the theory of stochastic processes which are continuous in time and discrete in space, i.e. processes describing systems which at every $t \geq 0$ are in one of countably many states. An interesting more recent paper which also has many references is one of G. E. H. Reuter [17]. In the simplest case of a birth-death process with constant rates the system is tridiagonal

$$x' = \begin{pmatrix} -\lambda_0 & \lambda_1 & & 0 \\ \lambda_0 & -(\mu_1+\lambda_1) & \lambda_2 & \\ & \mu_1 & -(\mu_2+\lambda_2) & \cdots \\ 0 & & \mu_2 & \cdots \end{pmatrix} x ,$$

$$x_i(0) = c_i \quad \text{with} \quad \sum_{i\geq 0} c_i = 1 , \tag{2}$$

the μ_i, λ_i are nonnegative and the unknown x_i are probabilities; therefore one looks for solutions such that $\sum_{i\geq 0} x_i(t) \equiv 1$ and thus l^1 is the natural Banach space for (1).

Also in the thirties, W. T. Reid considered linear boundary value problems for (1) in l^2. Recently, such problems where considered again for second order nonlinear systems by K.Schmitt and R. Thompson [18] and R. Thompson [19].

A paper of much influence has been that one of A. Tychonoff [20] from 1934, where he extended the Schauder fixed point theorem to locally convex spaces and applied this result to existence of solutions of the IVP (1), $x_i(0) = c_i$ in case the f_i are continuous with respect to the product topology on R^∞.

Until today, a large number of publications by Russian schools have been motivated by this paper and many of them are incorporated in the book of K. G. Valeev and O. A. Zautykov [21].

In recent years there have also appeared several models for chemical and biological problems that lead to countable systems. Reading these papers and also several of the "pure" mathematical ones we felt that it is high time to organize and to improve sometimes much of the existing material, and this has been done in the lecture notes [7] in 1977, where one can also find more details and references on the remarks given so far.

2. A good deal of the Russian papers mentioned above are concerned with stability of solutions to (1). Some references are given in [7], chapter VIII of the book [21] is devoted to this subject, and further references are included in the bibliography at the end of this paper. This year, A. Michel and R. K. Miller [12] have studied the question of stability too, and since the second author gave a lecture on this paper during this conference we may refer to his paper for the present state of the art.

3. We have already mentioned that only a few things are known about analytic solutions and about boundary value problems. The same may be said about existence of periodic solutions in case the f_i are periodic. In [4] we started to study this problem for $x' = f(t,x)$ in a general B-space, and in [5] we have indicated some existence theorems for countable systems of a special structure which can be obtained by means of Galerkin approximations. Since all results in general B-spaces X yield results for countable systems when X is a particular sequence space, let us also mention the recent paper of J. Mawhin/M. Willem [11] for $x' = f(t,x)$ in H-space, the lectures of J. Mawhin and M. Willem in these proceedings and a forthcoming paper of J. Prüss [16] for $x' = Ax + f(t,x)$ with an unbounded linear operator which contains also some concrete examples of periodic countable systems.

4. Another interesting specific problem for countable systems, where little is known although it is certainly the most

interesting one for people who are interested in numerical analysis, consists in the convergence of various truncation methods. At a first glance the most natural approach to (1) seems to be an N by N truncation, i.e. to consider the finite system

$$x_i^{N\prime} = f_i(t,x_1^N,\dots,x_N^N,0,\dots),\quad x_i^N(0) = c_i \quad \text{for } i \le N. \tag{3}$$

Concerning this method we have proved a rather general convergence theorem, Theore 7.3 in [7], for the linear system

$$x_i' = \sum_{j\ge 1} a_{ij}(t)x_j + b_i(t),\quad x_i(0) = c_i \quad \text{for } i \ge 0, \tag{4}$$

which can be extended to nonlinear systems via Lipschitz conditions on the f_i ; see Theorem 7.4 in [7]. Since we had requests from people working in applied stochastics interested in truncations for systems like (2), who didn't understand the proof, and since the ideas of proof may certainly be applied to other truncations, let us recall some facts. By a solution on $J = [0,\alpha]$ of (4) we understand a sequence of function $x_i \in C^1(J)$ which satisfies (4). Let

$$a_{ij} = \max_J |a_{ij}(t)|,\quad b_i = \max_J |b_i(t)|,\quad A = (a_{ij}) ,$$

and assume that the infinite matrix $\exp(A\alpha)$, defined by

$$(e^{A\alpha})_{ij} = \sum_{n>0} \frac{\alpha^n}{n!} (A^n)_{ij} \quad \text{for} \quad i,j \ge 0 ,$$

exists. Let

$$D_\alpha = \{y \in R^\infty : \sum_{j\ge 1} (e^{A\alpha})_{ij}|y_j| < \infty \quad \text{for every} \quad i \ge 1\}$$

and assume that $c = (c_i)$ and $b = (b_i)$ are in D_α . Then (4) may have many solutions, but there is one solution $\tilde{x}$ which is the limit of the successive approximations starting with $x^0 = c$. Now, it is possible to show that $x_i^N(t) \to \tilde{x}_i(t)$ as $N \to \infty$ uniformly on J for every $i \ge 0$.

However, in concrete models it is sometimes desirable that truncation preserves some properties of the original system, and this can not always be accomplished by the simple N by

N-truncation. For example, under reasonable conditions on the coefficients, a system like (2) has a solution satisfying $\sum_{i \geq 0} x_i(t) \equiv 1$ (an "honest" process) and one is interested in truncations such that the solution of the truncated system is honest too. Consider e.g. (2) with $c_o = 1$, $c_i = 0$ for $i \geq 1$ and λ_i and μ_i positive for all $i \geq 0$. Here the solution of the N by N truncation satisfies

$$(\sum_{i \leq N} x_i^N(t))' = -\mu_N x_N^N(t) \not\equiv 0$$

and therefore x^N is not honest, but the same truncation with μ_N replaced by 0 has an honest solution. To prove convergence, it is a very useful observation which can also be made in several quite different concrete examples (see [7]) that the finite s-stems are quasimonotone, i.e. increasing in the elements outside the main diagonal. Therefore the nonnegative initial values c_i for (2) imply nonnegativity of the solutions and $x_i^N(t) \leq x_i^{N+1}(t) \leq 1$ for $i \leq N$, and from this fact it is easy to obtain convergence if the λ_i and μ_i do not grow too fast. Another interesting example are systems for the moments of the solutions of Fokker-Planck equations, where the truncation should preserve moment inequalities etc.; an illustrative discussion of this problem can be found e.g. in R. Bellman/R. Wilcox [2] and R. Bellman/J. Richardson [3].

5. Our last topic is the existence of maximal and minimal solutions to the IVP

$$x_i' = f_i(t, x_1, x_2, \ldots), \quad x_i(0) = c_i \quad \text{for } i \geq 1. \tag{5}$$

In a forthcoming joint paper [8] with V. Lakshmikantham, we have considered the following situation. Let X be a real B-space, $K \subset X$ a cone, $J = [0,a] \subset R$ and $f: J \times K \mapsto X$ continuous and quasimonotone w.r. to K, i.e. let

$$\begin{aligned} &t \in J;\ x,y \in K \text{ and } x \leq y,\ x^* \in K^* \text{ and } x^*(x-y) = 0 \Rightarrow \\ &x^*(f(t,x) - f(t,y)) \leq 0 \end{aligned} \tag{6}$$

hold, where $x \leq y$ is defined by $y-x \in K$ and K^* is the set of continuous linear functionals $x^* \in X^*$ such that $x^*(x) \geq 0$ for all $x \in K$; see § 5 of [7] for more details. For simplicity of this exposition suppose also that $f(J \times K)$ is compact. We assume that $f(t,0) \in K$ on J . Then the IVP $x' = f(t,x)$, $x(0) = c \in K$ has at least one solution on J ; of course its range is in K since f is only defined on K ; see § 4 of [7]. There may be many solutions and then we look for a maximal $\bar{x}$ and a minimal solution $\underline{x}$ (w.r. to $\leq$). In case the interior $\mathring{K}$ of K is nonempty, e.g. for $X = l^\infty$ and the standard cone $K = \{x : x_i \geq 0 \text{ for all } i\}$, the existence of $\bar{x}$ is trivial; it may be obtained like in the case $X = \mathbb{R}^n$ and $K = \{x \in \mathbb{R}^n : x_i \geq 0 \text{ for all } i \leq n\}$ as the limit (as $p \to \infty$) of a decreasing sequence of solutions x^p of

$$x' = f(t,x) + \frac{1}{p} e \ , \quad x(0) = c + \frac{1}{p} e \ , \ e \in \mathring{K} \text{ fixed.} \tag{7}$$

The existence of $\underline{x}$ is an open problem; the usual trick ($-e$ instead of e) does not work since the IVP with $-e$ need not have a solution (in K!). The situation becomes nearly hopeless if $\mathring{K} = \phi$, since then we cannot apply the e-trick (7) and even the existence of $\bar{x}$ is doubtful. Unfortunately this situation occurs quite often in the sequence spaces of interest for (5), e.g. $X = (c_o)$, the space of all sequences tending to zero or $X = l^p$ with $p \geq 1$ and the standard cone K . Sometimes, however, we find such extremal solutions by means of Galerkin approximation (n by n-truncation). The result is as follows. Let X be a B-space with a Schauder-base (e_i) , i.e. every $x \in X$ has a unique norm-convergent expansion $x = \sum_{i \geq 1} x_i e_i$, and consider the standard cone $K = \{x : x_i \geq 0 \text{ for all } i\}$. Then the above conditions imply the existence of $\underline{x}$. But the only example where we have found $\bar{x}$ is $X = (c_o)$! This curiosity should stimulate further activity in this field.

In another forthcoming paper [9] with V. Lakshmikantham we replace the quasimonotonicity by conditions which have been introduced by M. Müller [13] for finite systems in 1926. The main result is a follows. Consider again a B-space with a Schauder base and the corresponding standard cone K which defines $\leq$. Let $J = [0,a]$; $v,w : J \mapsto X$ continuously differentiable and such that $v(t) \leq w(t)$ on J . $D = \{(t,z) : t \in J$

and $v(t) \leq z \leq w(t)\}$; $f : D \mapsto X$ continuous and such that

$$\begin{aligned} v_i'(t) &\leq f_i(t,z) \text{ whenever } (t,z) \in D \text{ and } z_i = v_i(t) \\ w_i'(t) &\geq f_i(t,z) \text{ whenever } (t,z) \in D \text{ and } z_i = w_i(t) \\ &\text{for } i \geq 1 \end{aligned} \tag{8}$$

holds, and let $(0,c) \in D$. Then $x' = f(t,x)$, $x(0) = c$ has a solution on J provided that either K is normal or f is compact (for simplicity).

K is called normal if there exists a $\delta > 0$ such that $x,y \in K$ and $|x| = |y| = 1$ implies $|x+y| \geq \delta$. In this case D is compact and therefore the continuity of f is sufficient. However, the paper contains an example of a standard cone which is not normal. This theorem can be proved by means of Galerkin approximations again. The autonomous version, with v and w constant, yields a funny fixed point theorem; see Theorem 3 in [6].

By this limited survey we hope to have shown that countable systems of ODEs is an interesting field, and we hope to obtain further papers or references on this subject, since we certainly missed many of them in the notes [7] and in the bibliography to follow, especially those which will be written in the future.

ACKNOWLEDGEMENT

Prof. W. Hahn (TU Graz) has been so kind to send us the references [23]-[27] on stability of countable systems. Prof. E. A. Coddington (UCLA) told us that he has written a paper on countable systems in l^2 in connection with V. Karman vortices (≈1951), and one of Prof. Y. Sibuya's (Univ. of Minnesota) students has considered analytic countable systems. Unfortunately, there was not enough time between conference and deadline to search for the precise references.

REFERENCES

1. Arley, N. and Borchsenius V., On the theory of infinite systems of differential equations and their application to the theory of stochastic processes and the perturbation theory of quantum mechanics. *Acta Math. 76*, 261-322 (1945).
2. Bellman, R. and Wilcox R., Truncation and preservation of moment properties for Fokker-Planck moment equations. *J. Math. Anal. Appl. 32*, 432-542 (1970).
3. Bellman, R. and Richardson J., Closure and preservation of moment properties. *J. Math. Anal. Appl. 23*, 639-644 (1968).
4. Deimling, K., Periodic solutions of differential equations in Banach spaces. *Manuscripta Math. 24*, 31-44 (1978).
5. Deimling, K., Cone-valued periodic solutions of ordinary differential equations. Proc. Conf. "Applied Nonlinear Analysis", Academic Press (to appear).
6. Deimling, K., Fixed points of condensing maps. Proc. Helsinki-Sympos. on Volterra Integral Equations. Springer Lecture Notes (to appear).
7. Deimling, K., Ordinary differential equations in Banach Spaces. Lect. Notes in Math. vol. 596, Springer Verlag, Berlin,(1977).
8. Deimling, K. and Lakshmikantham V., On existence of extremal solutions of differential equations in Banach spaces (preprint).
9. Deimling, K. and Lakshmikantham V., Existence and comparison theorems for differential equations in Banach spaces (preprint).
10. Hille, E., Classical analysis and functional analysis, selected papers (R.R. Kallman, ed.). The MIT-Press, Cambridge, (1975).
11. Mawhin, J. and Willem M., Periodic solutions of nonlinear differential equations in Hilbert space. Comunicazioni Convegno "Equadiff 78", pp. 323-332. Centro 2P, Firenze, (1978)
12. Michel, A. and Miller R. K., Stability theory for countably infinite systems of differential equations (preprint).
13. Müller, M., Über das Fundamentaltheorem in der Theorie der gewöhnlichen Differentialgleichungen. *Math. Z. 26*, 619-645 (1926).

14. Nachbin, L., Topology on spaces of holomorphic mappings. Ergeb. Math. Grenzgeb. Bd. 47, Springer Verlag, Berlin, (1969).
15. Noverraz, P., Pseudo-convexité, convexité polynomiale et domaines d'holomorphie en dimension infinie. North Holland Math. Studies Vol. 3, North Holland Publ. Comp., Amsterdam, (1973).
16. Prüss, J., Periodic solutions of semilinear equations of evolution (preprint).
17. Reuter, G. E. H., Denumerable Markov processes and the associated contraction semigroups on l^1. *Acta Math. 97*, 1-46 (1957).
18. Schmitt, K. and Thompson R., Boundary value problems for infinite systems of second order differential equations. *J. Diff. Eqs. 18*, 277-295 (1975).
19. Thompson, R., On extremal solutions to infinite dimensional nonlinear second order systems (preprint).
20. Tychnoff, A., Ein Fixpunktsatz. *Math. Ann. 111*, 767-776 (1935).
21. Valeev, K. G. and Zautykov O. A., Infinite Systems of differential equations. Izdat. "Nauka" Kazach. SSR, Alma-Ata 1974 (in Russian).
22. Wintner, A., Über die Differentialgleichungen der Himmelsmechanik. *Math. Ann. 96*, 284 (1927).
23. Goršin, S., On the stability of the solutions of a denumerable system of differential equations with constantly acting disturbances. *Izv. Akad. Nauk Kazach. SSR 60, Ser. Mat. Meh. 3*, 32-38 (1949). On Lyapunov's second method. Ibid. *97 (4)*, 42-50 (1959). Some criteria of stability with constant disturbances. Ibid. *97 (4)*, 51-56 (1950). On stability with a countable number of perturbations in a critical case. Ibid. *4 (8)*, 38-42 (1956). Some problems of stability in the large for constantly acting perturbations in normed linear spaces. *Diff. Urav. 4*, 631-638; Transl. in "Diff. Eqs.".
24. Harasahal, V., On stability in the first approximation of the solutions of denumerable systems of differential equations. *Izv. Akad. Naik. Kazach. SSR 60 (3)*, 77-84 (1949).
25. Makarov, J. P., New criteria of stability according to

Lyapunov in the case of an infinite triangular matrix. *Math. Sbornik N. S. 30 (72)*, 53-58 (1952).

26. Resetov, M.R., On the stability of the solutions of a denumerable system of differential equations, the linear parts of which have triangular form. *Izv. Akad. Nauk. Kazah. SSR 60 (3)*, 39-76 (1949).

27. Slyusarchuk, V. E., Stability of solutions of an infinite system of differential equations. *Diff. Eds. 12*, 1414-1419 (1976); original in *Diff. Urav. 12*, 2019-2026 (1976).

THE ROLE OF THE STRUCTURAL OPERATOR AND THE QUOTIENT SPACE STRUCTURE IN THE THEORY OF HEREDITARY DIFFERENTIAL EQUATIONS

M. C. Delfour[1,2]

C.R.M., Université de Montréal
Montréal, Québec, Canada

I. INTRODUCTION

The object of this paper is to illustrate the role of a certain *structural operator* (later denoted by F) in the theory and control of linear *hereditary differential systems*. The paper only contains a description of a small part of the available results. Complete results and detailed proofs can be found in Delfour-Manitius [2],[3],[1], A. Manitius [1],[2], Bernier-Manitius [1], R. B. Vinter [1],[2] and Delfour-Lee-Manitius [1].

In Section II, we review the theory of autonomous linear systems in the product space setting. R. B. Vinter [1],[2] has shown that systems of the form

$$\dot{x}(t) = L(x_t) \tag{1.1}$$

[1]This work has been supported in part by National Research Council of Canada Grant A-8730 and a FCAC Grant from Québec Ministry of Education.

[2]Present address: Centre de Recherches Mathématiques, Université de Montréal, C.P. 6128, Montréal, Québec, Canada H3C 3J7

RECENT ADVANCES IN DIFFERENTIAL EQUATIONS

ISBN 0-12-186280-1

(cfr. equations (2.5), (2.12) and (2.8)) generate strongly continuous semigroups of operators on the product space of class C_0 when $L : C[-h,0] \to \mathbb{R}^n$ is a continuous linear map; this already indicated that the three extraconditions imposed by Borisovic-Trubabin [1] were always verified for such L's. An alternate proof of this fact is given here. Moreover the results of R. B. Vinter [1],[2] are extended to systems characterized by continuous linear maps $L : W^{1,p}[-h,0] \to \mathbb{R}^n$, $1 \leq p < \infty$. This provides the answer to the following question raised by R. B. Vinter and J. Zabczyk: what is the largest family of maps L for which system (1.1) induces a strongly continuous semigroup of operators on the product space of class C_0.

In Section III, we introduce the *structural operator* F and give the extension of the *intertwining theorem* of Bernier-Manitius [1] to systems characterized by maps L which are continuous from $C[-h,0]$ to $\mathbb{R}^n$. We indicate the connection of the operator F with the pairing and pseudo adjoint semigroup of J. K. Hale and S. Shimanov; we also show how it enters in the definition of the spectral projection operator and affects the convergence of spectral projection series.

In Section IV, we show that the semigroup on the product space naturally induces a C_0 strongly continuous semigroup on the quotient space of the product space by the kernel of the structural operator F. This has some interesting connections with the concepts of F-completeness, F-controllability and F-observability as introduced by A. Manitius [1] (see also Delfour-Manitius [1]). There are other applications of the operator F to control theory. For instance savings in computation can be achieved by F-*reduction* of the Riccati equation (cf. Delfour-Lee-Manitius [1]).

Notation and Terminology

Let $\mathbb{R}$ and $\mathbb{C}$ denote the fields of real and complex numbers, respectively. For each integer $n \geq 1$, let $\mathbb{R}^n$ (resp. $\mathbb{C}^n$) denote the n-fold cartesian product of $\mathbb{R}$ (resp. $\mathbb{C}$). Given a real number $h > 0$, $C[-h,0]$ denotes the Banach space of all continuous maps $\phi : [-h,0] \to \mathbb{R}^n$ endowed with the sup norm, $L^p[-h,0]$ the Banach space of all Lebesgue measurable maps which are p-integrable $(1 \leq p < \infty)$ or essentially bounded $(p = \infty)$, $W^{1,p}[-h,0]$ the Sobolev space of $L^p[-h,0]$-maps with a distributional derivative in $L^p[-h,0]$ and $L^1_{loc}[0,\infty[$ the Fréchet space of all maps $x : [0,\infty[\to \mathbb{R}^n$ whose restriction to $[0,T]$ belongs to $L^1[0,T]$ for all $T > 0$. Given a continuous linear map $U : E \to X$ from a Banach space E into another Banach space X, its dual operator will be denoted $U : X' \to E'$ (X' and E' are the respective topological duals of X and E). The transposed of an $n \times n$ real matrix M will be written M^T. The orthogonal complement of a subspace S of a Banach space will be denoted by $S^{\perp}$.

II. REVIEW OF AUTONOMOUS LINEAR SYSTEMS

We first consider hereditary or dealy differential equations of the form

$$\begin{aligned} \dot{x}(t) &= \int_{-h}^{0} d\eta(\theta)x(t+\theta) + f(t) \quad \text{in } [0,\infty[, \\ x(\theta) &= \phi(\theta) \quad \text{in } [-h,0], \end{aligned} \tag{2.1}$$

where $x(t) \in \mathbb{R}^n$ ($n \geq 1$, an integer), η is an $n \times n$ matrix of functions of bounded variation, ϕ is a continuous function $[-h,0] \to \mathbb{R}^n$ (denoted $\phi \in C[-h,0]$) and $f : [0,\infty[\to \mathbb{R}^n$ is a function which is at least locally integrable (that is, $f \in L^1_{loc}[0,\infty[$). As an example we can choose

$$\eta(\theta) = A_0 \chi_{[0]}(\theta) + \sum_{i=1}^{N} A_i \chi_{]\theta_i,0]}(\theta) + \int_{-h}^{0} A_{01}(\alpha)d\alpha, \tag{2.2}$$

where A_i, $i = 0,\dots,N$ are $n \times n$ matrices, $A_{01} : [-h,0] \to \mathbb{R}^{n^2}$ is a $n \times n$ matrix of functions in $L^1[-h,0]$, $N \geq 0$ is an integer and θ_i are delays chosen as follows

$$-h = \theta_N < \theta_{N-1} < \dots < \theta_{i+1} < \dots < \theta_0 = 0. \qquad (2.3$$

For η given by expression (2.2), system (2.1) is of the form

$$\dot{x}(t) = \sum_{i=0}^{N} A_i x(t+\theta_i) + \int_{-h}^{0} A_{01}(\theta)x(t+\theta)d\theta + f(t). \tag{2.4}$$

A complete theory of systems of the form (2.1) is available (cf. Bellman-Cooke [1], J. K. Hale [1],[2],[3] and their bibliographies).

II.1. The notion of state.

For researchers motivated by control and filtering problems, an important notion is the one of *state* of the control system. It seems that N. N. Krasovskii [1],[2] was the first to define the state of system (2.1) at time t as the piece of trajectory x between time $t-h$ and time t or, equivalently, as the element x_t of $C[-h,0]$,

$$x_t(\theta) = x(t+\theta), \quad -h \le \theta \le 0. \tag{2.5}$$

It is well known that the evolution of x_t can be described by a semigroup of bounded linear operators $\mathcal{S} = \{S(t) : t \ge 0\}$ of class C_0 defined on $C[-h,0]$. Define the map $L : C[-h,0] \to \mathbb{R}^n$ as

$$L\phi = \int_{-h}^{0} d\eta(\theta)\phi(\theta)d\theta. \tag{2.6}$$

The infintesimal generator A of $\mathcal{S}$ and its domain $\mathcal{D}(A)$ are given by

$$(A\phi)(\theta) = \left\{\begin{array}{ll} L\phi & , \theta = 0 \\ \frac{d\phi}{d\theta}(\theta), & -h\le\theta<0 \end{array}\right\}, \quad \mathcal{D}(A) = \{\phi \in C^1[-h,0] : L\phi = \dot{\phi}(0)\}, \tag{2.7}$$

where $C^1[-h,0] = \{\phi \in C[-h,0] : \dot{\phi} \in C[-h,0]\}$. For control problems this formulation has some limitations. Technically speaking, for optimization problems, it is preferable to work with a Hilbert space rather than with the non-reflexive Banach space $C[-h,0]$. It is also extremely convenient to use the formula

$$x_t = S(t)\phi + \int_0^t S(t-s)\tilde{f}(s)ds, \tag{2.8}$$

where $\tilde{f}(s) : [-h,0] \to \mathbb{R}^n$ is defined as

$$\tilde{f}(s)(\theta) = \begin{Bmatrix} f(s), & \theta = 0 \\ 0, & -h \le \theta < 0 \end{Bmatrix}. \tag{2.9}$$

Unfortunately formula (2.8) does not hold since $\tilde{f}(s)$ does not belong to $C[-h,0]$. We cannot set $f = 0$ since, for control problems, f is precisely the *control variable*. So this motivated a reformulation of the non homogeneous Cauchy problem in a space big enough to accomodate $\tilde{f}(s)$.

II.2. The product space approach.

The right space turned out to be the product space $\mathbb{R}^n \times L^p[-h,0]$. It was introduced almost simultaneously by several authors in different contexts. To the author's knowledge they can be found in the work of Coleman-Mizel [1], [2] on the theory of materials with fading memory, in the work of M. Artola [1], [2], [3], [4], [5] on parabolic partial differential equations with delays and in the work of Borisovic-Turbabin |1| who wanted to make sense of formula (2.8) for the linear non homogeneous system

$$\dot{x}(t) = L(x_t) + f(t). \tag{2.10}$$

At the same time, but motivated by optimal control and filtering problems for non autonomous systems, Delfour-Mitter [1] (cf. also M. C. Delfour [1], [2], Delfour-McCalla-Mitter [1], R. B. Vinter [4], Bensoussan-Delfour-Mitter [1] and their bibliography) were also led to the introduction and the "exploitation" of the product space. Perhaps not as intuitive as one would have liked it to be, the product space has, however, become an extremely useful tool in the study of stability, control and filtering of hereditary systems.

The basic idea is very simple. Since the continuity is only used to make sense of the initial point $x(0) = \phi(0)$, we shall disconnect the initial value $x(0) = \phi^0$ in $\mathbb{R}^n$ from the initial function $\phi^1 : [-h,0] \to \mathbb{R}^n$ which is required in order to make sense of the right hand side of the differential equation (2.1). In this way it becomes possible to choose ϕ^1 in a space as big as $L^1[-h,0]$. An initial condition is then an element ϕ of the product space:

$$\phi = (\phi^0, \phi^1) \in \mathbb{R}^n \times L^p[-h,0], \quad 1 \leq p < \infty. \tag{2.11}$$

For some time, it was thought that the product space setting was limited to η's of the form (2.2) or to continuous linear maps $L : C[-h,0] \to \mathbb{R}^n$ for which the three conditions of Borisovic-Turbabin [1] are verified. Recently R. B. Vinter [1],[2] has shown that the result is true for all continuous linear maps $L : C[-h,0] \to \mathbb{R}^n$. Therefore the three conditions of Borisovic-Turbarin [1] are always satisfied and can be dropped. In Section III we shall sketch an alternate proof to the one of R. B. Vinter [1],[2] (1). We shall also see, in Section II.3 that everything can be extended to systems of the form

$$\dot{x}(t) = L(x_t), \quad \phi \in \mathbb{R}^n \times L^p[-h,0], \quad 1 \leq p < \infty, \tag{2.12}$$

for arbitrary continuous linear maps

$$L : W^{1,p}[-h,0] \to \mathbb{R}^n, \tag{2.13}$$

provided that the solution is interpreted in an appropriate integral sense.

Analogously to Section II.1 the *state* of system (2.1) in $\mathbb{R}^n \times L^p[-h,0]$ is defined as the pair $\tilde{x}(t) = (x(t), x_t)$

$$\tilde{x}(t) = (\tilde{x}(t)^0, \tilde{x}(t)^1), \; \tilde{x}(t)^0 = x(t), \; \tilde{x}(t)^1 = x_t. \tag{2.14}$$

It is readily seen that $\tilde{x}(t)$ induces a strongly continuous semigroup $\{S(t) : t \geq 0\}$ of class C^0 on the product space $M^p = \mathbb{R}^n \times L^p[-h,0]$, $1 \leq p < \infty$, and that formula (2.8) now

---

(1) In fact, the author is pleased to acknowledge discussions with R. B. Vinter about this alternate proof.

makes sense for all f in $L^1_{loc}[0,\infty[$. Here the infinitesimal generator A of the semigroup takes the simpler form

$$(A\phi)^0 = L\phi, \qquad (A\phi^1)(\theta) = \frac{d\phi}{d\theta}(\theta), \tag{2.15}$$

where the domain $\mathcal{D}(A)$ of A,

$$\mathcal{D}(A) = \{(\phi(0),\phi) : \phi \in W^{1,p}[-h,0]\}, \tag{2.16}$$

is independent of the map L.

II.3. The largest class of hereditary systems defining a C_0 semigroup on the product space.

For some time, R. B. Vinter and J. Zabczyk have been interested in characterizing the largest class of maps L for which system (2.12) induces a C_0 strongly continuous semigroup on the product space. Their obvious candidate was the family $\mathcal{L}$ of all continuous linear maps $L : W^{1,p}[-h,0] \to \mathbb{R}^n$ since the very special form (2.15) of the infinitesimal generator indicates that the "largest class" of L's cannot be larger than $\mathcal{L}$. In fact it turns out to be exactly $\mathcal{L}$. To see this associate with L two $n \times n$ matrices $A_1(\theta)$ and $A_2(\theta)$ of functions in $L^q[-h,0]$, $p^{-1}+q^{-1} = 1$,

$$L\phi = \int_{-h}^{0} [A_1(\theta)\phi(\theta) + A_2(\theta)\dot{\phi}(\theta)]d\theta \tag{2.17}$$

(this representation of L always exists but is not unique). For each initial condition ϕ in $W^{1,p}[-h,0]$, the differential equation is of the form

$$\begin{aligned} \dot{x}(t) &= \int_{-h}^{0} [A_1(\theta)x(t+\theta) + A_2(\theta)\dot{x}(t+\theta)]d\theta + f(t), \quad t \geq 0 \\ x(\theta) &= \phi(\theta), \quad -h \leq \theta \leq 0. \end{aligned} \tag{2.18}$$

By integrating both sides of (2.18) and changing the order of integration in the term in $\dot{x}(t+\theta)$ we obtain the following integral equation

$$x(t) = \phi(0) + \int_0^t ds \int_{-h}^0 d\theta \; A_1(\theta)x(s+\theta)$$
$$+ \int_{-h}^0 d\theta \; A_2(\theta) \int_0^t ds\dot{x}(s+\theta) + \int_0^t f(s)ds \tag{2.19}$$

which can be further transformed to

$$\begin{cases} x(t) = \phi(0) + \int_0^t ds \int_{-h}^0 d\theta \; A_1(\theta)x(s+\theta) \\ \qquad + \int_{-h}^0 d\theta \; A_2(\theta)\left[x(t+\theta)-x(\theta)\right] + \int_0^t f(s)ds \\ x(\theta) = \phi(\theta) \quad \text{in } [-h,0]. \end{cases} \tag{2.20}$$

This last integral equation can be further generalized to

$$x(t) = \phi^0 + \int_{-h}^0 d\theta \left[A_1(\theta)\int_0^t ds \begin{cases} s(s+\theta), & s+\theta \geq 0 \\ \phi^1(s+\theta), & \text{oth.} \end{cases}\right.$$
$$\left. + A_2(\theta) \begin{cases} x(t+\theta)-\phi^1(\theta), & t+\theta \geq 0 \\ \phi^1(t+\theta)-\phi^1(\theta), & \text{oth.} \end{cases}\right] + \int_0^t f(s)ds \tag{2.21}$$

This equation has a unique continuous solution which is continuous with respect to the data $\phi = (\phi^0,\phi^1)$ in M^p. Hence by introducing the state $\tilde{x}(t) = (x(t),x_t)$, we obtain a strongly continuous semigroup of class C_0 on M^p. When ϕ belongs to $W^{1,p}[-h,0]$ we can show that x belongs to $W^{1,1}_{loc}[0,\infty[$ and that equation (2.18) is satisfied.

This last result considerably extends the class of systems (2.1) but not enough to include functional differential equations of the neutral type. This latter class generates a strongly continuous semigroup of class C_0 on $C[-h,0]$, but again formula (2.8) does not make sense since $\tilde{f}(s) \notin C[-h,0]$. In fact, it is doubtful that a formula (2.8) will ever hold, even if a "big enough space" can be constructed. Non homogeneous functional differential equations of the neutral type seem to be analogous to parabolic partial differential equations with non homogeneous Dirichlet conditions. To make sense of L^2-boundary conditions, it is necessary to use a "generalized Green formula" and a weak formulation (cf. Lions-Magenes [1]).

III. THE STRUCTURAL OPERATOR

In the remaining part of this paper we shall limit ourselves to systems of the form

$$\begin{aligned} \dot{x}(t) &= \int_{[-h,0]} d\eta(\theta)x(t+\theta), \quad t \geq 0 \\ \tilde{x}(0) &= \phi = (\phi^0,\phi^1) \in \mathbb{R}^n \times L^p[-h,0], \quad 1 \leq p < \infty. \end{aligned} \tag{3.1}$$

For simplicity, we shall denote by M^p the product space $\mathbb{R}^n \times L^p[-h,0]$ and identify an element ϕ of M^p with the pair (ϕ^0,ϕ^1). Proofs of the main results can be found in Delfour-Manitius [2],[3].

III.1. Definition of the structural operator.

Pick ϕ in $C[-h,0]$ and rewrite (3.1) separating terms in $x : [0,T] \to \mathbb{R}^n$ from those in $\phi : [-h,0] \to \mathbb{R}^n$:

$$\begin{cases} \dot{x}(t) = \begin{cases} \displaystyle\int_{[-t,0]} d\eta(\theta)x(t+\theta) + \int_{[-h,-t[} d\eta(\theta)\phi(t+\theta), & 0\leq t\leq h \\ \displaystyle\int_{[-h,0]} d\eta(\theta)x(t+\theta), & t > h \end{cases} \\ x(0) = \phi(0). \end{cases} \tag{3.2}$$

Since the term containing ϕ is zero for $t > h$, we introduce the function $H\phi : [-h,0] \to \mathbb{R}^n$ defined as

$$(H\phi)(\alpha) = \int_{[-h,\alpha[} d\eta(\theta)\phi(\theta-\alpha). \tag{3.3}$$

For ϕ in $C[-h,0]$, the function $H\phi$ belongs to $BV[-h,0]$ and the following inequality holds

$$|(H\phi)(\alpha)| \leq \int_{[-h,\alpha[} |d\eta(\theta)||\phi(\theta-\alpha)| \leq V(\eta,[-h,0[)\cdot\|\phi\|_C. \tag{3.4}$$

Hence $H\phi$ can be viewed as an element of $L^p[-h,0]$ for all p, $1 \leq p < \infty$.

Theorem 3.1. Let η be a $n \times n$ matrix of functions in $BV[-h,0]$ and p, $1 \leq p < \infty$. Then
(i) H can be extended to a continuous linear map (still denoted H) $L^p[-h,0] \to L^p[-h,0]$.

Remark 3.1. Notice that as $L^p[-h,0]$ functions, $H\phi$ is indistinguishable from the map

$$\alpha \to \int_{[-h,\alpha]} d\eta(\theta)\phi(\theta-\alpha) : [-h,0] \to \mathbb{R}^n . \tag{3.5}$$

Remark 3.2. If we separate the terms in x from those in ϕ^1 in (3.1) we see that the differential equation still makes sense for ϕ^1 in $L^p[-h,0]$

$$\begin{cases} \dot{x}(t) = (\mathcal{A}x)(t) + \begin{cases} (H\phi^1)(-t), & 0 \leq t \leq h \\ 0 & , \quad t > h \end{cases} \\ x(0) = \phi^0 . \end{cases} \tag{3.6}$$

Remark 3.3. For all $T \geq h$

$$x = 0 \quad \text{in} \quad [0,T] \Leftrightarrow \phi^0 = 0 \quad \text{and} \quad H\phi^1 = 0. \tag{3.7}$$

Definition 3.2. The structural operator is the map $F : M^p \to M^p$ constructed from the operator H as follows

$$F\phi = F(\phi^0,\phi^1) = (\phi^0, H\phi^1). \tag{3.8}$$

Remark 3.4. For all $T \geq h$

$$x = 0 \quad \text{in} \quad [0,T] \Leftrightarrow \phi \in \text{Ker } F; \tag{3.9}$$

hence the map $[\phi] \to x : M^2/\text{Ker } F \to H^1[0,T]$ is a continuous linear *injection*.

III.2. Hereditary pairing.

It is convenient (but not necessary) to assure that $p = 2$. We associate with F the bilinear form

$$<\psi,\phi> = ((\psi,F\phi)), \tag{3.10}$$

where $(\ ,\ )$ is the inner product in M^2,

$$((\psi,\xi)) = \psi^0 \cdot \xi^0 + (\psi^1,\xi^1)_2, \tag{3.11}$$

and $\psi^0 \cdot \xi^0$ and $(\psi^1,\xi^1)_2$ are the respective inner products in $\mathbb{R}^n$ and $L^2[-h,0]$.
Introduce the subspace

$$C = \{(\phi(0),\phi) : \phi \in C[-h,0]\} \tag{3.12}$$

of M^2. For all ϕ and ψ in C

$$<\psi,\phi> = \psi(0)\cdot\phi(0) + \int_{-h}^{0} d\alpha\psi(\alpha) \cdot \int_{[-h,\alpha[} d\eta(\theta)\phi(\theta-\alpha). \tag{3.13}$$

By using Fubini's theorem the last identity can be rewritten as

$$<\psi,\phi> = \psi(0)\cdot\phi(0) + \int_{-h}^{0} d\xi \int_{\theta}^{0} \psi(\theta-\xi)\cdot d\eta(\theta)\phi(\xi). \tag{3.14}$$

But this is precisely the well known pairing introduced by J. K. Hale and S. Shimanov except for the fact that ψ is in $C[-h,0]$ instead of $C[0,h]$.

The dual operators H^* and F^* of H and F will also naturally arise in the study of "adjoint systems". It is fundamental to notice that H^* and F^* have the same structure as H and F. The dual operator F^* of F is of the form

$$F^*\psi = (\psi^0,H^*\psi^1) \tag{3.15}$$

and by direct computation, it can be seen that

$$(H^*\psi)(\alpha) = \int_{[-h,\alpha[} d\eta^\tau(\theta)\psi(\theta-\alpha). \tag{3.16}$$

Therefore H^* is identical to H up to a transposition of the matrix $\eta(\theta)$. It is naturally associated with the analogue of equation (3.1) which can be formally written as

$$\begin{aligned} \dot{p}(t) &= \int_{[-h,0]} d\eta^\tau(\theta)p(t+\theta), \quad t \geq 0 \\ p(0) &= \psi^0, \; p(\theta) = \psi^1(\theta) \text{ in } [-h,0[, \; \psi = (\psi^0,\psi^1) \in M^2. \end{aligned} \tag{3.17}$$

Remark 3.5. For all $T \geq h$

$$p = 0 \quad \text{in} \quad [0,T] \Leftrightarrow \psi \in \text{Ker } F^*; \tag{3.18}$$

the map $[\psi] \to p : M^2/\text{Ker } F^* \to H^1[0,T]$ is a continuous linear injection.

Remark 3.6. Results are available to characterize the null space and range of F. Situations where F or F^* are injective, surjective or have a dense image in M^2 are of special interest. See Delfour-Manitius [2],[3] for further details and additional results.

III.3. Adjoint semigroups and intertwining properties of the structural operator.

Identify the elements of the topological dual of M^2 with those of M^2. The adjoint semigroup $\mathcal{S}^* = \{S(t)^* : t \geq 0\}$ is also a C_0 semigroup of operators on M^2. Its infinitesimal generator A^* has been characterized by R. B. Vinter [1] and Burns-Herdman [1],[2]

$$\begin{aligned} &\mathcal{D}(A^*) = \{(\psi^0,\psi^1) \mid \psi^1 = H^*\hat{\psi}^0 + g, \text{ for some } g \in H^1[-h,0], \; g(-h) = 0\} \\ &[A^*\psi]^0 = L^\tau\hat{\psi}^0 + g(0), \quad [A^*\psi]^1 = -\dot{g} \end{aligned} \tag{3.19}$$

where $\hat{\psi}^0$ denotes the constant function in $[-h,0]$ which is equal to ψ^0

$$\hat{\psi}^0(\theta) = \psi^0, \quad -h \le \theta \le 0, \text{ and } \quad L^\tau \psi = \int_{-h}^{0} d\eta^\tau(\theta)\psi(\theta). \tag{3.20}$$

Analogously to the construction of the semigroup $\{S(t)\}$ from system (3.1) we can construct from the solution of system (3.19) a semigroup $\mathcal{S}^\tau = \{S^\tau(t) : t \ge 0\}$ of operators on M^2 of class C_0. We shall refer to $\{S^\tau(t)\}$ as the "*transposed semigroup*".Note that $\mathcal{D}(A^\tau) = \mathcal{D}(A)$. The following theorem provides the essential connection between the adjoint and the transposed semigroups and between their respective infinitesimal generators. It is an extension of the results of Bernier-Manitius [1, Thm. 5.4] to the case of an arbitrary $n \times n$ matrix η of bounded variation.

Theorem 3.3. (i) $F^*\mathcal{D}(A^\tau) \subset \mathcal{D}(A^*)$. (ii) $A^*F^* = F^*A^\tau$ on $\mathcal{D}(A^\tau)$. (iii) For all $t \ge 0$, $S(t)^*F^* = F^*S^\tau(t)$.

The following Corollary to Theorem 3.2 gives the classical "duality relationship" (cf. J. K. Hale [3, §7.3]) between the semigroup $\{S(t)\}$ and the transposed semigroup $\{S^\tau(t)\}$ through the hereditary pairing (3.10).

Corollary 3.4. (i) $\forall\ \phi,\psi \in \mathcal{D}(A) = \mathcal{D}(A^\tau)$, $\langle\phi,A\psi\rangle = \langle A^\tau\phi,\psi\rangle$. (ii) $\forall\ \phi,\psi \in M^2$, $\langle\phi,S(t)\psi\rangle = \langle S^\tau(t)\phi,\psi\rangle$.

Remark 3.7. Since F and F^* differ from each other by a transposition of $\eta(\cdot)$, we also have

$$S^\tau(t)^*F = FS(t), \quad \forall\ t \ge 0, \tag{3.21}$$

where $S^\tau(t)^*$ is the adjoint of $S^\tau(t)$.

Remark 3.8. From Remark 3.4 it follows that

$$S(h)\phi = 0 \Leftrightarrow F\phi = 0. \tag{3.22}$$

In general, $S(t)\phi = 0$ for some $t > h$ does not necessarily imply that ϕ belongs to Ker F (cf. D. Henry [1]).

III.4. Spectral projections.

The product space framework provides an alternate approach to finite dimensional spectral projections since M^2 is now a

Hilbert space. The results are obviously the same as the ones already existing in the non-reflexive Banach space $C[-h,0]$.

In the $C[-h,0]$ framework, it is difficult to work directly with the dual semigroup of the semigroup $\{S(t)\}$ which is now defined on the dual of the non-reflexive Banach space $C[-h,0]$. This was certainly part of the motivation for the introduction of the hereditary pairing (3.10) and the construction of the transposed semigroup $\{S^{\tau}(t)\}$.

In the M^2 framework, standard spectral theory in a Hilbert space (or even in a reflexive Banach space for M^p, $1 \leq p < \infty$) (cf. A. E. Taylor [1]) can be used involving the semigroup $\{S(t)\}$ and the adjoint semigroup $\{S^*(t)\}$. In a second step the intertwining property of the semigroup $S(t)^*$ and $S^{\tau}(t)$ with respect to F^* are used to recover the classical characterization of eigensubspaces and eigenprojections. Similar results are obtained for the spectral theory of the quotient semigroup $\{[S(t)]\}$. All of this more and more suggests that the right space to study the convergence of spectral projection series is the quotient space $M^2/\text{Ker } F$ and not $C[-h,0]$ or M^2 (see equation (3.25) below).

Let $R(\lambda,A)$ denote the resolvent of A, $\sigma(A)$ its spectrum and $\rho(A)$ the complement of $\sigma(A)$. Following A. E. Taylor [1, p. 306], for each pole λ of $R(\lambda,A)$ of order m, define the (canonical) projection operator P_λ and spaces ${}_{\lambda}$ and $\mathcal{Q}_\lambda$

$$P_\lambda \phi = \frac{1}{2\pi i} \oint_{\Gamma_\lambda} R(\mu,A)\phi \, d\mu, \tag{3.23}$$

$$\mathcal{M}_\lambda = \text{Ker}(I\lambda - A)^m, \quad \mathcal{Q}_\lambda = \text{Im}(I\lambda - A)^m ;$$

it is also known that

$$M^2 = \mathcal{M}_\lambda \oplus \mathcal{Q}_\lambda \quad \text{and} \quad \mathcal{Q}_\lambda = (\text{Ker}(I\bar{\lambda} - A^*)^m)^\perp . \tag{3.24}$$

<u>Theorem 3.5</u>. (i) For all λ in $\sigma(A)$

$$P_\lambda \phi = \mathcal{E}_\lambda F \phi, \quad \mathcal{E}_\lambda \phi = \frac{1}{2\pi i} \oint_{\Gamma_\lambda} E_\lambda \Delta^{-1}(\lambda) E_\lambda^* \phi \, d\lambda, \tag{3.25}$$

where $E_\lambda : C^n \to M^2$ is defined by

$$(E_\lambda x)^0 = x, \quad (E_\lambda x)^1(\theta) = e^{\lambda\theta}x, \tag{3.26}$$

E_λ^* is the dual of E_λ,

$$E_\lambda^* \phi = \phi^0 + \int_{-h}^{0} e^{\bar{\lambda}\theta}\phi^1(\theta)d\theta, \quad (\bar{\lambda} \text{ complex conjugate of } \lambda) \tag{3.27}$$

and $\Delta(\lambda)$ is the $n \times n$ complex matrix

$$\Delta(\lambda) = \lambda I - LE_\lambda, \quad (I \quad n\times n \text{ identity matrix}). \tag{3.28}$$

(ii) For all $\ell = 1,\ldots,m$

$$\text{Ker}(I\lambda - A^*)^\ell = F^*\text{Ker}(I\lambda - A^\tau)^\ell; \tag{3.29}$$

(iii) $\text{Ker } P_\lambda = \{\phi \mid \langle\psi,\phi\rangle = 0, \quad \phi \in \text{Ker}(I\bar{\lambda} - A^\tau)^m\}$;

(iv) if $\{\phi_1,\ldots,\phi_d\}$ in $\mathcal{M}_\lambda$ and $\{\psi_1,\ldots,\psi_d\}$ in $\text{Ker}(I\bar{\lambda} - A^\tau)^m$ are bases for which

$$\langle\psi_i,\phi_j\rangle = \langle(F^*\psi_i,\phi_j)\rangle = \delta_{ij} \tag{3.30}$$

then the operator

$$\hat{P}_\lambda\phi = \sum_{i=1}^{d} \langle\psi_i,\phi\rangle \; \phi_i \tag{3.31}$$

coincides with P_λ.

IV. QUOTIENT SPACE AND QUOTIENT SEMIGROUP

We have seen (cf. Remark 3.4) that two initial functions are distinguishable by the solution x only modulo $\text{Ker } F$. This naturally suggests to go to the quotient space $M^2/\text{Ker } F$. The canonical surjection is written

$$\phi \to [\phi] : M^2 \to M^2/\text{Ker } F. \tag{4.1}$$

IV.1. Definitions and basic results.

Proposition 4.1. (i) $\phi \in \text{Ker } F \Rightarrow \forall t \geq 0,\ S(t)\phi \in \text{Ker } F$.
(ii) If we define the mapping $[S(t)] : M^2/\text{Ker } F \to M^2/\text{Ker } F$ as

$$[S(t)][\phi] = [S(t)\phi], \tag{4.2}$$

then $[S(t)]$ is a continuous linear map.

Theorem 4.3. The family $[\mathcal{S}] = \{[S(t)] : t \geq 0\}$ is a C_0-semigroup of operators on the quotient space $M^2/\text{Ker } F$. Its domain coincides with $[\mathcal{D}(A)]$ and its infinitesimal generator $[A]$ is equal to

$$[A]|\phi| = [A\phi], \quad \phi \in [\phi] \cap \mathcal{D}(A). \tag{4.3}$$

Furthermore the hereditary pairing $\langle\cdot,\cdot\rangle$ generates a new pairing $\langle\!\langle , \rangle\!\rangle$ on $M^2/\text{Ker } F \times M^2/\text{Ker } F$ given by

$$\langle\!\langle [\psi], [\phi] \rangle\!\rangle = \langle\psi,\phi\rangle, \psi \in [\psi]^*, \ \phi \in [\phi], \tag{4.4}$$

which separates points. We also obtain the analogue of Corollary 3.3:

Proposition 4.3. For all $[\psi]^*$ in $M^2/\text{Ker } F^*$ and $[\phi]$ in $M^2/\text{Ker } F$

$$\langle\!\langle [\psi]^*, [S(t)][\phi] \rangle\!\rangle = \langle\!\langle [S^{\sim}(t)][\psi]^*, [\phi] \rangle\!\rangle . \tag{4.5}$$

In introducing the semigroup $\{S(t)\}$ on M^2 we have introduced some redundancies in the system. For instance

$$\begin{bmatrix} \dot{x}_1(t) \\ \dot{x}_2(t) \end{bmatrix} = \begin{bmatrix} 1 & 0 \\ 0 & 0 \end{bmatrix} \begin{bmatrix} x_1(t) \\ x_2(t) \end{bmatrix} + \begin{bmatrix} 0 & 0 \\ 0 & 1 \end{bmatrix} \begin{bmatrix} x_1(t-1) \\ x_2(t-1) \end{bmatrix} \tag{4.6}$$

only need the value of $x_1(0)$ and $x_2(\theta)$ in $[-1,0]$. In going to M^2 it is also necessary to provide $x_1(\theta)$ in $[-1,0]$. However for this system

$$\text{Ker } F = (0 \times L^2[-h,0]) \times (0,0) \in (\mathbb{R} \times L^2[-h,0])^2 \tag{4.7}$$

and by going to the quotient

$$M^2/\text{Ker } F \simeq \mathbb{R} \times (\mathbb{R} \times L^2[-h,0]). \tag{4.8}$$

In other words the quotient space is completely equivalent to the original system (4.6) and does not introduce any unnecessary part in the state. It seems that $[\tilde{x}(t)] = [S(t)][\phi]$ is what is closest to describe the real (minimal) state of the system.

IV.2. Application to the problem of completeness of eigenfunctions.

Let $\mathcal{M}$ (resp. $[\mathcal{M}]$) denote the span of all eigensubspaces $\mathcal{M}_\lambda$ of A (resp. $[\mathcal{M}_\lambda]$ of $[A]$) in M^2 (resp. $M^2/\text{Ker } F$):

$$\mathcal{M} = \underset{\lambda \in \sigma(A)}{\text{span}} \mathcal{M}_\lambda, \quad [\mathcal{M}] = \underset{\lambda \in \sigma([A])}{\text{span}} [\mathcal{M}_\lambda] \tag{4.9}$$

Definition 4.4. (i) The system of generalized eigenfunctions of A (resp. $[A]$) is *complete* if

$$\overline{\mathcal{M}} = M^2 \quad (\text{resp. } \overline{[\mathcal{M}]} = M^2/\text{Ker } F). \tag{4.10}$$

(ii) (A. Manitius [1]) The system of generalized eigenfunctions of A if *F-complete* if

$$\overline{\underset{\lambda \in \sigma(A)}{\text{span}} F \mathcal{M}_\lambda} = \overline{\text{Im } F}. \tag{4.11}$$

Notice that when $H = 0$, that is when there are no delays in the system the completeness for $[A]$ in $M^2/\text{Ker } F$ and the concept of F-completeness for A in M^2 both reduce to the usual completeness in $\mathbb{R}^n$. In fact this is a special case of the more general result.

Theorem 4.5. (i) $\overline{[\mathcal{M}]} = M^2/\text{Ker } F \Rightarrow$ F-completeness.
(ii) When $\text{Im } F$ is closed, the two concepts coincide.

Proof. Denote by $j : M^2 \to M^2/\text{Ker } F$, the canonical surjection. There exists a unique continuous linear injection $[F]$ such that

$$[F] : M^2/\text{Ker } F \to \overline{\text{Im } F} \quad \text{and} \quad [F]j = F.$$

Moreover since the generalized eigensubspaces are finite dimensional

$$[\mathcal{M}] = j\mathcal{M}.$$

(i) Let $\overline{[\mathcal{M}]} = M^2/\text{Ker } F$. Then

$$\overline{j\mathcal{M}} = \overline{[\mathcal{M}]} = M^2/\text{Ker}\, F = [F]\overline{j\mathcal{M}} = [F]M^2/\text{Ker } F = \text{Im } F.$$

But

$$[F]\overline{j\mathcal{M}} \subset \overline{[F]j\mathcal{M}} = \overline{F\mathcal{M}}$$

and necessarily $\text{Im } F \subset \overline{F\mathcal{M}}$.

(ii) If we assume that $\text{Im } F = \overline{\text{Im } F}$, then $[F]$ is a continuous linear bijection and for all sets A $[F]\bar{A} = \overline{[F]A}$. Consider $[F]M^2/\text{Ker } F = \text{Im } F \subset \overline{F\mathcal{M}} = \overline{[F]j\mathcal{M}} = \overline{[F][\mathcal{M}]}$. But we know that $\overline{[F][\mathcal{M}]} = [F]\overline{[\mathcal{M}]}$ and

$$[F]M^2/\text{Ker } F \subset [F]\overline{[\mathcal{M}]} \Rightarrow M^2/\text{Ker } F \subset \overline{[\mathcal{M}]}.$$

This Theorem 4.5 shows that the concept of F-completeness introduced by A. Manitius [1] (see also Delfour-Manitius [1]) is in general weaker that the concept of completeness in the quotient space. However they coincide when $\text{Im } F$ is closed. This later condition is satisfied for difference-differential equations.

IV.3. Application to control theory.

Consider now the system

$$\dot{x}(t) = \int_{-h}^{0} d\eta(\theta)x(t+\theta) + Bu(t), \tag{4.12}$$

where B is an $n \times m$ matrix. Defining $\tilde{B} : U = \mathbb{R}^m \to R^2$ as

$$\tilde{B}u = (Bu,0), \tag{4.13}$$

we can rewrite (4.12) in the more compact form

$$\dot{\tilde{x}}(t) = A\tilde{x}(t) + \tilde{B}u(t), \quad t \geq 0, \quad \tilde{x}(0) = \phi, \tag{4.14}$$

or more precisely

$$\tilde{x}(t) = S(t)\phi + \int_0^t S(t-s)\tilde{B}u(s)ds. \tag{4.15}$$

Define the set K_t of reachable states from the origin at time t

$$K_t = \{\int_0^t S(t-s)\tilde{B}v(s)ds : v \in L^2(0,t;U)\}$$
$$K_\infty = \bigcup_{t>0} K_t. \tag{4.16}$$

Definition 4.6. (i) $(A,\tilde{B})$ is M^2-controllable if $\overline{K_\infty} = M^2$. (ii) (A. Manitius [1]) $(A,\tilde{B})$ is F-controllable if $\overline{FK_\infty} = \overline{\text{Im } F}$.

In turns out that for the study of stabilizability of hereditary systems, the concept of F-controllability is the right one; the concept of M^2-controllability is too strong. As for the problem of completeness, this concept is related to its counterpart for the quotient space.

Notice that since the L^2-component of $\tilde{B}$ is zero, the new operator $[\tilde{B}] : U \to M^2/\text{Ker } F$ is well defined as

$$[\tilde{B}]u = [\tilde{B}u]. \tag{4.17}$$

Introduce the sets $[K]_t$ of reachable states from the origin at time t

$$[K]_t = \{\int_0^t [S(t-s)][\tilde{B}]v(s)ds : v \in L^2(0,t;U)\}$$
$$[K]_\infty = \bigcup_{t>0} [K]_t. \tag{4.18}$$

Notice that for all t

$$[K]_t = jK_t, \quad [K]_\infty = jK_\infty. \tag{4.19}$$

Theorem 4.7. (i) $\overline{[K]_\infty} = M^2/\text{Ker } F \Rightarrow (A,\tilde{B})$ is F-controllable. (ii) When Im F is closed, the two concepts coincide.

Proof. (i) We use the same notation as in the proof of Theorem 4.5. In view of (4.19)

$$\overline{jK_\infty} = \overline{[K]_\infty} = M^2/\text{Ker } F \Rightarrow [F]\overline{jK_\infty} = [F]M^2/\text{Ker } F = \text{Im } F.$$

But

$$\overline{FK_\infty} = \overline{[F]jK_\infty} \supset [F]\overline{jK_\infty} = \text{Im } F.$$

(ii) If $(A,\tilde{B})$ is F-controllable, then

$$\text{Im } F \subset \overline{FK_\infty} \Rightarrow [F]\ M^2/\text{Ker } F \subset \overline{[F]jK_\infty}.$$

Since Im F is closed

$$[F]M^2/\text{Ker } F \subset \overline{[F]jK_\infty} = [F]\overline{jK_\infty} = [F]\overline{[K]_\infty}$$

and $\overline{[K]_\infty} = M^2/\text{Ker } F$.

Again the concept of F-controllability introduced by A. Manitius [1] is, in general, weaker than the corresponding quotient space controllability. However they coincide when Im F is closed.

REFERENCES

Artola, M. [1], Equations paraboliques à retardement, *C.R. Acad. Sc. Paris 264*, 668-671 (1967).

[2], Sur les perturbations des équations d'évolution, application á des problèmes de retard, *Ann. Scient. Ec. Norm. Sup., 4ième série, t. 2*, 137-253 (1969).

[3], Sur une équation d'évolution du premier ordre à argument retardé, *C. R. Acad. Sc. Paris, Série A, 268*, 1540-1543 (1969).

[4], Sur une équation d'évolution du premier ordre à argument retardé, to appear.

Bellman, R., and Cooke, K. L. [1], Differential-difference equations, Academic Press, New York, (1963).

Bensoussan, A., Delfour, M. C., and Mitter, S. K. [1], Optimal filtering for linear stochastic hereditary differential systems, Proceedings of the 1972 IEEE Conference on Decision and Control and 11th Symposium on Adaptive Processes, pp. 378-380, New York, (1972).

Bernier, C. [1], Etude des semi-groupes d'opérations associés aux équations linéaires retardées, Mémoire de maîtrise, Université de Montréal, (1976).

Bernier, C., and Manitius, A. [1], On semigroups in $\mathbb{R}^n \times L^p$ corresponding to differential equations with delays, *Canadian Journal of Mathematics*, (1978), to appear.

Borisovic, J. G., and Turbabin, A. S. [1], On the Cauchy problem for linear non-homogeneous differential equations with retarded argument, *Soviet Math. Doklady 10*, 401-405 (1969).

Burns, J. A., and Herdman, T. L. [1], Adjoint semigroup theory for a Volterra integrodifferential system, *Bulletin of the American Mathematical Society 81*, 1099-1102 (1975).

[2], Adjoint semigroup theory for a class of functional differential equations, *SIAM J. Math. Anal. 7*, 729-745 (1976).

Coleman, B. D., and Mizel, V. J. [1], Norms and semi-groups in the theory of fading memory, *Arch. Rational Mech. Anal. 23*, 87-123 (1966).

[2], Stability of functional differential equations, *Arch. Rational Mech. Anal. 30*, 173-196 (1968).

Delfour, M. C. [1], The linear quadratic optimal control problem for hereditary differential systems: theory and numerical solution, *J. Applied Mathematics and Optimization 3*, 101-162 (1977).

[2], State theory of linear hereditary differential systems, *J. Math. Anal. and Appl. 60*, 8-35 (1977).

Delfour, M. C., Lee, E. B., and Manitius, A. [1], F-reduction of the operator Riccati equation for hereditary differential-systems, *Automatica 14*, (1978).

Delfour, M. C., and Manitius, A. [1], Control Systems with delays: areas of applications and present status of the linear theory, *in* "New trends in systems analysis" (A.Bensoussan and J. L. Lions, eds.), pp. 420-437. Springer Verlag, New York, (1978).

[2], The structural operator F and its role in the theory of retarded systems I, *J. Math. Anal. Appl.* (1979), to appear.

[3], The structural operator F and its role in the theory of retarded systems II, *J. Math. Anal. Appl.* (1979), to appear.

Delfour, M. C., McCalla, C., and Mitter, S. K. [1], Stability and the infinite-time quadratic cost problem for linear hereditary differential systems, *SIAM J. Control 13*, 48-88 (1975).

Delfour, M. C., and Mitter, S. K. [1], Controllability, observability and optimal feedback control of affine hereditary differential systems, *SIAM J. Control 10*, 298-328 (1972).

Hale, J. K. [1], Linear functional-differential equations with constant coefficients, *Contributions to Differential Equations 2*, 291-319 (1963).

[2], Functional Differential Equations, Springer-Verlag, New York, (1971).

[3], Theory of Functional Differential Equations, Springer Verlag, New York, (1977).

Henry, D. [1], Small solutions of linear autonomous functional differential equations, *J. Differential Equations 9*, 55-56 (1970).

Jones, G. S. [1], Hereditary structure in differential equations, *Math. Systems Theory 1*, 263-278 (1967).

Krasovskii, N. N. [1], On the analytic construction of an optimal control in a system with time lags, *Prikl. Mat. Mech. 26*, 39-51 (1962). (English transl. *J. Appl. Math. Mech.*, 50-67 (1962)).

[2], Optimal processes in systems with time lag, Proc. Second IFAC Congress [in Russian], Vol. I, Izd-vo "Nauka", (1964).

Lions, J. L., and Magenes, E. [1], Problèmes aux limites non homogènes et applications. Vol. 1 et 2, Dunod, (1968);

Vol. 3, Dunod, (1969), Paris.

Manitius, A. [1], Controllability, observability and stabilizability of retarded systems, Proc. 1976 IEEE Conference on Decision and Control, pp. 752-758, IEEE Publications, New York, (1976).

[2], Completeness and F-completeness of eigenfunctions associated with retarded functional differential equations, Rapport CRM-755, Montréal, Nov. 1977.

Shimanov, S. N. [1], On the theory of linear differential equations with after-effect, *Differentsial'nye Uravneniya 1*, 102-116 (1965). (English transl. *Differential Equations 1*, 76-86 (1965)).

Taylor, A. E. [1], Functional Analysis, John Wiley and Sons, New York, (1958).

Vinter, R. B. [1], Semigroups on product spaces with applications to initial value problems with non-local boundary conditions, *in* "Control of distributed parameter systems" (S.P. Banks and A.J. Pritchard, eds.), Pergamon Press, Oxford, (1978).

[2], On a problem of Zabczyk concerning semigroups generated by operators with non-local boundary conditions, Publications 77/8 (1977), Department of Computing and Control, Imperial College of Science and Technology, London.

[3], On the evolution of the state of a linear differential delay equation in M^2: properties of the generator, *J. Inst. Math. Appl.* (1978), to appear.

[4], A representation of solutions to stochastic delay equations, *J. Appl. Math. Optimization*, in press.

Zabczyk, J. [1], On semigroups corresponding to non-local boundary conditions with applications to system theory, Report 49, Control Theory Centre, University of Warwich, Coventry, (1976).

[2], On decomposition of generators, *SIAM J. Control and Optimization 16*, 523-534 (1978).

DEGENERATE EVOLUTION EQUATIONS AND SINGULAR OPTIMAL CONTROL

Angelo Favini[1,2]

Istituto di Matematica Generale e Finanziaria
Università di Bologna
Bologna, Italy

In a recent paper [2], S. L. Campbell, C. D. Meyer Jr. and M. J. Rose applied the Razin inverse of a $n \times n$ matrix to solve the equation $B\dot{x} + Ax = f$, where A, B are $n \times n$ matrices that may be singular.

In the paper [3], S. L. Campbell uses these results to handle optimal control problems relative to $\dot{x} = Ax + Bu$, with singular matrices in the quadratic cost functional $J(x,u) = \frac{1}{2} \int_{t_0}^{t_1} |\langle Hx,x\rangle + \langle Qu,u\rangle| dt$. In fact, Q is a positive semi-definite $m \times m$ matrix, also not invertible. Campbell establishes that the system

$$\frac{d}{dt} \begin{bmatrix} 1 & 0 & 0 \\ 0 & 1 & 0 \\ 0 & 0 & 0 \end{bmatrix} \begin{bmatrix} \lambda(t) \\ x(t) \\ u(t) \end{bmatrix} = - \begin{bmatrix} A^* & H & 0 \\ 0 & -A & -B \\ B^* & 0 & Q \end{bmatrix} \begin{bmatrix} \lambda(t) \\ x(t) \\ u(t) \end{bmatrix} \tag{1}$$

provides both necessary and sufficient conditions for optimi-

---

[1] Work supported by the C.N.R. (G.N.A.F.A.)

[2] Present address: Istituto di Matematica Generale e Finanziaria, P.za Scaravilli 2, 40126 Bologna, Italy.

ISBN 0-12-186280-1

zation.

In this paper we want to give some results about the abstract equation in a Banach space: $\mathcal{B}\dot{x} = -\mathcal{A}x + f$, where $\mathcal{A}$, $\mathcal{B}$ are suitable linear closed operators and $\mathcal{B}$ may be not invertible.

1. Let X be a complex Banach space, $\mathcal{B}$ a linear closed operator in X, with domain $\mathcal{D}(\mathcal{B})$ everywhere dense in X. Suppose f a strongly continuous function from $\bar{R}_+ = [0, +\infty[$ into X. Then we say that $x = x(t)$, $t \in R^+$, is a strong solution of

$$\frac{d}{dt}\mathcal{B}x(t) = -x(t) + f(t), \quad t \in R^+, \quad x(0) = x_o \in \mathcal{D}(\mathcal{B}) \qquad (2)$$

if $x \in C^{(o)}(R^+;X)$, $x(t) \in \mathcal{D}(\mathcal{B})$ for all $t \in R^+$, $t \mapsto \mathcal{B}x(t) \in C^{(1)}(R^+;X)$, x satisfies the equation in (2) and the initial condition in the sense $||\mathcal{B}x(t) - \mathcal{B}x_o|| \xrightarrow[t \to o+]{} 0$.

If $x \in C^{(o)}(\bar{R}_+;X) \cap C^{(1)}(R^+;X)$, $\frac{dx(t)}{dt} = \dot{x}(t) \in \mathcal{D}(\mathcal{B})$ $\forall\, t \in R^+$, and $\dot{x}(t) = -x(t) + f(t)$, $t \in R^+$, $||x(t) - x_o|| \xrightarrow[t \to o+]{} 0$, then x is said to be a classical solution of (2).

Suppose X a Hilbert space. If $f \in L^2(O,T;X)$, $T > 0$, $x_o \in X$, then $x :]0,T[\to X$ is a weak solution of (2) when $t \mapsto \langle x(t), \mathcal{B}^* y\rangle$ is absolutely continuous for any $y \in \mathcal{D}(\mathcal{B}^*)$, $\frac{d}{dt}\langle x(t), \mathcal{B}^* y\rangle = -\langle x(t), y\rangle + \langle f(t), y\rangle$ a.e. in $]0,T[$, and $\lim_{t \to o+} \langle x(t), \mathcal{B}^* y\rangle = \langle x_o, \mathcal{B}^* y\rangle$; here $\langle , \rangle$ denotes the inner product in X; see [1], pp. 204-205.

Remark 1.1. If $\mathcal{B} \in L(X,X)$ and x is a classical solution of (2), then x is a strong solution of (2), too.

In this paragraph we assume that $\lambda + \mathcal{B}$ has a bounded inverse for all $\lambda \in C^+$ and $||(\lambda + \mathcal{B})^{-1}|| \leq M \cdot |\lambda|^{-1}$, $\lambda \in C^+$.

It follows that $(\lambda + \mathcal{B})^{-1}$ exists for any λ in a sector Σ / $\mathrm{Re}\,\lambda > \alpha\,|\mathrm{Im}\,\lambda|$, α a negative number, and $||(\lambda + \mathcal{B})^{-1}|| \leq M'$ $|\lambda|^{-1}$, $\lambda \in \Sigma$. Let Γ be a regular (infinite) curve in Σ, agreeing with the rays $\mathrm{Re}\,\lambda = 1 + \alpha\,|\mathrm{Im}\,\lambda|$, $|\lambda| > R$, avoiding the origin to the right, going from $\infty \cdot \exp(-i\vartheta)$ to $\infty \cdot \exp(i\vartheta)$, $\vartheta = \pi + \mathrm{arctg}\,\alpha^{-1}$. Define:

$$W_k(t) = (2\pi i)^{-1} \cdot \int_\Gamma e^{\lambda t}\lambda^{-(k+1)} (\lambda\mathcal{B}+1)^{-1}d\lambda \;, \; k \in \mathbb{N} \cup \{0\}, \; t\in R^+,$$

$$U(t) = 1 - W_o(t) = (2\pi i)^{-1} \int_\Gamma e^{\lambda t}\, \mathcal{B}(\lambda\mathcal{B}+1)^{-1}d\lambda \;, \quad t \in R^+ .$$

We then have:

Proposition 1.1. If (2) has a strong (respectively, classical, weak) solution, then this solution is the one.

Further, the following results about existence hold:

Proposition 1.2. Assume that $x_o \in \mathcal{D}(\mathcal{B})$ and $f \in C^{(2)}(\bar{R}_+;X)$. Then $U(t)x_o - W_o(t)f(0) + W_1(t)\, f'(0) + \int_o^t W_1(t-s)\, f^{(2)}(s)\, ds$ is the unique strong solution of (2).

We note that $W_o(t)x = (2\pi i)^{-1} \int_\Gamma e^\xi\, \xi^{-1} (\xi t^{-1}\mathcal{B}+1)^{-1}x\, d\xi$, $x \in X$, implies $\|W_o(t)\| \leq$ Const, $t \in R^+$. If $k \in \mathbb{N}$, then $\|W_k(t)\, x\| \xrightarrow[t \to o+]{} 0$; on the other hand, $U(t)x_o + W_o(t)\, f(0) = x_o + W_o(t)\left[f(0) - x_o\right]$ and $W_o(t)\mathcal{B}\, y = ty - W_1(t)y \xrightarrow[t \to o+]{} 0$. That implies:

Proposition 1.3. If $x_o \in \overline{\mathcal{R}(\mathcal{B})}$, $f \in C^{(3)}(\bar{R}_+;X)$ and $f(0) \in \overline{\mathcal{R}(\mathcal{B})}$, then $U(t)x_o + \sum_{k=0}^{2} W_k(t)\, f^{(k)}(0) + \int_o^t W_2(t-s)\, f^{(3)}(s)\, ds$, $t \in R^+$, is the classical solution of (2).

As regards weak solutions, we can demonstrate

Proposition 1.4. Let X be a Hilbert space and $f \in H^1(0,T;X)$. Then for all $x_o \in X$, problem (2) has the weak solution $U(t)x_o + W_o(t)f(0) + \int_o^t W_o(t-s)\, f'(s)\, ds$.

Remark 1.2. We have

$$\mathcal{B}\, W_k(t) = \frac{t^{k+1}}{(k+1)!} - W_{k+1}(t), \quad k = 0,1,\ldots,$$

$$\frac{d}{dt} W_{k+1}(t)x = W_k(t)x \;, \quad k = 0,1,\ldots, \quad x \in X \;,$$

$$\|W_k(t)\| \leq M_k t^k \;, \quad t \in R^+ \;, \quad k = 0,1,\ldots .$$

These facts imply that if $f = \mathcal{B}\, g$, it is possible to weaken the hypothesis on the regularity of f .

2. Consider the differential equation in C^2 :

$$\frac{d}{dt}\begin{bmatrix}0 & 1\\ 0 & 0\end{bmatrix}\begin{bmatrix}x(t)\\ y(t)\end{bmatrix} = -\begin{bmatrix}x(t)\\ y(t)\end{bmatrix} + \begin{bmatrix}f(t)\\ g(t)\end{bmatrix}, \qquad (3)$$

where f, g are given continuous functions from $\bar{R}_+$ into C . If $f = g = 0$ then (3) has the unique solution $(0,0)$. If $g \in C^{(1)}(\bar{R}_+; C)$, we have $x(t) = f(t) - \dot{g}(t)$, $y(t) = g(t)$; and thus one cannot assign arbitrary initial conditions or non-homogeneous parts. Note that in our example $\|(\lambda\mathcal{B} + 1)^{-1}\| \leq C(1+|\lambda|)$, $\lambda \in C^+$.

We then want to handle (2) in the general case in which $\|(\lambda\mathcal{B}+1)^{-1}\| \leq C|\lambda|^k$, $\mathrm{Re}\,\lambda \geq \alpha_o > \alpha$, $k \in \mathbb{N}$. As regards this problem, we can prove that if (2) has strong or classical solutions, then these are unique. Moreover, the following existence result holds:

<u>Proposition 2.1</u>. Suppose that $x_o = \mathcal{B}^{k+1} y_o$, $y_o \in \mathcal{D}(\mathcal{B}^{k+2})$, $f \in C^{(k+2)}(\bar{R}_+;X)$, $f^{(j)}(0) = \mathcal{B}^{k+1-j} z_j$, $j = 0,1,\ldots,k$, (respectively, $x_o = \mathcal{B}^{k+2} y_o$, $y_o \in \mathcal{D}(\mathcal{B}^{k+2})$, $f \in C^{(k+3)}(\bar{R}_+; X)$, $f^{(j)}(0) = \mathcal{B}^{k+2-j} z_j$, $j = 0,1,\ldots,k+1$).

Then (2) has a strong (respectively, classical) solution.

<u>Remark 2.1</u>. Concerning (3), we have $\mathcal{B}^2 = 0$, and thus $x_o = 0 = y_o$.

Let γ be $\{\lambda = 1+is , s \in R\}$. Our general result implies that if $f, g \in C^{(3)}(\bar{R}_+;C)$, $f(0) = f'(0) = g(0) = 0$, then

$$x(t) = (2\pi i)^{-1}\{-\int_\gamma \lambda^{-1} e^{\lambda t} g'(0)d\lambda + \int_\gamma \lambda^{-3} e^{\lambda t}|f''(0)-\lambda g''(0)|d\lambda + \int_0^t\int_\gamma \lambda^{-3} e^{\lambda(t-s)}|f^{(3)}(s) - \lambda g^{(3)}(s)|d\lambda ds\} = f(t) - g'(t) ,$$

$$y(t) = (2\pi i)^{-1}\{\int_\gamma \lambda^{-2} e^{t} g'(0)\, d\lambda + \int_\gamma \lambda^{-3} e^{\lambda t} g''(0)\, d\lambda + \int_0^t\int_\gamma \lambda^{-3} e^{\lambda(t-s)} g^{(3)}(s)\, d\lambda\, ds\} = g(t) .$$

<u>Remark 2.2</u>. If $-\mathcal{B}$ generates a contraction semi-group, then $\|(\lambda\mathcal{B} +1)^{-1}\| \leq |\lambda|(\mathrm{Re}\;)^{-1}$, $\lambda \in C^+$, and Proposition 2.1 applies with $k = 1$.

Remark 2.3. Let X, Y be complex Banach spaces, A, B linear closed operators from Y into X. Let f be a continuous function from $\bar{R}_+$ into X.

We say that $x : R^+ \mapsto Y$ is a strong solution of

$$B\dot{x}(t) = -Ax(t) + f(t) , \quad t \in R^+ , \quad x(0) = x_o , \tag{4}$$

if $x \in C^{(o)}(R^+;Y)$, $x(t) \in \mathcal{D}(A) \cap \mathcal{D}(B)$ for all $t \in R^+$, $Ax(t) \in C^{(o)}(R^+;X)$, $Bx(t) \in C^{(1)}(R^+;X)$ and $\frac{d}{dt} Bx(t) + Ax(t) = f(t)$, $t \in R^+$, $\|Bx(t) - Bx_o\|_X \xrightarrow[t \to o+]{} 0$, $x_o \in \mathcal{D}(B)$.

If we assume that $(\lambda B + A)^{-1}$ exists for sufficiently large positive λ and $\limsup_{\lambda \to \infty} \lambda^{-1} \ln\| A(\lambda B + A)^{-1}\| = 0$, then (4) has at most one strong solution; in fact, we can suppose, without loss of generality, that $\mathcal{D}(A) \subseteq \mathcal{D}(B)$ and that A is invertible. This implies that x is a strong solution of (4) iff $x = A^{-1}y$, and y is a strong solution of (2), with $\mathcal{B} = BA^{-1}$. Hence:

Proposition 2.2. Assume that $\|A(\lambda B + A)^{-1}\| \leq M$, $\lambda \in \bar{C}_+$, $x_o \in \mathcal{D}(A)$ and $f \in C^{(2)}(R_+;X)$; then (4) has a unique strong solution.

If $\|A(\lambda B + A)^{-1}\| \leq C(1 + |\lambda|)^k$, $\lambda \in \bar{C}_+$, $k \in \mathbb{N}$, then the conditions $x_o = (A^{-1}B)^{k+1}y_o$, $y_o \in \mathcal{D}(A)$, $f \in C^{(k+2)}(\bar{R}_+;X)$, $f^{(j)}(0) = B(A^{-1}B)^{k-j} z_j$, $z_j \in \mathcal{D}(A)$, $j = 0,1,\ldots,k$, ensure that a strong solution of (4) exists.

Application 1. Suppose that A, C are linear closed operators, B, D, E, F are bounded operators acting between suitable Banach spaces, such that $Q_\mu = F - E\,(\mu+A)^{-1}B(\mu-C)^{-1}D$ is well-defined and has a bounded inverse for all $\mu \in C^+$. Further,

$$\|(\mu+A)^{-1}\| \leq M(\mathrm{Re}\mu)^{-1} , \qquad \|(\mu-C)^{-1}\| \leq M(\mathrm{Re}\mu)^{-1} ,$$

$$\|Q_\mu^{-1}\| \leq C(1+|\mu|)^k , \quad k \in \mathbb{N} \cup \{0\}, \quad \mu \in C^+ .$$

F may be not invertible.

Under these assumptions we can apply our results to the

$$\dot{\lambda} = -A\lambda+Bx, \quad \dot{x} = Cx+Du, \quad 0 = E\lambda+F\ddot{u}, \quad x(0)=x_o, \ \lambda(0) = \lambda_o. \quad (5)$$

Note that the system considered in the introduction is a particular case of (5).

In control theory one is often concerned with a initial-final value problem, where $x(0)$ and $\lambda(T)$, $T > 0$, or $x(0)$, $x(T)$ are given. If we have the general solution of (5) (grosso modo, if $C = G$, $A = G$ and G generates a group of linear operators), then the question is to find λ_o such that the corresponding $\lambda(T)$ or $x(T)$ coincide with the given λ_T or x_T.

Application 2. Consider the strongly elliptic operators on the bounded open set Ω of R^n with a C^∞ boundary, defined by $A(x,D) = \sum_{|\alpha|\leq 2m} a_\alpha(x) D^\alpha$. $B(x,D) = \sum_{|\beta|\leq 2p} b_\beta(x) D^\beta$; assume that $A(x,D)$ is uniformly strongly elliptic and let A, B be the operators defined in L^q with domains $H^{2m,q}(\Omega,\{B_j\})$, $H^{2p,q}(\Omega,\{C_j\})$, respectively, (see [6], p. 75). We can show that if the conditions in [6], p. 101, hold with the $k = 0$, and $p < m$, then for all $u \in C^{2m}(\bar{\Omega})$,

$$\|Bu\|_{L^q} \leq \varepsilon \|Au\|_{L^q} + C\,\varepsilon^{-p(m-p)^{-1}} \|u\|_{L^q}, \quad q\geq 1, \ 0<\varepsilon<\varepsilon_o \quad (6)$$

(6) permits to estimate $\|A(\lambda B+A)^{-1}\|$ and thus to handle equations of the type $\frac{\partial}{\partial t} B(x,D)\, u(t,x) = - A(x,D)\, u(t,x) + f(t,x)$, $(t,x) \in R^+ \times \Omega$, (another approach is given in [4]).

Application 3. Suppose $\alpha(x) \geq 0$ a C^∞ function on $\bar{\Omega}$, Ω a bounded regular domain of R^n. For example, $\alpha(x) = \mathrm{dist}\,(x, \partial\Omega)$. With the notations of Application 2, suppose $\mathrm{Re}\,\langle Au,u\rangle \geq \lambda_o \|u\|^2_{L^2(\Omega)}$, $\lambda_o > 0$, $u \in \mathcal{D}(A) = H^{2m}(\Omega) \cap H^m_o(\Omega)$. Define $B : L^2(\Omega) \to L^2(\Omega)$ by $(Bu)(x) = \alpha(x)\, u(x)$. Then our results suit well to treat problem $\frac{\partial}{\partial t}(\alpha(x) u(t,x)) + A(x,D)\, u(t,x) = f(t,x)$, $t \in R^+$, $x \in \Omega$, $\|u(t,.) - u_o\|_{L^2\sqrt{\alpha}} \xrightarrow[t \to o+]{} 0$, where $L^2\sqrt{\alpha} = \{u$ measurable from Ω into C such that $\|u\|^2_{L^2\sqrt{\alpha}} = \int_\Omega \alpha(x)|u(x)|^2 dx < +\infty\}$.

In the paper [5] we consider the same problem in the spaces $X = L^2_{1/\sqrt{\alpha}}$, $Y = L^2_{\sqrt{\alpha}}$.

REFERENCES

1. Balakrishnan, A. V., Applied Functional Analysis, Springer Verlag, Berlin, (1976).
2. Campbell, S. L., Meyer C. D., and Rose, N. J., *SIAM J. Appl. Math. 31*, 411-425 (1976).
3. Campbell, S. L., *SIAM J. Control and Optimization 14*, 1092--1106 (1976).
4. Carroll, R. W. and Showalter, R. E., Singular and Degenerate Cauchy Problems, ed. Academic Press, London, (1976).
5. Favini, A., *Rend. Sem. Mat. Univ. Padova 52*, 243-263 (1974).
6. Friedman, A., Partial Differential Equations, ed. Holt, Rinehart and Winston, (1969).

COMMUTATIVE LINEAR DIFFERENTIAL OPERATORS

Wolfgang Hahn[1]

Technische Universität Graz
Institut für Mathematik I
Graz, Austria

I. INTRODUCTION

Let us consider linear differential operators

$$P : p_o(x)D^m + p_1(x)D^{m-1} + \ldots + p_{m-1}(x)D + p_m(x) \qquad (1.1)$$

with $D := d/dx$. We can define a composition: if $Q = q_o(x)D^n +$ $+ \ldots$ is an other operator we put

$$PQ : = p_o(x)D^mQ + p_1(x)D^{m-1}Q + \ldots + p_m(x)Q . \qquad (1.2)$$

Usually, the operators are not commutative. We have $D(qD) \neq$ $\neq (qD)D$ unless q is a constant. On the other hand, certain operators are commutative, e.g. polynomials of the same operator P, i.e. linear combinations of powers of P with constant coefficients. Moreover, there are operators which are "nontrivially" commutative, e.g.

$$P = D^2 - 2x^{-2}, \quad Q = D^3 - 3x^{-2} - 9x^{-3} .$$

---

[1]Present address: Technische Universität Graz, Institut für Mathematik I, Graz, Austria.

ISBN 0-12-186280-1

We are interested in such operators.

Comparing the coefficients of equal powers of D in the equation $PQ = QP$ we obtain a sequence of differential equations for the coefficients p_i, q_j whose first runs

$$mp_o q_o' = nq_o p_o' \tag{1.3}$$

whence $p_o = a^m$, $q_o = a^n$ with suitable a. (The occurring constant can be put equal one). Introducing the variable y by $dy = a^{-1}dx$ we obtain the operators in the form

$$P = D^m + \ldots, \quad Q = D^n + \ldots \tag{1.4}$$

which will be assumed in the following.

The further differential equations are extremely intricate for small m and n already.

The problem of commutative operators has been dealt with some fifty years ago by *Burchnall* and *Chaundy* (Commutative Ordinary Differential Operators. Proc. London Math. Soc. (2) 21, 420-440 (1923)). The developped some formalisms and gave the complete solution for the case $m = 2$. We shall show how the problem can be handled with algebraic methods, but the complete solution is still far off.

II. NOTATIONS

Small latin letters $a, b, \ldots, p, q, \ldots, u, v, \ldots$ (apart from x) denote functions of x, the only formal supposition being that the required derivations are possible. Sometimes such letters, especially m and n, will denote exponents and indices, but there will be no confusion. Underlined small letter $\underline{u}$, $\underline{v}$ denote vectors, i.e. $\underline{u} = \mathrm{col}(u_1, \ldots, u_m)$. Capital latin letters $P, Q, \ldots$ are differential operators of type (1.1) whereas underlined capital letters $\underline{A}$, $\underline{B}$, ... denote quadratic matrices with constant coefficients. Small greek letters $\alpha, \beta, \ldots$ are constants and capital greek letters $\Phi(.)$, $\psi(.)$ polynomials of their argument.

An operator can act upon a vector: $P\underline{u}$ is another vector. The vector equation $P\underline{u} = \underline{A}\,\underline{u}$ can be transformed into a scalar equation. Let $\Phi_A(.)$ be the minimal polynomial of the matrix $\underline{A}$. Then we get

$$\Phi_A(P)\underline{u} = \Phi_A(A)\underline{u} = 0$$

and the corresponding scalar equation

$$\Phi_A(P)u = 0 \,. \tag{2.1}$$

Its order is m. degree of $\Phi_A(.)$ where $m = \sigma(P)$ means the order of the operator P.

A function which satisfies the differential equation $Pu = 0$ will be called a "solution of P".

III. THE RING OF OPERATORS

The operators form a euclidean ring, i.e. there exists a division with remainder. If $\sigma(P) \leq \sigma(Q)$ one can find operators L and R uniquely which satisfy

$$Q = LP + R, \qquad \sigma(R) < \sigma(P). \tag{3.1}$$

Repeated application permits to construct the greatest common divisor (g.c.d.) of P and Q which we denote, as usual, by (P,Q). If its order τ differs from zero, the differential equations

$$Pu = 0, \qquad Qv = 0 \tag{3.2}$$

have τ linearly independent solutions in common. If $\tau = 0$, P and Q have no common divisor and no common solution. The operator in (2.1) admits a decomposition

$$\Phi_A(P) = \prod_i (P-\alpha_i)^{k_i} \tag{3.3}$$

where the numbers α_i are the eigenvalues of $\underline{A}$ and the k_i

their multiplicities. The factors of the product are by pairs commutative and have no common divisors.

An operator Q is called to be a right hand factor (r.h.f.) of P if PQ is divisible by P. We then have an operator S such that $PQ = SP$. The following statements hold for the r.h.f. of a given P.

a) They form a subring in the ring of all operators as easily seen.

b) There exist operators R with $\sigma(R) < \sigma(P)$ in the subring. For, if Q is in the subring and $\sigma(Q) \geq \sigma(P)$ we take R from (3.1). Since $PR = P(Q - LP) = (S - PL)P$, R is in the subring too.

c) We can coordinate a matrix of type (m,m) to each operator of the subring. Consider a fundamental system

$$u_1, u_2, \ldots, u_m \tag{3.4}$$

of solutions of P and put $\underline{u} = \mathrm{col}\,(u_1,\ldots,u_m)$. Since $SP = PQ$ we have $PQ\underline{u} = P(Q\underline{u}) = 0$, i.e. $Q\underline{u}$ is a solution vector of P, therefore $Q\underline{u} = \underline{A}\,\underline{u}$ with suitable $\underline{A}$. $\underline{A}$ depends upon the special system (3.4) but obviously all those matrices belong to the same similarity class and we can assume that $\underline{A}$ is given in the Jordan normal form.

If $\underline{A}$ is singular, P and Q have common solutions. Their number equals $\sigma(P) - \mathrm{rk}\,\underline{A}$.

Example. Suppose P is a polynomial of a first order operator K with single roots,

$$P = \prod_{i=1}^{m} (K - \gamma_i),$$

and Q is another polynomial of K. Since the solutions of P satisfy equations $Ku_i = \gamma_i u_i$ one has

$$Qu_i = \psi(K)u_i = \psi(\gamma_i)u_i.$$

The matrix $\underline{A}$ is a diagonal matrix with $a_{ii} = \psi(\gamma_i)$. If multiple roots occur, $\underline{A}$ contains elementary divisors of higher order.

d) If two operators Q and Q_1 of the subring are congruent mod P, i.e. if $Q = KP + Q_1$, they correspond to the same matrix $\underline{A}$.

e) If $Q\underline{u} = \underline{A}\,\underline{u}$ we have according to (2.1) $\Phi_A(Q)\underline{u} = 0$ whence we conclude that the polynomial is divisible by P,

$$\Phi_A(Q) = \prod_i (Q - \alpha_i)^{k_i} = KP. \tag{3.5}$$

f) The numbers α_i, the eigenvalues of $\underline{A}$, can be found without knowledge of the solutions u_i of P. If α is a single eigenvalue of $\underline{A}$, the order of $(P,Q-\alpha)$ is one. We carry through the euclidean algorithm with α as a parameter. The procedure must break off before the order of the last remainder has become zero. This condition yields and equation for α. If α belongs to an elementary divisor of order two, the order of $(P,Q-\alpha)$ is one and the order of $(P,(Q-\alpha)^2)$ is two. We have to construct the g.c.d. $(P,(Q-\alpha)^2)$ etc. In each case we obtain an equation for α which is equivalent to the characteristic equation of $\underline{A}$.

IV. COMMUTATIVE OPERATORS

a) We now assume $PQ = QP$. Each operator is r.h.f. of the other. There are matrices $\underline{A}$ and $\underline{B}$ such that

$$Q\underline{u} = \underline{A}\,\underline{u}, \quad P\underline{v} = \underline{B}\,\underline{v}. \tag{4.1}$$

If $(P,Q) = 1$, these relations are necessary and sufficient for commutativity. For in this case PQ and QP have the same fundamental system of solutions, namely $u_1,\ldots,u_m$; $v_1,\ldots,v_n$. If $\underline{A}$ and $\underline{B}$ are singular we conclude from (4.1) that PQ and QP have a common divisor of order $m + \mathrm{rk}\,\underline{B} = n + \mathrm{rk}\,\underline{A}$. If $\mathrm{rk}\,\underline{A} = m-1$ and if P and Q are in the standard form (1.7), then (4.1) implies $PQ = QP$, since in this case the order of $PQ - QP$ is no greater than $m + n - 2$ whereas this operators has $m + n - 1$ linear independent solutions, i.e. is identically zero.

b) If $n \geq m$ and $Q = LP + R$, we have $Q\underline{u} = R\underline{u} = \underline{A}\,\underline{u}$.

If $\sigma(R) = 0$, R must be a constant, $R = \rho$, whence

$$Q - \rho = LP$$

Since $P(Q-\rho) = (Q-\rho)P$, we follow $PLP = LPP$ and $PL = LP$. In the same way we conclude $KP = PK$ for the operators in (3.5) because $\Phi_A(Q)P = P\Phi_A(Q)$.

c) Polynomials $P = \Phi(K)$ and $Q = \psi(K)$ of the same operator K are commutative. Elimination of K yields an algebraic relation

$$\Omega(P,Q) = 0. \tag{4.2}$$

Such a relation holds for each commutative pair P, Q. In order to find it out we start with the equation

$$(Q-\delta)\underline{u} = (\underline{A} - \delta\underline{I})\underline{u}$$

where δ denotes a parameter. If δ is an eigenvalue of $\underline{A}$ the matrix $\underline{A} - \delta\underline{I}$ is singular. Let $\underline{v}_\delta$ be the solution vector of $Q-\delta$. This operator is commutative with P. Therefore exists $\underline{B}_\delta$ with $P\underline{v}_\delta = \underline{B}_\delta\underline{v}_\delta$. We can choose the number γ such that

$$(P-\gamma)\underline{v}_\delta = (\underline{B}_\delta - \gamma\underline{I})\underline{v}_\delta$$

and $\underline{B}_\delta - \gamma\underline{I}$ singular. Similarily we define $\underline{u}_\gamma$ and $\underline{A}_\gamma$ by

$$(P-\gamma)\underline{u}_\gamma = 0, \qquad Q\underline{u}_\gamma = \underline{A}_\gamma\underline{u}_\gamma.$$

If α_γ is an eigenvalue of $\underline{A}_\gamma$, then $\underline{A}_\gamma - \alpha_\gamma\underline{I}$ is singular, and the operators $Q - \alpha_\gamma$ and $P-\gamma$ have a common divisor. We state: given any γ, we have at least one δ such, that the matrices $\underline{A}_\gamma - \delta\underline{I}$ and $\underline{B}_\delta - \gamma\underline{I}$ are singular and $P-\gamma$ and $Q-\delta$ have a common divisor. We carry through the euclidean algoritm for $P-\gamma$ and $Q-\delta$, considering γ and δ as parameters, and write down the condition for $(P-\gamma, Q-\delta) > 1$. We obtain an algebraic relation

$$\Omega(\gamma, \delta) = 0. \tag{4.3}$$

Let the common solutions of $P - \gamma$ and $Q - \delta$ be denote by $y_{\gamma\delta}$. The relation

$$\Omega(P,Q)y_{\gamma\delta} = \Omega(\gamma,\delta)y_{\gamma\delta} = 0$$

holds for an infinite number of pairs. The operator must vanish identically, and we follow (4.2).

d) If $(Q - \delta)\underline{v}_\delta = 0$, we have $P\underline{v}_\delta = \underline{B}_\delta \underline{v}_\delta$ and $\det(\underline{B}_\delta - \beta_\delta \underline{I}) = 0$. If $(P - \gamma)\underline{u}_\gamma = 0$, we have $Q\underline{u}_\gamma = \underline{A}_\gamma \underline{u}_\gamma$ and $\det(\underline{A}_\gamma - \alpha_\gamma \underline{I}) = 0$. All these conditions guarantee $(P - \gamma, Q - \delta) > 1$. The equation $\det(\underline{B}_\delta - \gamma\underline{I}) = 0$ is equivalent to $\det(\underline{A}_\gamma - \delta\underline{I}) = 0$. We write these relations in the form

$$(\gamma - \beta_1(\delta))(\gamma - \beta_2(\delta)) \ldots (\gamma - \beta_n(\delta)) = 0,$$

$$(\gamma - \delta_1)(\gamma - \delta_2) \ldots (\gamma - \delta_n) = 0.$$

The product of the $\beta_i(\delta)$ is $\det \underline{B}_\delta$, apart from the sign, and the product of the δ_j equals $\det(\underline{A}_o - \delta\underline{I})$. We obtain

$$\det(\underline{A}_o - \delta\underline{I}) = \pm \det \underline{B}_\delta$$

and

$$\det \underline{A}_o = \det \underline{A} = \pm \det \underline{B}_o = \pm \det \underline{\dot{B}} .$$

As a matter of fact, we have for the matrices in (4.1)

$$\det \underline{A} = (-1)^{mn} \det \underline{B} \tag{4.4}$$

as can be shown by the following

Example. Let P and Q be polynomials of a first order operator K with single roots, $P = \prod_{i=1}^{m}(K - \gamma_i)$, $Q = \prod_{j=1}^{n}(K - \delta_j)$. We have as in 3c

$$Qu_i = \prod_{j=1}^{n}(\gamma_i - \delta_j)u_i; \quad Pv_j = \prod_{i=1}^{m}(\delta_j - \gamma_i)v_j$$

and

$$\det \underline{A} = \prod_{i,j}(\gamma_i - \delta_j) = (-1)^{mn} \det \underline{B} = (-1)^{mn} \prod_{i,j}(\delta_j - \gamma_i),$$

The product is the resultant of the two polynomials.

e) If two polynomials of the operators P and Q are commutative, the operators are commutative too. The way of proving this statement can be illustrated as follows.

Let P^2Q and PQ^2 be commutative and assume that $PQ \neq QP$, i.e. $PQ = QP + R$, $R \neq 0$. Since $P^2QPQ^2 = PQ^2P^2Q$ we have

$$PQPQ = Q^2P^2 .$$

We replace PQ by $QP + R$ repeatedly and get

$$PQPQ = Q(QP + R)P + RQP + QPR + R^2 = Q^2P^2$$

whence

$$QRP + RQP + QPR + R^2 = 0.$$

This equation is an identity. The coefficient of the highest power of D must vanish. This coefficient is $3p_oq_or_o$, and we follow $r_o = 0$. Continuation of the argument yields $R = 0$.

In the general case

$$\Phi(P,Q)\psi(P,Q) - \psi(P,Q)\Phi(P,Q) = 0$$

one rewrites the left hand side expression in the form

$$\sum p_{ij}q_{rs} \; (P^iQ^jP^rQ^s - P^rQ^sP^iQ^j)$$

and transforms the first product in each bracket into $P^rQ^sP^iQ^j + \Omega(P,Q,R)$. The expression $\Omega(P,Q,R)$ must vanish identically, and since the coefficients of the highest powers of D have the same sign we conclude $r_o = 0$ as above etc.

We learn from this statement that the relation (4.2) is characteristic for commutative operators. We have $P\Omega(P,Q) = = \Omega(P,Q)P = 0$ whence $PQ = QP$.

f) If n is divisible by m, $n = m'm$, the order of $R := Q - P^{m'}$ is smaller than n. Since P and R are commutative the problem is reduced in a sense: the sum of the orders is smaller. In the case $m = 1$ the order of Q can be reduced repeatedly, and Q proves to be a polynomial in P.

In connection with the foregoing consideration we can state: An operator which is commutative with the polynomial of a first order operator is a polynomial of this operator.

V. THE CASE m = 2

Regarding 4 f we can restrict ourselves to odd values of n. We have $P\underline{v} = \underline{B}\,\underline{v}$ for $Q\underline{v} = 0$ and according to (3.5) and 4 b

$$\Phi_B(P) = LQ = QL \tag{5.1}$$

where $\Phi_B(.)$ is the minimal polynomial of B. The order of L is at most n. If it equals n, we put $L = Q + Q_1$ where $PQ_1 = Q_1P$, $\sigma(Q_1) < n$. At any rate there exists an operator of order less than n which is commutative with P, and the problem is reduced apart from the case $Q_1 = \rho$ where

$$\Phi_B(P) = Q(Q + \rho).$$

The left hand side is of the form $\psi(P)P - \det \underline{B}$. Equation (3.5) runs

$$\Phi_A(Q) = Q^2 - (\alpha_1 + \alpha_2)Q + \det \underline{A} = KP.$$

We substract the second equation from the first and consider the equality of the determinants. We get

$$(\rho + \alpha_1 + \alpha_2)Q = (\psi(P) - K)P$$

and follow that either P is a divisor of Q or the two equations are identical and equivalent to

$$Q^2 = \Phi_B(P). \tag{5.2}$$

In the first case one has $Q = Q_1P$ and $Q_1PP = PQ_1P$, i.e. Q_1 and P are commutative, and the problem is reduced.

We can summarize: Each problem $PQ = QP$, $o(P) = 2$, can be reduced to an equation (5.2) either with the given Q or with a Q of lower order.

Burchnall and *Chaundy* l.c. could show that the coefficients can be represented by means of hyperelliptic functions. If $m = 2$, $n = 3$, the corresponding equation can be treated with the aid of the Weierstrass p-function.

VI. A SPECIAL PAIR OF COMMUTATIVE OPERATORS

Let us assume that the commutative operators P and Q are of type (1.4) and that, in addition, there exist two non-constant functions f and g such that fP and gQ are commutative too. Regarding (1.3) we can put $f = a^m$, $g = a^n$. P and fP have the same solutions just as Q and gQ. Therefore

$$Q\underline{u} = \underline{A}\,\underline{u} \quad \text{and} \quad a^n Q\underline{u} = \underline{A}_1\underline{u} \tag{6.1}$$

whence

$$a^n\underline{A}\,\underline{u} = \underline{A}_1\underline{u} \quad \text{and} \quad \det(\underline{A}_1 - a^n\underline{A}) = 0. \tag{6.2}$$

The latter equation cannot hold but for singular $\underline{A}$ and $\underline{A}_1$. Certain functions u_i can be expressed by the functions $a^n u_j$. If $\underline{A}$ has a single eigenvalue zero we can express $m-1$ functions u_i by the $a^n u_j$, and the operators P and (Pa^n) have $m-1$ common solutions, namely

$$a^n u_1 = u_2,\ a^n u_2 = u_3,\ \ldots,\ a^n u_{m-1} = u_m.$$

$\underline{A}$ is of diagonal type, and $\underline{A}_1$ has the only eigenvalue zero with an elementary divisor of order m. If we put

$$fP = a^m P = \prod_{i=1}^{m} (aD + b + \gamma_i),$$

we get

$$(aD+b+\gamma_i)(a^{(i-1)n}u_1) = 0, \quad i = 1,2,\dots,m-1,$$

and

$$a(a^{(i-1)n})' + (\gamma_i - \gamma_1)a^{(i-1)n} = 0,$$

$$((i-1)n-1)a' + (\gamma_i - \gamma_1) = 0$$

and finally

$$a = x+\delta, \quad \gamma_i = (i-1)n.$$

We can assume $\delta = 0$ and obtain the operators

$$P = x^{-m} \prod_{i=1}^{m} (xD+b+(i-1)n),$$
$$Q = x^{-n} \prod_{j=1}^{n} (xD+b+(j-1)m). \tag{6.3}$$

x^mP and x^nQ are commutative as polynomials of a first order operator. The commutativity of P and Q follows from 4 a because the rank of A is $m-1$. It is easily seen that

$$P^n = Q^m. \tag{6.4}$$

For $m = 2$, $n = 3$ one has

$$P = D(D+2x^{-1}), \quad Q = D(D+3x^{-1}).$$

Here is

$$x^2P = (xD+2)(xD-1), \quad x^3Q = xD(xD-2)(xD+2).$$

If the commutative operators $P = P_1F$ and $Q = Q_1F$ have a common factor F, the operators FP_1 and FQ_1 are also commutative. The pair FP_1, FQ_1 originates by "transference" of the factor F. (cf. *Burchnall* and *Chaundy* l.c.).

The operators (6.3) have the common factor $xD+b := H$. Regarding the relation $Hx^{-k} = x^{-k}(H-k)$ we can find out the result of the transference of the common factor H. We obtain the operators

$$P_1 := x^{-m}(H-m) \prod_{i=2}^{m} (H-(i-1)n),$$

$$Q_1 := x^{-n}(H-n) \prod_{j=2}^{n} (H-(j-1)m)$$

which have the common factors $H-m$ and $H-n$. Repeated transference originates two operators with three common factors. After $m-1$ transferences the operators P_{m-1} and Q_{m-1} are in hand, and P_{m-1} is a divisor of Q_{m-1}. Further transferences do not change this state: P_r divides Q_r for $r \geq m-1$. The relation

$$P_r^n = Q_r^m \tag{6.5}$$

holds for all $r \geq 0$.

Our example shows that there are nontrivially commutative operators of arbitrary order.

APPROXIMATIONS OF DELAYS BY ORDINARY DIFFERENTIAL EQUATIONS

A. Halanay[1]

Faculty of Mathematics
Bucharest 1, Romania

Vl. Răsvan[2]

I.C.P.E.T. - Bucharest, Romania

I. INTRODUCTION. STATE OF THE ART

In the last three or four years the problem of approximating functional differential equations (FDE) by a sequence of ordinary differential equations (ODE) has got a renewed attention [2], [8], [10], [11], [12]. We say "renewed" because such type of approximation had been pointed out nearly 15 years ago in connection with stability or optimal control problems for time lag control systems [9], [15].

However, it must be remarked that the approximation of FDE by ODE occured in the analogue computation of the dynamical systems described by other types of equations than ODE. The

---

[1]Present address: Faculty of Mathematics, 14 str. Academiei, Bucharest 1, Romania.

[2]Present address: I.C.P.E.T. - Bucharest, Romania.

ISBN 0-12-186280-1

simplest class of such systems is the class of linear time lag systems, and the transfer functions of such systems contain the transcendental function $e^{-h\sigma}$. In the analogue computation use was made of a certain approximation in the complex domain of this exponential by rational functions. Such approximations are the Padé approximation [7] or the following approximation [13]

$$e^{-h\sigma} \approx \frac{1}{(1 + \frac{h}{N}\sigma)^N} \tag{1.1}$$

One can see that the right hand side (RHS) of (1.1) converges to the left hand side (LHS) when $N \to \infty$, uniformly on compacts. This complex domain approach served as a heuristic argument to the approximation of time lag systems by a certain ODE of higher order; N. N. Krasovskii [9] and Yu. M. Repin [15] gave rigorous convergence arguments in the time domain.

Later, H. T. Banks and J. A. Burns [2] considered the type of approximation of time lag systems generated by (1.1) in the more general framework of the approximation of an operator defined on an infinite dimensional space by a sequence of operators defined on finite dimensional projections of this space. In this case the problem is, as it was pointed out by F.Kappel and W. Schappacher [8], to approximate the translation operator acting on the solutions of the systems with deviating argument. This is also the significance of the complex domain formal relation (1.1).

The approximation results of Kappel and Schappacher were extended by K. Kunisch [10],[11] to the case of neutral FDE.

What is to be pointed out in all these results is that, due to technical reasons, they are true only for spaces of sufficiently smooth functions. In fact they are obtained for the case of Lipschitz functions. For delay-differential equations (DDE) this is not quite a restriction because the solution is smoothed in time. For NFDE this requirement is accomplished if the initial data are Lipschitz.

The aim of the present paper is to obtain approximation results for new classes of time lag systems, particularly for those occurring from mixed initial boundary value problems for hyperbolic partial differential equations (PDE), pointing out also the connection between these approximation results and the method of lines for hyperbolic PDE. The sense of the approximation will allow to weaken the assumptions on the initial data.

PART I. COUPLED DIFFERENTIAL AND DIFFERENCE EQUATIONS

II. A PRELIMINARY RESULT

In the following we shall obtain what was called "the approximation of the pure delay" [15]. In fact this result gives a significance to the formal relation (1.1). The result is the following:

Lemma. Consider the system of linear differential equations

$$\begin{aligned} \dot{y}_1 &= \frac{N}{h}[x(t) - y_1] \\ \dot{y}_j &= \frac{N}{h}[y_{j-1} - y_j] \qquad j = 2,\dots,N \end{aligned} \tag{2.1}$$

with the initial conditions

$$y_j(t_o) = \frac{N}{h}\int_{t_o - j\frac{h}{N}}^{t_o-(j-1)\frac{h}{N}} x(s)\,ds \qquad j = 1,\dots,N \tag{2.2}$$

where $x \in L^1(t_o-h, t_o+T; R^n)$ and t_o, T are arbitrary ($T > 0$). Then the following results are true:

$$\int_{t_o}^{t_o+T} |y_j(s) - x(s - j\tfrac{h}{N})|\,ds \le \gamma_1(\tfrac{h}{N}), \qquad j = 1,\dots,N \tag{2.3}$$

$$\sum_1^N \int_{-j\frac{h}{N}}^{-(j-1)\frac{h}{N}} |y_j(t) - x(t+s)|\,ds \le \gamma_2(\tfrac{h}{N}), \qquad t_o \le t \le t_o+T \tag{2.4}$$

If $x \in C(t_o-h, t_o+T; R^n)$ then

$$|y_j(t) - x(t - j\frac{h}{N})| \leq \gamma_3(\frac{h}{N}), \qquad j = 1,\ldots,N \tag{2.5}$$

for $t_o \leq t \leq t_o+T$, and this is true also if the initial conditions (2.2) are replaced by

$$y_j(t_o) = x(t_o - j\frac{h}{N}), \qquad j = 1,\ldots,N \tag{2.2'}$$

All these estimates are not dependent on t_o. Here $\gamma_i(r)$ are monotonically increasing and $\lim_{r\to 0} \gamma_i(r) = 0$.

Proof. We define the following regularized function [3], [16]

$$\xi_\delta(t) = \frac{1}{\delta}\int_{-\infty}^{\infty} \omega(\frac{t-s}{\delta})x(s)ds$$

where $x(t)$ is extended by 0 to the whole real axis and $\omega(r)$ is a regularization kernel with the following properties:

i) $\omega(r) > 0$ for $|r| < 1$ and $\omega(r) \equiv 0$ for $|r| \geq 1$

ii) $\omega(r) = \omega(-r)$

iii) $\omega \in C^\infty(-\infty,\infty)$

iv) $\int_{-\infty}^{\infty} \omega(r)dr = \int_{-1}^{1} \omega(r)dr = 1$

An example of such regularization kernel is the following

$$\omega(r) = \begin{cases} C_o e^{\frac{r^2}{r^2-1}} & \text{for } r < 1 \\ 0 & \text{for } r \geq 1 \end{cases}$$

It is obvious that $\xi_\delta \in C^\infty(-\infty,\infty)$.

The proof of the Lemma will be performed in several steps.

1. Define $z_j(t)$ to be the solutions of the system

$$\dot{z}_1 = \frac{N}{h}\left[\xi_\delta(t) - z_1\right]$$
$$\dot{z}_j = \frac{N}{h}\left[z_{j-1} - z_j\right] \qquad j = 2,\dots,N \tag{2.6}$$

with the initial conditions

$$z_j(t_o) = y_j(t_o) = \frac{N}{h}\int_{t_o - j\frac{h}{N}}^{t_o-(j-1)\frac{h}{N}} x(s)\,ds, \qquad j = 1,\dots,N \tag{2.7}$$

and $\hat{z}_j(t)$ to be the solutions of system (2.6) with the initial conditions

$$\hat{z}_j(t_o) = \xi_\delta(t_o - j\tfrac{h}{N}), \qquad j = 1,\dots,N \tag{2.8}$$

Consider the differences

$$\zeta_j^\delta(t) = \hat{z}_j(t) - \xi_\delta(t-j\tfrac{h}{N})$$

These differences are solutions of the following system

$$\dot{\zeta}_1^\delta(t) = \frac{N}{h}\left[\xi_\delta(t) - \xi_\delta(t-\tfrac{h}{N}) - \tfrac{h}{N}\,\dot{\xi}_\delta(t-\tfrac{h}{N}) - \zeta_1^\delta\right]$$

$$\dot{\zeta}_j^\delta(t) = \frac{N}{n}\left[\xi_\delta(t-(j-1)\tfrac{h}{N}) - \xi_\delta(t-j\tfrac{h}{N}) - \tfrac{h}{N}\,\dot{\xi}_\delta(t-j\tfrac{h}{N}) + \zeta_{j-1}^\delta - \zeta_j^\delta\right]$$

$$j = 2,\dots,N$$

with the initial conditions $\zeta_j^\delta(t_o) = 0 \quad (j = 1,\dots,N)$.

Using the following relation

$$\xi_\delta(t-(j-1)\tfrac{h}{N}) - \xi_\delta(t-j\tfrac{h}{N}) = \frac{h}{N}\int_{-1}^{0}\dot{\xi}_\delta(t-(j-1)\tfrac{h}{N}+\theta\tfrac{h}{N})\,d\theta$$

and the variation of constants formula, one finds

$$\zeta_1^\delta(t) = \int_{t_o}^{t} e^{-\frac{N}{h}(t-s)}\left\{\int_{-1}^{0}\left|\dot{\xi}_\delta(s+\theta\tfrac{h}{N}) - \dot{\xi}_\delta(s-\tfrac{h}{N})\right|d\theta\right\}ds$$

$$\zeta_j^\delta(t) = \int_{t_o}^{t} e^{-\frac{N}{h}(t-s)} \left\{ \int_{-1}^{0} |\dot{\xi}_\delta(s-(j-1)\frac{h}{N}+\theta\frac{h}{N}) - \dot{\xi}_\delta(s-j\frac{h}{N})|d\theta \right\} ds$$

$$+ \frac{N}{h} \int_{t_o}^{t} e^{-\frac{N}{h}(t-s)} \zeta_{j-1}^\delta(s)ds$$

It can be easily seen that

$$\sup_{|t_1-t_2|\le r} |\dot{\xi}_\delta(t_1) - \dot{\xi}_\delta(t_2)| = \sup_{|\lambda|\le 1} |\omega'(\lambda)|\frac{1}{\delta^2}\alpha_o(r)$$

where

$$\alpha_o(r) = \sup_{|t_1-t_2|\le r} \int_{-h}^{T} |x(t+t_1) - x(t+t_2)|dt$$

It is obvious that $\alpha_o(r)$ is monotonically increasing and $\lim_{r\to 0} \alpha_o(r) = 0$. Therefore

$$|\zeta_j^\delta(t)| = |\hat{z}_j(t) - \xi_\delta(t-j\frac{h}{N})| \le h \sup_{|\lambda|\le 1} |\omega'(\lambda)|\frac{1}{\delta^2}\alpha_o(\frac{h}{N}) \tag{2.9}$$

$$j = 1,\dots,N$$

We shall perform now a comparison between $\hat{z}_j(t)$ and $z_j(t)$ which are both solutions of (2.6), but with different initial conditions. By induction one gets

$$|\hat{z}_j(t) - z_j(t)| \le e^{-\frac{N}{h}(t-t_o)} \sum_1^j |z_{j-k+1}(t_o)$$

$$- z_{j-k+1}(t_o)| \frac{1}{(k-1)!}\left[\frac{N}{h}(t-t_o)\right]^{k-1} \tag{2.10}$$

It can be easily shown that the following estimate is true

$$\sum_1^N |\hat{z}_j(t_o) - z_j(t_o)| \leq \frac{N}{h}\Big[h \sup_{|\lambda|\leq 1} \omega(\lambda)\ \frac{1}{\delta}\ \alpha_o(\tfrac{h}{N}) + \int_{t_o-h}^{t_o} |\xi_\delta(s) - x(s)|ds\Big] \tag{2.11}$$

We define now the functions $u_j(t) = y_j(t) - z_j(t)$.
It can be seen that $u_j(t)$ are the solutions of the system

$$\dot{u}_1 = \frac{N}{h}\ [x(t) - \xi_\delta(t) - u_1]$$

$$\dot{u}_j = \frac{N}{h}\ [u_{j-1} - u_j] \qquad j \geq 2,\quad u_j(t_o) = 0,\quad j = 1,\dots,N$$

Applying Fubini's theorem one gets, by induction

$$|u_j(t)| \leq \frac{N}{h} \int_{t_o}^{t} \frac{1}{(j-1)!}\Big[\frac{N}{h}(t-s)\Big]^{j-1} e^{-\frac{N}{h}(t-s)} |\xi_\delta(s) - x(s)|ds \tag{2.12}$$

2. We shall obtain now the estimate (2.3)

$$\int_{t_o}^{t} |y_j(s) - x(s-j\tfrac{h}{N})|ds \leq \int_{t_o}^{t} |\hat{z}_j(s) - \xi_\delta(s-j\tfrac{h}{N})|ds + \int_{t_o}^{t} |z_j(s) - \hat{z}_j(s)|ds$$

$$+ \int_{t_o}^{t} |\xi_\delta(s-j\tfrac{h}{N}) - x(s-j\tfrac{h}{N})|ds + \int_{t_o}^{t} |u_j(s)|ds$$

$$\leq \int_{t_o}^{t_o+T} |\hat{z}_j(s) - \xi_\delta(s-j\tfrac{h}{N})|ds + \int_{t_o}^{t} |z_j(s) - \hat{z}_j(s)|ds$$

$$+ \int_{t_o-h}^{t_o+T} |\xi_\delta(s) - x(s)|ds + \int_{t_o}^{t} |u_j(s)|ds$$

Taking into account the estimates obtained above one gets:

$$\int_{t_o}^{t} |y_j(s)-x(s-j\frac{h}{N})|ds \leq T\ h \sup_{|\lambda|\leq 1} |\omega'(\lambda)| \frac{1}{\delta^2} \alpha_o(\frac{h}{N})$$

$$+ \sum_{1}^{j} |\hat{z}_{j-k+1}(t_o) - z_{j-k+1}(t_o)| \int_{o}^{t-t_o} \frac{1}{(k-1)!} (\frac{N}{h}s)^{k-1} e^{-\frac{N}{h}s} ds$$

$$+ \int_{t_o-h}^{t_o+T} |\xi_\delta(s) - x(s)|ds +$$

$$+ \frac{N}{h} \int_{t_o}^{t} \left\{ \int_{o}^{s} \frac{1}{(j-1)!} \left[\frac{N}{h}(s-\sigma)\right]^{j-1} e^{-\frac{N}{h}(s-\sigma)} |\xi_\delta(\sigma)-x(\sigma)|d\sigma \right| ds$$

In the last integral we use Fubini's theorem, the estimate (2.11) and also the estimate

$$\frac{N}{h} \int_{o}^{t-t_o} \frac{1}{(k-1)!} (\frac{N}{h}s)^{k-1} e^{-\frac{N}{h}s} ds \leq \int_{o}^{\infty} \frac{\lambda^{k-1}}{(k-1)!} e^{-\lambda} d\lambda = 1$$

Therefore

$$\int_{t_o}^{t} |y_j(s)-x(s-j\frac{h}{N})|ds \leq T\ h \sup_{|\lambda|\leq 1} |\omega'(\lambda)| \frac{1}{\delta^2} \alpha_o(\frac{h}{N})$$

$$+ h \sup_{|\lambda|\leq 1} \omega(\lambda) \frac{1}{\delta} \alpha_o(\frac{h}{N}) + 2 \int_{-\infty}^{\infty} |\xi_\delta(s) - x(s)|ds$$

The last integral can be estimated using the approximation properties of regularized functions [3],[16]. By straightforward computation it follows

$$\int_{-\infty}^{\infty} |\xi_\delta(t) - x(t)|dt \leq 2 \sup_{|\lambda|\leq 1} \omega(\lambda)\alpha_o(2\delta)$$

where $\alpha_o(r)$ is the same as above. Therefore, by choosing $\delta = \sqrt[4]{\alpha_o(\frac{h}{N})}$ one gets

$$\int_{t_o}^{t} |y_j(s) - x(s-j\tfrac{h}{N})|ds \leq T\ h \sup_{|\lambda|\leq 1} |\omega'(\lambda)| \sqrt{\alpha_o(\tfrac{h}{N})}$$

$$+ \sup_{|\lambda|\leq 1} \omega(\lambda) \left[(\alpha_o(\tfrac{h}{N}))^{3/4} + 4\ \alpha_o\ (2\sqrt[4]{\alpha_o(\tfrac{h}{N})})\right]$$

and the estimate (2.3) is proved.

3. We shall obtain now the estimate (2.4). We have

$$\sum_1^N \int_{-j\frac{h}{N}}^{-(j-1)\frac{h}{N}} |y_j(t) - x(t+s)|ds \leq \frac{h}{N} \sum_1^N |\hat{z}_j(t) - \xi_\delta(t-j\tfrac{h}{N})|$$

$$+ \sum_1^N \int_{-j\frac{h}{N}}^{-(j-1)\frac{h}{N}} |\xi_\delta(t-j\tfrac{h}{N}) - \xi_\delta(t+s)|ds$$

$$+ \int_{-h}^{o} |\xi_\delta(t+s) - x(t+s)|ds + \frac{h}{N} \sum_1^N |u_j(t)|$$

Using again the estimates (2.9)-(2.12) we have, with the same choice for δ as above

$$\sum_1^N \int_{-j\frac{h}{N}}^{-(j-1)\frac{h}{N}} |y_j(t) - x(t+s)|ds \leq h^2 \sup_{|\lambda|\leq 1} |\omega'(\lambda)| \sqrt{\alpha_o(\tfrac{h}{N})}$$

$$+ 2 \sup_{|\lambda|\leq 1} \omega(\lambda) \left[h(\alpha_o(\tfrac{h}{N}))^{3/4} + 2\alpha_o(2\sqrt[4]{\alpha_o(\tfrac{h}{N})}\right]$$

what proves the estimate (2.4).

4. Suppose now that $x \in C(t_o-h,\ t_o+T;\ R^n)$. Instead of (2.9) the following estimate can be easily found

$$|\hat{z}_j(t) - \xi_\delta(t-j\tfrac{h}{N})| \le 2\cdot h \sup_{|\lambda|\le 1} |\omega'(\lambda)|\frac{1}{\delta}\alpha_1(\tfrac{h}{N}), \qquad j = 1,\dots,N \tag{2.9'}$$

The comparison between $\hat{z}_j(t_o)$ and $z_j(t_o)$ will give:

$$|\hat{z}_j(t_o) - z_j(t_o)| \le 2[\alpha_1(\tfrac{h}{N}) + \alpha_1(2\delta)] \sup_{|\lambda|\le 1} \omega(\lambda) \tag{2.11'}$$

The estimates (2.9') and (2.11') were obtained using the following properties of $\xi_\delta(t)$ for the case of continuous $x(t)$:

$$\sup_{|t_1-t_2|\le r} |\xi_\delta(t_1) - \xi_\delta(t_2)| = 2 \sup_{|\lambda|\le 1} \omega(\lambda)\alpha_1(r)$$

$$\sup_{|t_1-t_2|\le r} |\dot{\xi}_\delta(t_1) - \dot{\xi}_\delta(t_2)| = 2 \sup_{|\lambda|\le 1} |\omega'(\lambda)|\frac{1}{\delta}\alpha_1(r)$$

$$|\xi_\delta(t) - x(t)| \le 2 \sup_{|\lambda|\le 1} \omega(\lambda)\alpha_1(2\delta)$$

where

$$\alpha_1(r) = \sup_{|t_1-t_2|\le r} |x(t_1) - x(t_2)|, \qquad t_o-h \le t_i \le t_o+T$$

(The extension of $x(t)$ outside of the interval is performed by continuity).

Using (2.11') one gets, by induction

$$|\hat{z}_j(t) - z_j(t)| \le 2[\alpha_1(\tfrac{h}{N}) + \alpha_1(2\delta)] \sup_{|\lambda|\le 1} \omega(\lambda) \tag{2.10'}$$

If $z_j(t_o)$ are given by (2.2') the above estimates become

$$|\hat{z}_j(t_o) - z_j(t_o)| \le 2 \sup_{|\lambda|\le 1} \omega(\lambda)\alpha_1(2\delta) \tag{2.11''}$$

$$|\hat{z}_j(t) - z_j(t)| \le 2 \sup_{|\lambda|\le 1} \omega(\lambda)\alpha_1(2\delta) \tag{2.10''}$$

The inequality (2.12) is still valid. Therefore

$$|u_j(t)| \le 2 \sup_{|\lambda|\le 1} \omega(\lambda)\alpha_1(2\delta), \qquad j = 1,\dots,N \tag{2.12'}$$

We can write now

$$|y_j(t) - x(t-j\tfrac{h}{N})| \le |\hat{z}_j(t) - \xi_\delta(t-j\tfrac{h}{N})| + |z_j(t) - \hat{z}_j(t)|$$

$$+ |\xi_\delta(t-j\tfrac{h}{N}) - x(t-j\tfrac{h}{N})| + |u_j(t)|$$

Using the estimates (2.9'), (2.10'), (2.12') and the approximation property of $\xi_\delta(t)$ for continuous $x(t)$ we get

$$|y_j(t) - x(t-j\tfrac{h}{N})| \le 2\,h \sup_{|\lambda|\le 1} |\omega'(\lambda)| \tfrac{1}{\delta}\, \alpha_1(\tfrac{h}{N})$$

$$+ 6 \sup_{|\lambda|\le 1} \omega(\lambda)\alpha_1(2\delta) + 2 \sup_{|\lambda|\le 1} \omega(\lambda)\alpha_1(\tfrac{h}{N})$$

$$j = 1,\dots,N; \quad t_o \le t \le t_o+T$$

By choosing $\delta = \sqrt{\alpha_1(\frac{h}{N})}$ the following estimate is obtained

$$|y_j(t) - x(t-j\tfrac{h}{N})| \le 2\,h \sup_{|\lambda|\le 1} |\omega'(\lambda)| \sqrt{\alpha_1(\tfrac{h}{N})}$$

$$+ 6 \sup_{|\lambda|\le 1} \omega(\lambda)\alpha_1(2\sqrt{\alpha_1(\tfrac{h}{N})}) + 2 \sup_{|\lambda|\le 1} \omega(\lambda)\alpha_1(\tfrac{h}{N})$$

$$j = 1,\dots,N; \quad t_o \le t \le t_o+T$$

If (2.2) are replaced by (2.2') then (2.10') is replaced by (2.10") and the last term in the RHS of the above estimate is missing. The estimate (2.5) has been obtained in both cases what ends the proof.

Remark. If the restriction of $x(t)$ to the interval $[t_o-h, t_o]$ is continuous, then (2.3) and (2.4) remain true when (2.2) are replaced by (2.2').

III. THE APPROXIMATION OF LINEAR DIFFERENCE EQUATIONS

Consider the difference equation:

$$x(t) = Ax(t-h) + g(t) \tag{3.1}$$

with the initial condition $x_{t_o} = \varphi$, where $\varphi \in L^1(-h,0;R^n)$, and with $g \in L^1(t_o, t_o+T; R^n)$. Here $x_t(s) = x(t+s)$, $-h \leq s \leq 0$, as usually.

It is obvious that this equation has a unique integrable solution which can be constructed by steps. This solution satisfies (3.1) a.e. and in the following we shall mean by solution the class of integrable functions which satisfy (3.1) a.e. with φ as initial conditions.

Define $y_j(t)$ to be the solutions of the following system of ODE:

$$\dot{y}_1 = \frac{N}{h}[Ay_N + g(t) - y_1]$$
$$\dot{y}_j = \frac{N}{h}[y_{j-1} - y_j] \qquad j = 2,\dots,N \tag{3.2}$$

with the initial conditions

$$y_j(t_o) = \frac{N}{h}\int_{-j\frac{h}{N}}^{-(j-1)\frac{h}{N}} \varphi(s)ds \qquad j = 1,\dots,N \tag{3.3}$$

This system is called the approximating system of (3.1) due to the following approximation result:

Theorem 1. Consider the difference equation (3.1) with the initial condition as above and with $g(t)$ integrable on (t_o, t_o+T). Suppose that the eigenvalues of A are inside the unit circle.

If $y_j(t)$ are solutions of (3.2) with the initial conditions (3.3), then:

$$\int_{t_o}^{t_o+T} |y_j(s) - x(s-j\frac{h}{N})|ds \leq \gamma_4(\frac{h}{N}) \qquad j = 1,\dots,N \tag{3.4}$$

$$\sum_1^N \int_{-j\frac{h}{N}}^{-(j-1)\frac{h}{N}} |y_j(t) - x(t+s)|ds \leq \gamma_5(\frac{h}{N}) \qquad t_o \leq t \leq t_o+T \tag{3.5}$$

If $\varphi \in C(-h,0; R^n)$, $g(t)$ is continuous on $[t_o, t_o+T]$ and if the following condition of "matching" is fulfilled

$$\varphi(0) = A\varphi(-h) + g(t_o) \tag{3.6}$$

then

$$|y_j(t) - x(t-j\tfrac{h}{N})| \le \gamma_6(\tfrac{h}{N}) \qquad j = 1,\dots,N \quad t_o \le t \le t_o+T \tag{3.7}$$

and this is true also if the initial conditions (3.3) are replaced by the following ones:

$$y_j(t_o) = \varphi(-j\tfrac{h}{N}) \qquad j = 1,\dots,N \tag{3.3'}$$

The functions $\gamma_i(r)$ are of the same type as those of the Lemma and the estimates are independent of t_o.

Proof. Taking into account (3.1), (3.2) can be written as follows:

$$\begin{aligned} \dot{y}_1 &= \frac{N}{h}\left[A(y_N - x(t-h)) + x(t) - y_1\right] \\ \dot{y}_j &= \frac{N}{h}\left[y_{j-1} - y_j\right] \qquad j = 2,\dots,N \end{aligned} \tag{3.2'}$$

1. We define $u_j(t)$ to be the solutions of the system:

$$\begin{aligned} \dot{u}_1 &= \frac{N}{h}\left[x(t) - u_1\right] \\ \dot{u}_j &= \frac{N}{h}\left[u_{j-1} - u_j\right] \qquad j = 2,\dots,N \end{aligned} \tag{3.8}$$

with the initial conditions

$$u_j(t_o) = y_j(t_o) = \frac{N}{h}\int_{-j\frac{h}{N}}^{-(j-1)\frac{h}{N}} \varphi(s)\,ds$$

If we extend $x(t)$ on (t_o-h, t_o) by $\varphi(t)$ this extended function belongs to $L^1(t_o-h, t_o+T; R^n)$. The application of the Lemma gives:

$$\int_{t_o}^{t_o+T} |u_j(s) - x(s-j\tfrac{h}{N})|\,ds \le \gamma_1(\tfrac{h}{N}) \qquad j = 1,\dots,N \tag{3.9}$$

$$\sum_{1}^{N} \int_{-j\frac{h}{N}}^{-(j-1)\frac{h}{N}} |u_j(t) - x(t+s)|ds \le \gamma_2(\tfrac{h}{N}) \qquad t_o \le t \le t_o+T \tag{3.10}$$

Define now $v_j(t) \equiv y_j(t) - u_j(t)$. It can be seen that $v_j(t)$ are solutions of the following system:

$$\dot{v}_1 = \frac{N}{h}\left[Av_N + A(u_N(t) - x(t-h)) - v_1\right]$$

$$\dot{v}_j = \frac{N}{h}\left[v_{j-1} - v_j\right] \qquad j \ge 2; \quad v_j(t_o) = 0, \quad j = 1,\ldots,N$$

Using the variations of constants formula and applying Fubini's theorem we get, by induction:

$$|v_j(t)| \le |A|\frac{N}{h}\int_{t_o}^{t} \frac{1}{(j-1)!}\left[\frac{N}{h}(t-s)\right]^{j-1} e^{-\frac{N}{h}(t-s)}\Big[|v_N(s)|$$
$$+ |u_N(s) - x(s-h)|\Big]ds$$

2. We shall obtain now the estimate (3.4)

$$\int_{t_o}^{t} |y_j(s) - x(s-j\tfrac{h}{N})|ds \le \int_{t_o}^{t} |u_j(s) - x(s-j\tfrac{h}{N})|ds + \int_{t_o}^{t} |v_j(s)|ds$$

Using the estimate for $v_j(t)$ we get, after a change of order of integration

$$\int_{t_o}^{t} |v_j(s)|ds \le |A|\int_{t_o}^{t}\left[\int_{0}^{\frac{N}{h}(t-s)} \frac{\lambda^{j-1}}{(j-1)!} e^{-\lambda}d\lambda\right]\Big[|v_N(s)|$$
$$+ |u_N(s) - x(s-h)|\Big]ds$$

Therefore

$$\int_{t_o}^{t} |v_j(s)|ds \le |A|\int_{t_o}^{t} |v_N(s)|ds + |A|\int_{t_o}^{t} |u_N(s) - x(s-h)|ds$$

Due to the spectral condition upon the matrix A, it is possible to choose norms in order that $|A| < 1$. Therefore:

$$\int_{t_o}^{t} |v_N(s)|ds \le \frac{|A|}{1-|A|}\int_{t_o}^{t} |u_N(s) - x(s-h)|ds$$

and

$$\int_{t_o}^{t} |v_j(s)|ds \le \frac{|A|}{1-|A|}\int_{t_o}^{t} |u_N(s) - x(s-h)|ds \tag{3.11}$$

Taking into account (3.9) and (3.11) we get finally:

$$\int_{t_o}^{t_o+T} |y_j(s) - x(s-j\tfrac{h}{N})|ds \le \frac{1}{1-|A|}\gamma_1(\tfrac{h}{N}) \qquad j = 1,\dots,N$$

3. We shall prove now (3.5).

$$\sum_1^N \int_{-j\frac{h}{N}}^{-(j-1)\frac{h}{N}} |y_j(t) - x(t+s)|ds \le \sum_1^N \int_{-j\frac{h}{N}}^{-(j-1)\frac{h}{N}} |u_j(t) - x(t+s)|ds + \frac{h}{N}\sum_1^N |v_j(t)|$$

Using again the estimate for $v_j(t)$ we get:

$$\frac{h}{N}\sum_1^N |v_j(t)| \le |A|\int_{t_o}^{t}\left\{\sum_1^N \frac{1}{(j-1)!}\left[\frac{N}{h}(t-s)\right]^{j-1}\right\} e^{-\frac{N}{h}(t-s)}\Big[|v_N(s)|$$

$$+ |u_N(s) - x(s-h)|\Big]ds \le |A|\int_{t_o}^{t}|v_N(s)|ds$$

$$+ |A|\int_{t_o}^{t}|u_N(s) - x(s-h)|ds$$

Taking into account (3.9), (3.10) and (3.11) it follows:

$$\sum_1^N \int_{-j\frac{h}{N}}^{-(j-1)\frac{h}{N}} |y_j(t) - x(t+s)|ds \leq \frac{|A|}{1-|A|} \gamma_1(\tfrac{h}{N}) + \gamma_2(\tfrac{h}{N})$$

4. Suppose now that $\varphi \in C(-h,0;\ R^n)$, $g \in C(t_o-h,t_o+T;\ R^n)$ and that (3.6) is fulfilled. From the construction by steps it follows that $x(t)$ is continuous on $[t_o-h,t_o+T]$. Defining $u_j(t)$ and $v_j(t)$ as in the previous cases we have:

$$|u_j(t) - x(t-j\tfrac{h}{N})| \leq \gamma_3(\tfrac{h}{N}) \qquad t_o \leq t \leq t_o+T, \quad j = 1,\dots,N \tag{3.12}$$

This estimate follows from the Lemma for both types of initial conditions (3.3) or (3.3').

From the estimate for $v_j(t)$ we get:

$$|v_j(t)| \leq |A|\Big[\sup_{t_o\leq s\leq t}|v_N(s)| + \sup_{t_o\leq s\leq t}|u_N(s) - x(s-h)|\Big]\frac{N}{h}\int_{t_o}^{t}\frac{1}{(j-1)!}\Big[\frac{N}{h}(t-s)\Big]^{j-1}e^{-\frac{N}{h}(t-s)}ds$$

Using (3.12) and the upper bound for the integral it follows that

$$|v_j(t)| \leq |A| \sup_{t_o\leq s\leq t}|v_N(s)| + |A|\gamma_3(\tfrac{h}{N})$$

This inequality is true for any $t \in [t_o,t_o+T]$ and for any j. Therefore it is true for $\tilde{t} \in [t_o,t_o+T]$ such that

$$|v_N(\tilde{t})| = \sup_{t_o\leq t\leq t_o+T}|v_N(s)|$$

The existence of such t is obvious. Taking also into account that

$$\sup_{t_o\leq s\leq \tilde{t}}|v_N(s)| \leq \sup_{t_o\leq s\leq t_o+T}|v_N(s)|$$

we have

$$|v_N(\tilde{t})| \leq |A|\,|v_N(\tilde{t})| + |A|\gamma_3(\tfrac{h}{N})$$

If norms are chosen again in order that $|A| < 1$, then

$$|v_N(\tilde{t})| \leq \frac{|A|}{1-|A|}\gamma_3(\tfrac{h}{N})$$

and, therefore

$$|v_j(t)| \leq \frac{|A|}{1-|A|}\gamma_3(\tfrac{h}{N}) \qquad j = 1,\dots,N; \quad t_o \leq t \leq t_o+T$$

Taking into account (3.12) we get

$$|y_j(t) - x(t-j\tfrac{h}{N})| \leq \frac{1}{1-|A|}\gamma_3(\tfrac{h}{N}) \qquad j=1,\dots,N; \quad t_o \leq t \leq t_o+T$$

what ends the proof of Theorem 1.

Remarks 1. It can be seen that, after using (3.1) to give (3.2) the form (3.2'), no reference is made any longer throughout the proof to difference equation (3.1). In fact the results of Theorem 1 can be viewed as concerning system (3.2'), where $x(t)$ is a L^1 function extended by φ for $t_o-h \leq t \leq t_o$.

2. All the estimates of Theorem 1 depend in fact on $\alpha_o(r)$, where

$$\alpha_o(r) = \sup_{|t_1-t_2|\leq r} \int_{-h}^{T} |x(t_1+s) - x(t_2+s)|ds$$

This dependence is of the following type

$$\gamma_i(\tfrac{h}{N}) = \tilde{\gamma}_i(\alpha_o(\tfrac{h}{N}))$$

where $\gamma_i(r)$ and $\tilde{\gamma}_i(r)$, as well as $\alpha_o(r)$, are continuous, monotonically increasing and tend to 0 when $r \to 0$. In fact the estimates (i.e. the convergence of the approximation) depend on $\alpha_o(r)$ - a function which is a measure of the smoothness of the approximated solution.

3. The proof of Theorem 1 can be performed in the same way as above if the difference equation (3.1) is replaced by the nonlinear difference equation:

$$x(t) = f(t,x(t-h)) \tag{3.13}$$

where $f(t,\cdot)$ is globally Lipschitzian with the Lipschitz constant smaller than 1. This condition is quite restrictive.

4. As in the case of the Lemma, if $\varphi \in C(-h,0;\ R^n)$ then (3.4) and (3.5) remain true when (3.3) are replaced by (3.3').

IV. THE APPROXIMATION OF COUPLED DIFFERENTIAL AND DIFFERENCE EQUATIONS

Consider the following system:

$$\begin{aligned} \dot{x}(t) &= f(t,x(t),y(t-h)) \\ y(t) &= g(t,x(t)) + A_0 y(t-h) \end{aligned} \tag{4.1}$$

with the initial conditions $x(t_0) = x_0$, $y_{t_0} = \varphi$, where $\varphi \in L^1(-h,0;\ R^m)$. We denote as usually $y_t(s) = y(t+s)$, $-h \le s \le 0$.

Assuming that $f(t,x,y)$ is globally Lipschitzian with respect to x and y uniformly for $t_0 \le t \le t_0+T$ and that $g(t,x)$ is uniformly globally Lipschitz with respect to x for $t_0 \le t \le t_0+T$, the solution of (4.1) has global existence and can be constructed by steps. The solution is defined by the couple $(x(t),y(t))$ where $x(t)$ is absolutely continuous and $y(t)$ is integrable; the difference equation of the system is satisfied a.e.

Define $y_j(t)$ $(j = 0,1,\ldots,N)$ to be the solution of the following system of ODE:

$$\dot{y}_o = f(t, y_o, y_N)$$
$$\dot{y}_1 = \frac{N}{h}[g(t, y_o) + A_o y_N - y_1] \qquad (4.2)$$
$$\dot{y}_j = \frac{N}{h}[y_{j-1} - y_j] \qquad j = 2, \ldots, N$$

with the initial conditions

$$y_o(t_o) = x_o; \quad y_j(t_o) = \frac{N}{h} \int_{-j\frac{h}{N}}^{-(j-1)\frac{h}{N}} \varphi(s)\,ds, \quad j = 1, \ldots, N \qquad (4.3)$$

Theorem 2. Consider the coupled system (4.1) with the initial conditions as above. Suppose that the eigenvalues of A_o are inside the unit circle and

$$|f(t, x_2, y_2) - f(t, x_1, y_1)| \leq L_1|x_2 - x_1| + L_2|y_2 - y_1|$$

$$|g(t, x_2) - g(t, x_1)| \leq L_3|x_2 - x_1|$$

for any $x \in R^n$, $y \in R^m$. Here L_k are the Lipschitz constants.

If $y_j(t)$ are solutions of (4.2) with the initial conditions (4.3), then

$$|y_o(t) - x(t)| \leq \gamma_7(\tfrac{h}{N}) \qquad t_o \leq t \leq t_o + T \qquad (4.4)$$

$$\int_{t_o}^{t_o+T} |y_j(s) - y(s - j\tfrac{h}{N})|\,ds \leq \gamma_8(\tfrac{h}{N}) \qquad j = 1, \ldots, N \qquad (4.5)$$

$$\sum_1^N \int_{-j\frac{h}{N}}^{-(j-1)\frac{h}{N}} |y_j(t) - y(t+s)|\,ds \leq \gamma_9(\tfrac{h}{N}) \qquad t_o \leq t \leq t_o + T \qquad (4.6)$$

If $\varphi \in C(-h, 0;\ R^m)$ and if the following condition of "matching" is fulfilled

$$\varphi(0) = g(t_o, x_o) + A_o \varphi(-h) \qquad (4.7)$$

then

$$|y_j(t) - y(t-j\frac{h}{N})| \le \gamma_{10}(\frac{h}{N}), \qquad j = 1,\dots,N \quad t_o \le t \le t_o+T \tag{4.8}$$

and this is true also if (4.3) are replaced by

$$y_o(t_o) = x_o, \quad y_j(t_o) = \varphi(-j\frac{h}{N}), \qquad j = 1,\dots,N \tag{4.3'}$$

Here $\gamma_i(r)$ have the properties from the previous cases and are independent of t_o.

Proof. Taking into account the second equation of (4.1) system (4.2) can be written as follows:

$$\begin{aligned}
\dot{y}_o &= f(t,y_o,y_N) \\
\dot{y}_1 &= \frac{N}{h}[y(t) + A_o(y_N - y(t-h)) + g(t,y_o) - g(t,x(t)) - y_1] \\
\dot{y}_j &= \frac{N}{h}[y_{j-1} - y_j]
\end{aligned} \tag{4.2'}$$

We define $u_j(t)$ to be solutions of the system

$$\begin{aligned}
\dot{u}_1 &= \frac{N}{h}[y(t) + A_o(u_N - y(t-h)) - u_1] \\
\dot{u}_j &= \frac{N}{h}[u_{j-1} - u_j] \qquad j = 2,\dots,N
\end{aligned} \tag{4.9}$$

with the initial conditions

$$u_j(t_o) = y_j(t_o) = \frac{N}{h}\int_{-j\frac{h}{N}}^{-(j-1)\frac{h}{N}} \varphi(s)ds$$

System (4.9) is of the same type as (3.2). Applying Theorem 1 we get

$$\int_{t_o}^{t_o+T} |u_j(s) - y(s-j\frac{h}{N})|ds \le \gamma_4(\frac{h}{N}) \qquad j = 1,\dots,N \tag{4.10}$$

$$\sum_{1}^{N} \int_{-j\frac{h}{N}}^{-(j-1)\frac{h}{N}} |u_j(t) - y(t+s)|ds \leq \gamma_5(\tfrac{h}{N}) \qquad t_o \leq t \leq t_o+T \tag{4.11}$$

where $\gamma_4(r)$ and $\gamma_5(r)$ are independent of t_o.

Define now $v_j(t)$ to be solutions of the system

$$\begin{aligned}
\dot{v}_1 &= \tfrac{N}{h}[A_o v_N + g(t,y_o(t)) - g(t,x(t)) - v_1] \\
\dot{v}_j &= \tfrac{N}{h}[v_{j-1} - v_j] \\
v_j(t_o) &= 0
\end{aligned} \tag{4.12}$$

It can be seen that $v_j(t) \equiv u_j(t) + v_j(t)$ $(j = 1,...,N)$. As in the case of Theorem 1 we get the estimate

$$|v_j(t)| \leq \frac{N}{h} \int_0^t \frac{1}{(j-1)!}\left[\frac{N}{h}(t-s)\right]^{j-1} e^{-\frac{N}{h}(t-s)} \Big[|A_o||v_N(s)| + L_3|y_o(s) - x(s)|\Big]ds$$

1. We shall obtain now the estimates (4.4) and (4.5).

$$\int_{t_o}^{t} |y_j(s) - y(s-j\tfrac{h}{N})|ds \leq \int_{t_o}^{t} |u_j(s) - y(s-j\tfrac{h}{N})|ds + \int_{t_o}^{t} |v_j(s)|ds$$

Taking into account the estimate for $|v_j(t)|$ we get

$$\int_{t_o}^{t} |v_j(s)|ds \leq |A_o| \int_{t_o}^{t} |v_N(s)|ds + L_3 \int_{t_o}^{t} |y_o(s) - x(s)|ds$$

Due to the spectral condition upon A_o it follows

$$\int_{t_o}^{t} |v_N(s)|ds \leq \frac{L_3}{1 - |A_o|} \int_{t_o}^{t} |y_o(s) - x(s)|ds \tag{4.13}$$

and, therefore

$$\int_{t_o}^{t} |v_j(s)|ds \le \frac{L_3}{1 - |A_o|} \int_{t_o}^{t} |y_o(s) - x(s)|ds \tag{4.14}$$

Denoting

$$z_o(t) = y_o(t) - x(t)$$

we find

$$|z_o(t)| \le L_1 \int_{t_o}^{t} |z_o(s)|ds + L_2 \int_{t_o}^{t} |v_N(s)|ds + L_2 \int_{t_o}^{t} |u_N(s) - y(s-h)|ds$$

Taking into account (4.10) and (4.13) it follows

$$|z_o(t)| \le (L_1 + \frac{L_2L_3}{1 - |A_o|}) \int_{t_o}^{t} |z_o(s)|ds + L_2\gamma_4(\frac{h}{N})$$

We can apply now the Gronwall lemma and get

$$|z_o(t)| = |y_o(t) - x(t)| \le L_2 \exp\left[(L_1 + \frac{L_2L_3}{1- A_o})T\right]\gamma_4(\frac{h}{N}) = \gamma_7(\frac{h}{N})$$

what proves (4.4). Using (4.10), (4.14) and the above estimate we obtain

$$\int_{t_o}^{t_o+T} |y_j(s) - y(s-j\frac{h}{N})|ds \le \left\{1 + \frac{L_2L_3}{1-|A_o|} T \exp\left[(L_1 + \frac{L_2L_3}{1-|A_o|})T\right]\right\}\gamma_4(\frac{h}{N})$$

what proves (4.5).

2. We shall obtain now the estimate (4.6)

$$\sum_{1}^{N}\int_{-j\frac{h}{N}}^{-(j-1)\frac{h}{N}} |y_j(t) - y(t+s)|ds \leq \sum_{1}^{N}\int_{-j\frac{h}{N}}^{-(j-1)\frac{h}{N}} |u_j(t) - y(t+s)|ds + \frac{h}{N}\sum_{1}^{N}|v_j(t)|$$

Using again the estimate for $|v_j(t)|$ it follows

$$\frac{h}{N}\sum_{1}^{N}|v_j(t)| \leq |A_o|\int_{t_o}^{t}|v_N(s)|ds + L_3\int_{t_o}^{t}|y_o(s) - x(s)|ds$$

Taking into account (4.13) and the estimate for $|y_o(t) - x(t)|$ we get

$$\frac{h}{N}\sum_{1}^{N}|v_j(t)| \leq \frac{L_2L_3}{1-|A_o|}\, T\exp\left[(L_1 + \frac{L_2L_3}{1-|A_o|})T\right]\gamma_4(\frac{h}{N})$$

and, using also (4.11):

$$\sum_{1}^{N}\int_{-j\frac{h}{N}}^{-(j-1)\frac{h}{N}} |v_j(t) - y(t+s)|ds \leq \gamma_9(\frac{h}{N})$$

what proves (4.6).

3. Suppose now that $\varphi \in C(-h,0;R^m$ and that (4.7) holds. It is obvious that $y(t)$ is continuous. Defining $u_j(t)$ and $v_j(t)$ as above, system (4.9) is again of the same type as (3.2') but with continuous $y(t)$. Using the 3-d part of Theorem 1 we get

$$|u_j(t) - y(t-j\frac{h}{N})| \leq \gamma_6(\frac{h}{N}) \qquad j=1,\dots,N;\ t_o \leq t \leq t_o+T \qquad (4.15)$$

and this is true for both types of initial conditions - (4.3) or (4.3'). From the estimate for $v_j(t)$ it follows:

$$|v_j(t)| \leq |A_o| \sup_{t_o\leq s\leq t}|v_N(s)| + L_3 \sup_{t_o\leq s\leq t}|y_o(s) - x(s)|$$

Taking into account (4.4) and doing as in the 3-d case of Theorem 1 we find

$$|v_j(t)| \le \frac{L_3}{1 - |A_o|} \gamma_7(\frac{h}{N}) \qquad t_o \le t \le t_o+T, \quad j = 1,\dots,N$$

and, therefore

$$|v_j(t) - y(t-j\frac{h}{N})| \le \gamma_6(\frac{h}{N}) + \frac{L_3}{1 - |A_o|} \gamma_7(\frac{h}{N})$$

what ends the proof of Theorem 2.

Remark. As in the previous cases, if $\varphi \in C(-h,0;\ R^m)$ then the average approximations (4.5) and (4.6) still hold if (4.3) are replaced by (4.3').

As a straightforward application of the above result we shall consider the following neutral FDE:

$$\dot{y}(t) - A_o\dot{y}(t-h) = f(t,y(t),y(t-h)) \tag{4.16}$$

with the initial condition $y_{t_o} = \varphi$, where $\varphi : [-h,0] \to R^n$ is absolutely continuous. If $f(t,y,z)$ is uniformly globally Lipschitz with respect to y and z, global existence is ensured for solutions of (4.16) and these solutions can be constructed by steps. Denoting

$$x(t) = y(t) - A_oy(t-h) \tag{4.17}$$

the couple $(x(t),y(t))$ is the solution of the following system

$$\begin{aligned} \dot{x}(t) &= \tilde{f}(t,x(t),y(t-h)) \\ y(t) &= x(t) + A_oy(t-h) \end{aligned} \tag{4.18}$$

with the initial conditions

$$x(t_o) = y(t_o) - A_oy(t_o-h) = \varphi(0) - A_o\ \ (-h); \quad y_{t_o} = \varphi \tag{4.19}$$

Here $\tilde{f}(t,x,y) = f(t,x+A_oy,y)$ and it can be easily seen that $\tilde{f}(t,x,y)$ is uniformly globally Lipschitz with respect to x and y.

The solution $(x(t), y(t))$ is obviously absolutely continuous (relation (4.19) is exactly the "matching" condition (4.7) for this particular system) and $y(t)$ coincides with the solution of (4.16).

Define $y_j(t)$ $(j = 0,1,\ldots,N)$ to be the solution of the system

$$\begin{aligned} \dot{y}_o &= f(t, y_o, y_N) \\ \dot{y}_1 &= \frac{N}{h}[y_o + A_o y_N - y_1] \\ \dot{y}_j &= \frac{N}{h}[y_{j-1} - y_j] \qquad j = 2,\ldots,N \end{aligned} \tag{4.20}$$

with the initial conditions:

$$\begin{aligned} y_o(t_o) &= x(t_o) = \varphi(0) - A_o \varphi(-h) \\ y_j(t_o) &= \varphi(-j\tfrac{h}{N}), \qquad j = 1,\ldots,N \end{aligned} \tag{4.21}$$

We can state now the approximation result:

Theorem 3. Consider the neutral FDE (4.10) with the initial condition $\varphi : [-h, 0] \to R^n$ being absolutely continuous. Suppose that the eigenvalues of A_o are inside the unit circle and

$$|f(t,x_2,y_2) - f(t,x_1,y_1)| \leq L_1|x_2 - x_1| + L_2|y_2 - y_1|$$

for any $x,y \in R^n$.

Then system (4.20) with the initial conditions (4.21) approximates (4.16) in the sense that

$$|y_o(t) + A_o y_N(t) - y(t)| \leq \gamma_{11}(\tfrac{h}{N}) \qquad t_o \leq t \leq t_o+T \tag{4.22}$$

The proof of this Theorem is straightforward. Indeed, system (4.20) approximates (4.18) in the sense of Theorem 2. Therefore, applying Theorem 2 we get

$$|y_o(t) - x(t)| \leq \gamma_7(\tfrac{h}{N})$$

$$|y_j(t) - y(t-j\tfrac{h}{N})| \leq \gamma_{10}(\tfrac{h}{N}) \qquad j = 1,\ldots,N$$

and

$$|y_o(t) + A_o y_N(t) - y(t)| \leq \gamma_7(\frac{h}{N}) + |A_o|\gamma_{10}(\frac{h}{N})$$

what ends the proof.

Remark. The estimate (4.22) gives in fact the natural sense for approximating NFDE (4.16) by ODE. Indeed, if we make the substitution

$$z_o = y_o + A_o y_N \tag{4.23}$$

the approximating system (4.20) becomes

$$\begin{aligned} &\dot{z}_o - A_o\dot{y}_N = f(t,z_o,y_N) \\ &\dot{y}_1 = \frac{N}{h}[z_o - y_1] \\ &\dot{y}_j = \frac{N}{h}[y_{j-1} - y_j] \qquad j = 2,\dots,N \end{aligned} \tag{4.20'}$$

with the initial conditions

$$z_o(t_o) = \varphi(0); \qquad y_j(t_o) = \varphi(-j\frac{h}{N}), \qquad j = 1,\dots,N \tag{4.21'}$$

It is obvious that system (4.20') is the system one gets by natural approximation of (4.16). Using (4.23) (4.22) becomes

$$|z_o(t) - y(t)| \leq \gamma_{11}(\frac{h}{N}) \qquad t_o \leq t \leq t_o+T \tag{4.22'}$$

which again is the natural sense of the approximation.

However system (4.20) is more suitable for computer processing because it has fewer equations containing small parameters. Indeed, if (4.20') is written in Cauchy form, we have

$$\begin{aligned} &\dot{z}_o = f(t,z_o,y_N) + \frac{N}{h}A_o(y_{N-1} - y_N) \\ &\dot{y}_j = \frac{N}{h}[y_{j-1} - y_j] \qquad j = 2,\dots,N \\ &\dot{y}_1 = \frac{N}{h}[z_o - y_1] \end{aligned} \tag{4.20''}$$

while the first equation of (4.20) does not contain small parameters.

Once again the concept of coupled differential and difference equations has shown to be more adequate for NFDE than the usual form (4.16).

PART II. APPLICATIONS TO THE FOUNDATIONS OF THE METHOD OF LINES

V. STATEMENT OF SOME MIXED INITIAL BOUNDARY VALUE PROBLEMS FOR HYPERBOLIC PARTIAL DIFFERENTIAL EQUATIONS

In the following we shall consider a linear hyperbolic problem defined by

$$\begin{aligned} &\frac{\partial v_1}{\partial t} + q_1 \frac{\partial v_2}{\partial \lambda} = 0 \\ &\frac{\partial v_2}{\partial t} + q_2 \frac{\partial v_1}{\partial \lambda} = 0 \qquad t \geq t_o, \quad 0 \leq \lambda \leq \ell, \quad q_i > 0 \quad (i=1,2) \end{aligned} \tag{5.1}$$

with the initial conditions

$$v_i(\lambda, t_o) = \omega_i(\lambda) \qquad 0 \leq \lambda \leq \ell, \quad i = 1,2 \tag{5.2}$$

and with the boundary conditions

$$\begin{aligned} &v_1(0,t) + \beta_1 v_2(0,t) = \varphi_1(t) \\ &v_2(\ell,t) - \beta_2 v_1(\ell,t) = \varphi_2(t) \end{aligned} \tag{5.3}$$

This is a classical mixed problem for hyperbolic PDE. In several engineering problems concerning control systems containing lossless water, gas or steam pipes, as well as in analysis of electrical networks containing lossless transmission lines [14, chapter V] the boundary conditions (5.3) are controlled by a system of ordinary differential equations which is itself controlled by the boundary conditions as follows:

$$\begin{aligned} v_1(0,t) + \beta_1 v_2(0,t) &= c_1'x(t) - \beta_3\,\varphi(c_o'x(t)) + \varphi_1(t) \\ v_2(\ell,t) - \beta_2 v_1(\ell,t) &= c_2'x(t) - \beta_4\,\varphi(c_o'x(t)) + \varphi_2(t) \\ \dot{x}(t) &= Ax(t) + b_{11}v_1(0,t) + b_{12}v_2(0,t) + b_{21}v_1(\ell,t) \\ &\quad + b_{22}v_2(\ell,t) - b_o\,\varphi(c_o'x(t)) + f(t) \end{aligned} \tag{5.3'}$$

Here an initial condition for the system of ODE must be added:

$$x(t_o) = x_o$$

Other types of applications lead to various types of boundary conditions. For instance, K. L. Cooke [4],[5] considered derivative boundary conditions

$$\begin{aligned} \sum_{o}^{L_1} a_{j1}\frac{d^j}{dt^j} v_1(0,t) + \sum_{o}^{K_2} a_{j2}\frac{d^j}{dt^j} v_2(0,t) &= \varphi_1(t) \\ \sum_{o}^{K_1} b_{j1}\frac{d^j}{dt^j} v_1(\ell,t) + \sum_{o}^{L_2} b_{j2}\frac{d^j}{dt^j} v_2(\ell,t) &= \varphi_2(t) \end{aligned} \tag{5.3"}$$

An even more general situation occurs when the boundary conditions are described by some nonlinear Volterra operators acting on the boundary of the definition domain of PDE [1].

The way of solving such mixed problems is the following [1],[4],[5],[14]. At the beginning a change of functions is performed

$$\begin{aligned} v_1(\lambda,t) &= u_1(\lambda,t) + u_2(\lambda,t) \\ v_2(\lambda,t) &= \sqrt{\frac{q_2}{q_1}}\left[u_1(\lambda,t) - u_2(\lambda,t)\right] \end{aligned} \tag{5.4}$$

Consequently system (5.1) gets the normal hyperbolic form:

$$\begin{aligned} \frac{\partial u_1}{\partial t} + c\,\frac{\partial u_1}{\partial \lambda} &= 0 \\ \frac{\partial u_2}{\partial t} - c\,\frac{\partial u_2}{\partial \lambda} &= 9 \qquad t \geq t_o,\quad 0 \leq \lambda \leq \ell;\quad c = \sqrt{q_1 q_2} \end{aligned} \tag{5.5}$$

and the initial and boundary conditions also get the corresponding form.

Performing then an integration along the characteristics, it can be proved a one-to-one correspondence between the solutions of the mixed problem and the solutions of some functional equations: difference equations for the boundary conditions (5.3), coupled differential and difference equations for (5.3'), delay-differential equations or NFDE for (5.3"), etc.

In the following we shall look for approximations of solutions of the problems formulated above.

VI. APPROXIMATION OF THE SOLUTIONS OF MIXED PROBLEMS IN NORMAL FORM. THE METHOD OF LINES

To make more clear the illustration of the method we shall consider first the mixed problem (5.1)-(5.3). Its normal form is the following:

$$\frac{\partial u_1}{\partial t} + c\,\frac{\partial u_1}{\partial \lambda} = 0; \qquad \frac{\partial u_2}{\partial t} - c\,\frac{\partial u_2}{\partial \lambda} = 0$$

$$u_1(0,t) + \alpha_1 u_2(0,t) = \psi_1(t)$$

$$u_2(\ell,t) + \alpha_2 u_1(\ell,t) = \psi_2(t) \tag{6.1}$$

$$u_1(\lambda,0) = \frac{1}{2}\left[\omega_1(\lambda) + \sqrt{\frac{q_1}{q_2}}\,\omega_2(\lambda)\right] = \tilde{\omega}_1(\lambda)$$

$$u_2(\lambda,0) = \frac{1}{2}\left[\omega_1(\lambda) - \sqrt{\frac{q_1}{q_2}}\,\omega_2(\lambda)\right] = \tilde{\omega}_2(\lambda)$$

where

$$c = \sqrt{q_1 q_2}, \quad \alpha_1 = \frac{1-\beta_1\sqrt{\dfrac{q_2}{q_1}}}{1+\beta_1\sqrt{\dfrac{q_2}{q_1}}}, \quad \alpha_2 = \frac{1-\beta_2\sqrt{\dfrac{q_1}{q_2}}}{1+\beta_2\sqrt{\dfrac{q_1}{q_2}}}$$

$$\psi_1(t) = \frac{\varphi_1(t)}{1+\beta_1\sqrt{\dfrac{q_1}{q_1}}}, \qquad \psi_2(t) = \frac{\sqrt{\dfrac{q_1}{q_2}}\,\varphi_2(t)}{1+\beta_2\sqrt{\dfrac{q_1}{q_2}}} \tag{6.2}$$

Integrating along the characteristics we get, after some rearrengement, the associated difference system

$$\eta_1(t) = -\alpha_1\eta_2(t - \frac{\ell}{c}) + \psi_1(t)$$

$$\eta_2(t) = -\alpha_2\eta_1(t - \frac{\ell}{c}) + \psi_2(t) \tag{6.3}$$

$$\eta_1^o(t) = \tilde{\omega}_1(-c(t-t_o)), \quad \eta_2^o(t) = \tilde{\omega}_2(\ell + c(t-t_o)); \quad t_o - \frac{\ell}{c} \leq t \leq t_o$$

The solution of (6.1) is given by the following representation formulae:

$$u_1(\lambda,t) = \eta_1(t - \frac{\lambda}{c}); \qquad u_2(\lambda,t) = \eta_2(t - \frac{\ell - \lambda}{c}) \tag{6.4}$$

In order to obtain classical solutions for (6.1) $\eta_i(t)$ must be differentiable (at least piece-wise). Consequently some smoothness assumptions are needed. In the following we shall consider $\omega_i \in C^1[0,\ell]$ and $\varphi_i \in C^1[t_o,t_o+T]$.

Taking into account the results of Part I, a way of obtaining approximations for the solutions of (6.1) is to approximate (6.3) by a system of ODE and to define an approximating solution of (6.1) based on the solution of the approximating system.

Using (3.2) and (3.3') we get the following system:

$$\dot{\zeta}_1^1 = \frac{Nc}{\ell}[-\alpha_1\zeta_2^N + \psi_1(t) - \zeta_1^1]$$

$$\dot{\zeta}_2^1 = \frac{Nc}{\ell}[-\alpha_2\zeta_1^N + \psi_2(t) - \zeta_2^1] \tag{6.5}$$

$$\dot{\zeta}_k^j = \frac{Nc}{\ell}[\zeta_k^{j-1} - \zeta_k^j] \qquad k = 1,2; \quad j = 2,\dots,N$$

with the initial conditions

$$\zeta_1^j(t_o) = \tilde{\omega}_1(-c(t_o - j\frac{\ell}{Nc} - t_o)) = \tilde{\omega}_1(j\frac{\ell}{N})$$

$$\zeta_2^j(t_o) = \tilde{\omega}_2(\ell + c(t_o - j\frac{\ell}{Nc} - t_o)) = \tilde{\omega}_2((N-j)\frac{\ell}{N}) \tag{6.6}$$

$$j = 1,\dots,N$$

In order to apply Theorem 1 we shall check the discrete stability of matrix A given by

$$A = \begin{bmatrix} 0 & -\alpha_1 \\ -\alpha_2 & 0 \end{bmatrix}$$

The stability condition is $|\alpha_1\alpha_2| < 1$ which leads to a corresponding relation between β_i and q_i. If this condition is fulfilled we can use Theorem 1 and find

$$\int_{t_o}^{t_o+T} |\zeta_k^j(s) - \eta_k(s - j\tfrac{\ell}{Nc})|ds \le \hat{\gamma}_4(\tfrac{\ell}{N}) \qquad j = 1,\ldots,N, \quad k = 1,2$$

To point out the significance of this estimate we shall use (6.4). It follows:

$$\int_{t_o}^{t_o+T} |\zeta_1^j(s) - u_1(j\tfrac{\ell}{N}, s)|ds \le \gamma_4(\tfrac{\ell}{N}) \qquad j = 1,\ldots,N \qquad (6.7a)$$

$$\int_{t_o}^{t_o+T} |\zeta_2^{N-j}(s) - u_2(j\tfrac{\ell}{N}, s)|ds \le \hat{\gamma}_4(\tfrac{\ell}{N}) \qquad j = 0,\ldots,N-1 \qquad (6.7b)$$

Now it is obvious that $\zeta_i^j(t)$ represent the average approximation of $u_1(j\frac{\ell}{N}, t)$ $(j = 1,\ldots,N)$ and $\zeta_2^{N-j}(t)$ are the approximation of $u_2(j\frac{\ell}{N}), t)$ $(j = 0,\ldots,N-1)$. It can be seen that for $u_1(0,t)$ and $u_2(\ell,t)$ we have no approximation but they are given by the boundary conditions.

To divide the interval $(0,\ell)$ in N equal parts, to perform discretization with respect to this space variable and to solve the system of ODE thus obtained in order to get an approximation of $u_k(j\frac{\ell}{N}, t)$ is usually called the method of lines. Relations (6.7) show that system (6.5), discovered by applying a method of approximation used for FDE corresponds to a suitable discretization of (6.1) with respect to λ.

Indeed, let us perform the usual Euler discretization, but according to the following rule:

$$\frac{\partial v}{\partial \lambda}(\lambda,t) \approx \frac{v(\lambda+h,t) - v(\lambda,t)}{h}$$

in the equation where $-c$ occurs, and

$$\frac{\partial v}{\partial \lambda}(\lambda,t) \approx \frac{v(\lambda,t) - v(\lambda-h,t)}{h}$$

in the equation where $+c$ occurs.

Taking into account that $u_1(0,t)$ and $u_2(\ell,t)$ are given explicitely by the boundary conditions, we write

$$\frac{\partial u_1}{\partial t}\left(j\frac{\ell}{N}, t\right) + c\,\frac{\partial u_1}{\partial \lambda}\left(j\frac{\ell}{N}, t\right) = 0 \qquad j = 1,\dots,N$$

$$\frac{\partial u_2}{\partial t}\left(j\frac{\partial}{N}, t\right) - c\,\frac{\partial u_2}{\partial \lambda}\left(j\frac{\ell}{N}, t\right) = 0 \qquad j = 0,\dots,N-1$$

Applying the discretization rule we can write (formally):

$$\frac{d}{dt}\,u_1\left(j\frac{\ell}{N},t\right) + \frac{Nc}{\ell}\left[u_1\left(j\frac{\ell}{N},t\right) - u_1\left((j-1)\frac{\ell}{N}),t\right)\right] = 0 \qquad j = 1,\dots,N$$

$$\frac{d}{dt}\,u_2\left(j\frac{\ell}{N},t\right) - \frac{Nc}{\ell}\left[u_2\left((j+1)\frac{\ell}{N},t\right) - u_2\left(j\frac{\ell}{N},t\right)\right] = 0 \qquad j = 0,\dots,N-1$$

Using the correspondence suggested by (6.7)

$$\zeta_1^j(t) \sim u_1\left(j\frac{\ell}{N}, t\right) \qquad j = 1,\dots,N$$

$$\zeta_2^{N-j}(t) \sim u_2\left(j\frac{\ell}{N}, t\right) \qquad j = 0,\dots,N-1$$

and expressing $u_1(0,t)$ and $u_2(\ell,t)$, which are not defined by differential equations, from the boundary conditions, we get:

$$\dot{\zeta}_1^1 = \frac{Nc}{\ell}\left[-\alpha_1\zeta_2^N + \psi_1(t) - \zeta_1^1\right]$$

$$\dot{\zeta}_2^1 = \frac{Nc}{\ell}\left[-\alpha_2\zeta_1^N + \psi_2(t) - \zeta_2^1\right]$$

$$\dot{\zeta}_k^j = \frac{Nc}{\ell}\left[\zeta_k^{j-1} - \zeta_k^j\right] \qquad k = 1,2; \quad j = 2,\dots,N$$

We rediscovered system (6.5). If the initial conditions are choosen according to (6.6), then (6.7) are true and these relations give the sense of the approximation via the method of lines.

Therefore we obtained the foundation of the method of lines for the mixed problem (6.1); the discretization is performed according to the rule mentioned above and the initial conditions for the system of ODE thus obtained must be chosen according to (6.6).

We must point out that this foundation remains true if initial conditions (6.6) are replaced by the other initial conditions that are "compatible" with Theorem 1

$$\zeta_1^j(t_0) = \frac{N}{\ell}\int_{(j-1)\frac{\ell}{N}}^{j\frac{\ell}{N}} \tilde{\omega}_1(s)\,ds \;; \qquad \zeta_2^{N-j}(t_0) = \frac{N}{\ell}\int_{(N-j)\frac{\ell}{N}}^{(N-j+1)\frac{\ell}{N}} \tilde{\omega}_2(s)\,ds \tag{6.8}$$

The initial conditions (6.6) and (6.8) can be used both for pointwise and average approximations. The pointwise approximation can be obtained when the "matching" condition of Theorem 1 is fulfilled. In our case this "matching" condition gives:

$$\eta_1^0(t_0) = -\alpha_1\eta_2^0\left(t_0 - \frac{\ell}{c}\right) + \psi_1(t_0)$$

$$\eta_2^0(t_0) = -\alpha_2\eta_1^0\left(t_0 - \frac{\ell}{c}\right) + \psi_2(t_0)$$

or

$$\tilde{\omega}_1(0) = -\alpha_1\tilde{\omega}_2(0) + \psi_1(t_0)$$

$$\tilde{\omega}_2(\ell) = -\alpha_2\tilde{\omega}_2(\ell) + \psi_2(t_0)$$

which are exactly the conditions that must be observed in order to have "matched" initial and boundary conditions. In this case we have for the method of lines:

$$|\zeta_1^j(t) - u_1(j\tfrac{\ell}{N}, t)| \le \hat{\gamma}_6(\tfrac{\ell}{N}) \qquad j = 1,\dots,N; \quad t_0 \le t \le t_0+T$$

$$|\zeta_2^{N-j}(t) - u_2(j\tfrac{\ell}{N}, t)| \le \hat{\gamma}_6(\tfrac{\ell}{N}) \qquad j = 0,\dots,N-1; \quad t_0 \le t \le t_0+T$$

The application of Theorem 1 leads also to the following

approximation result:

$$\sum_1^N \int_{-j\frac{\ell}{Nc}}^{(-j-1)\frac{\ell}{Nc}} |\zeta_k^j(t) - \eta_k(t+s)|ds \le \hat{\gamma}_5(\tfrac{\ell}{N}) \qquad t_o \le t \le t_o+T; \qquad k = 1,2$$

Under the first integral (for $k = 1$) we make the change of variable

$$s = -\frac{\lambda}{c}$$

and under the second (for $k = 2$) the following change:

$$s = \frac{\lambda-\ell}{c}$$

Therefore

$$\sum_1^N \int_{(j-1)\frac{\ell}{N}}^{j\frac{\ell}{N}} |\zeta_1^j(t) - \eta_1(t-\tfrac{\lambda}{c})|d\lambda \le \hat{\gamma}_5(\tfrac{\ell}{N})$$

$$\sum_1^N \int_{(j-1)\frac{\ell}{N}}^{j\frac{\ell}{N}} |\zeta_2^{N-j+1}(t) - \eta_2(t-\tfrac{\lambda-\ell}{c})|d\lambda \le \hat{\gamma}_5(\tfrac{\ell}{N})$$

Defining the following piece-wise constant (with respect to λ) functions

$$\begin{aligned} u_1^N(\lambda,t) &= \zeta_1^j(t) \qquad (j-1)\tfrac{\ell}{N} < \lambda < j\tfrac{\ell}{N}\ ; \qquad j = 1,\dots,N \\ u_2^N(\lambda,t) &= \zeta_2^{N-j+1}(t) \qquad (j-1)\tfrac{\ell}{N} < \lambda < j\tfrac{\ell}{N}\ ; \qquad j = 1,\dots,N \end{aligned} \tag{6.9}$$

and taking into account (6.4) we find that $u_k^N(\lambda,t)$ are an approximation for the solution of (6.1) in the following sense:

$$\int_o^\ell |u_k^N(\lambda,t) - u_k(\lambda,t)|d\lambda \le \hat{\gamma}_5(\tfrac{\ell}{N}), \qquad t_o \le t \le t_o+T; \quad k = 1,2$$

It must be pointed out that although the approximation (6.9) is also generated by (6.5) it is not the solution via the method of lines. It is however an average approximation of the solution by piece-wise constant (with respect to the space variable) functions which may be useful in applications. The fact that both types of average approximations (with respect to t - the method of lines - and with respect to λ) are generated by the same system of ODE is due to the representation formulae (6.4).

The approximation methods above (the method of lines and the other one) can be extended for more complicated boundary conditions, some of them being mentioned in the previous paragraph. Indeed, the rule of discretization applies to the equations (5.5) which are the same for all the cases. The discretization of the PDE generates the differential equations for $j \geq 2$ i.e. the approximation of the delay (or of the translation operator [8], [10], [11]). The other part of the approximating system is generated by the boundary conditions and it is mainly this part that gives the specific structure. For instance, the boundary conditions (5.3') lead to a system of coupled differential and difference equations. Therefore Theorem 2 may be used for the foundation of the method of lines. It will result always a pointwise approximation for $x(t)$ - the state vector of the system of ODE that controls the boundary conditions and average or pointwise approximation for $u_k(\lambda,t)$ - the solution of the PDE.

The derivative boundary conditions (5.3") are much more general than the previous ones. According to the values of K_1, K_2, L_1, L_2, such boundary conditions can generate difference equations, coupled differential and difference equations, neutral FDE or time lag equations. For all these types of equations there are known approximations by ODE, hence the method of lines has a rigorous foundation. Moreover, it seems not very difficult to extend the approximation results to even more complicated FDE and to get the extension of the method of lines for the corresponding boundary conditions of the mixed problems.

VII. THE METHOD OF LINES FOR HYPERBOLIC PARTIAL DIFFERENTIAL EQUATIONS (MIXED INITIAL BOUNDARY VALUE PROBLEMS)

For simplicity we shall consider again the mixed problem defined by (5.1)-(5.3). The problem is to find approximations for $v_k(j\frac{\ell}{N},t)$. The boundary conditions (5.3) allow us to obtain $v_1(0,t)$ and $v_2(\ell,t)$ if $v_2(0,t)$ and $v_1(\ell,t)$ (or their approximations) are known. Hence we need approximations for $v_1(j\frac{\ell}{N},t)$ $(j = 1,\dots,N)$ and $v_2(j\frac{\ell}{N},t)$ $(j = 0,\dots,N-1)$.

To solve the approximation problem formulated above we have at our disposal the results concerning the mixed problem in normal form. We write again these results

$$\begin{aligned} &\int_{t_o}^{t_o+T} |\zeta_1^j(s) - u_1(j\tfrac{\ell}{N},s)|ds \le \hat{\gamma}_4(\tfrac{\ell}{N}) \qquad j = 1,\dots,N \\ &\int_{t_o}^{t_o+T} |\zeta_2^{N-j}(s) - u_2(j\tfrac{\ell}{N},s)|ds \le \hat{\gamma}_4(\tfrac{\ell}{N}) \qquad j = 0,\dots,N-1 \end{aligned} \tag{7.1}$$

where $\zeta_k^j(t)$ are solutions of (6.5) with the initial conditions (6.6). From (5.4) we have:

$$\begin{aligned} &v_1(j\tfrac{\ell}{N},t) = u_1(j\tfrac{\ell}{N},t) + u_2(j\tfrac{\ell}{N},t) \qquad j = 1,\dots,N \\ &v_2(j\tfrac{\ell}{N},t) = \sqrt{\frac{q_2}{q_1}}\,[u_1(j\tfrac{\ell}{N},t) - u_2(j\tfrac{\ell}{N},t)] \qquad j = 0,\dots,N-1 \end{aligned} \tag{7.2}$$

The previous results give approximating functions for $u_1(j\frac{\ell}{N},t)$ $(j = 1,\dots,N)$ and $u_2(j\frac{\ell}{N},t)$ $(j = 0,\dots,N-1)$. However, to define approximating functions for $v_k(j\frac{\ell}{N},t)$ we need approximations for $u_1(0,t)$ and $u_2(\ell,t)$. These approximations can be defined from the boundary conditions of (6.1):

$$\zeta_1^o(t) = -\alpha_1\zeta_2^N(t) + \psi_1(t)$$

$$\zeta_2^o(t) = -\alpha_2\zeta_1^N(t) + \psi_2(t)$$

We are able now to define the approximating functions for $v_1(j\frac{\ell}{N},t)$ $(j = 1,\dots,N)$ and $v_2(j\frac{\ell}{N},t)$ $(j = 0,\dots,N-1)$:

$$\xi_1^j(t) = \zeta_1^j(t) + \zeta_2^{N-j}(t) \qquad j = 1,\dots,N-1$$

$$\xi_1^N(t) = \zeta_1^N(t) + \zeta_2^o(t) = (1-\alpha_2)\zeta_1^N(t) + \psi_2(t)$$

$$\xi_2^o(t) = \sqrt{\frac{q_2}{q_1}}\left[\zeta_1^o(t) - \zeta_2^N(t)\right] = \sqrt{\frac{q_2}{q_1}}\left[-(1+\alpha_1)\zeta_2^N(t)+\psi_1(t)\right] \tag{7.3}$$

$$\xi_2^j(t) = \sqrt{\frac{q_2}{q_1}}\left[\zeta_1^j(t) - \zeta_2^{N-j}(t)\right] \qquad j = 1,\dots,N-1$$

We must check first that these functions are really approximating for $v_k(j\frac{\ell}{N},t)$. Using (7.1), (7.2), (7.3) one finds immediately:

$$\int_{t_o}^{t_o+T} |\xi_1^j(s) - v_1(j\tfrac{\ell}{N},s)|ds \le 2\hat{\gamma}_4(\tfrac{\ell}{N}) \qquad j = 1,\dots,N-1$$

$$\int_{t_o}^{t_o+T} |\xi_1^N(s) - v_1(\ell,s)|ds \le \frac{2}{1+\beta_2\sqrt{\frac{q_1}{q_2}}}\,\hat{\gamma}_4(\tfrac{\ell}{N})$$

$$\int_{t_o}^{t_o+T} |\xi_2^o(s) - v_2(0,s)|ds \le \frac{2\sqrt{\frac{q_2}{q_1}}}{1+\beta_1\sqrt{\frac{q_2}{q_1}}}\,\hat{\gamma}_4(\tfrac{\ell}{N})$$

$$\int_{t_o}^{t_o+T} |\xi_2^j(s) - v_2(j\tfrac{\ell}{N},s)|ds \le 2\sqrt{\frac{q_2}{q_1}}\,\hat{\gamma}_4(\tfrac{\ell}{N}) \qquad j = 1,\dots,N-1$$

The next problem is to find the system of ODE that is satisfied by the approximating functions (7.3). Taking into account that $\xi_k^j(t)$ are solutions of (6.5) we find, after some lengthy but straightforward computation, the following system:

$$\dot{\xi}_1^1 = \frac{Nc}{2\ell}\left[\xi_1^2 - 2\xi_1^1 - \beta_1\xi_2^o + \varphi_1(t)\right] - \frac{N}{2\ell} q_1(\xi_2^2 - \xi_2^o)$$

$$\dot{\xi}_1^j = \frac{Nc}{2\ell}(\xi_1^{j-1} - 2\xi_1^j + \xi_1^j) - \frac{N}{2\ell} q_1(\xi_2^{j+1} - \xi_2^j) \qquad j = 2,\dots,N-2$$

$$\dot{\xi}_1^{N-1} = \frac{Nc}{2\ell}(\xi_1^{N-2} - 2\xi_1^{N-1} + \xi_1^N) - \frac{N}{2\ell} q_1\left[\beta_2\xi_1^N - \xi_2^{N-2} + \varphi_2(t)\right]$$

$$\left(1 + \beta_2\sqrt{\frac{q_1}{q_2}}\right)\dot{\xi}_1^N = \frac{Nc}{\ell}(\xi_1^{N-1} - \xi_1^N) - \frac{N}{\ell} q_1\left[\beta_2\xi_1^N - \xi_2^{N-1} + \varphi_2(t) + \frac{\ell}{Nc}\dot{\varphi}_2(t)\right]$$

$$\left(1 + \beta_1\sqrt{\frac{q_2}{q_1}}\right)\dot{\xi}_2^o = \frac{Nc}{\ell}(\xi_2^1 - \xi_2^o) - \frac{N}{\ell} q_2\left[\beta_1\xi_2^o + \xi_1^1 - \varphi_1(t) - \frac{\ell}{Nc}\dot{\varphi}_1(t)\right] \tag{7.4}$$

$$\dot{\xi}_2^1 = \frac{Nc}{2\ell}(\xi_2^o - 2\xi_2^1 + \xi_2^2) - \frac{N}{2\ell} q_2\left[\beta_1\xi_2^o + \xi_1^2 - \varphi_1(t)\right]$$

$$\dot{\xi}_2^j = \frac{Nc}{2\ell}(\xi_2^{j-1} - 2\xi_2^j + \xi_2^{j+1}) - \frac{N}{2\ell} q_2(\xi_1^{j+1} - \xi_1^j) \qquad j = 2,\dots,N-2$$

$$\dot{\xi}_2^{N-1} = \frac{Nc}{2\ell}\left[\xi_2^{N-2} - 2\xi_2^{N-1} + \beta_2\xi_1^N + \varphi_2(t)\right] - \frac{N}{2\ell} q_2(\xi_1^N - \xi_1^{N-2})$$

We must find now the initial conditions for $\xi_k^j(t)$. Using (7.3) and (6.6) we find

$$\xi_1^j(t_o) = \zeta_1^j(t_o) + \zeta_1^{N-j}(t_o) = \omega_1(j\tfrac{\ell}{N}) = v_1(j\tfrac{\ell}{N}, t_o)$$
$$j = 1,\ldots,N-1$$

$$\xi_1^N(t_o) = (1-\alpha_2)\zeta_1^N(t_o) + \psi_2(t_o)$$

$$= \omega_1(\ell) + \frac{\sqrt{\dfrac{q_1}{q_2}}}{1+\beta_2\sqrt{\dfrac{q_1}{q_2}}}\left[\omega_2(\ell) - \beta_2\omega_1(\ell) - \varphi_2(t_o)\right]$$

$$= v_1(\ell,t_o) + \frac{\sqrt{\dfrac{q_1}{q_2}}}{1+\beta_2\sqrt{\dfrac{q_1}{q_2}}}\left[v_2(\ell,t_o) - \beta_2 v_1(\ell,t_o) - \varphi_2(t_o)\right] \tag{7.5}$$

$$\xi_2^o(t_o) = \sqrt{\frac{q_2}{q_1}}\left[-(1+\alpha_1)\zeta_2^N(t_o) + \psi_1(t_o)\right] =$$

$$= \omega_2(0) - \frac{\sqrt{\dfrac{q_2}{q_1}}}{1+\beta_1\sqrt{\dfrac{q_2}{q_1}}}\left[\omega_1(0) + \beta_1\omega_2(0) - \varphi_1(t_o)\right]$$

$$= v_2(0,t_o) - \frac{\sqrt{\dfrac{q_2}{q_1}}}{1+\beta_1\sqrt{\dfrac{q_2}{q_1}}}\left[v_1(0,t_o) + \beta_1 v_2(0,t_o) - \varphi_1(t_o)\right]$$

$$\xi_2^j(t_o) = \sqrt{\frac{q_2}{q_1}}\left[\zeta_1^j(t_o) - \zeta_2^{N-j}(t_o)\right] = \omega_2(j\tfrac{\ell}{N}) = v_2(j\tfrac{\ell}{N}, t_o)$$
$$j = 1,\ldots,N-1$$

It is obvious that the initial conditions for the approximating functions (7.3) are identical with the initial conditions of the approximated functions, except $v_1(\ell,t)$ a,d $v_2(0,t)$. The coincidence for these last functions is ensured if

$$\omega_1(0) + \beta_1\omega_2(0) = \varphi_1(t_o)$$

$$\omega_2(\ell) - \beta_2\omega_1(\ell) = \varphi_2(t_o)$$

i.e. if initial and boundary values are "matched". In this case, as it has already been shown (see (3.7)), the approximation is pointwise:

$$|\xi_1^j(t) - v_1(j\frac{\ell}{N},t)| \le 2\hat{\gamma}_6(\frac{\ell}{N}) \qquad j = 1,\dots,N-1$$

$$|\xi_1^N(t) - v_1(\ell,t)| \le \frac{2}{1+\beta_2\sqrt{\frac{q_1}{q_2}}}\hat{\gamma}_6(\frac{\ell}{N})$$

$$|\xi_2^o(t) - v_2(0,t)| \le \frac{2\sqrt{\frac{q_2}{q_1}}}{1+\beta_1\sqrt{\frac{q_2}{q_1}}}\hat{\gamma}_6(\frac{\ell}{N})$$

$$|\xi_2^j(t) - v_2(j\frac{\ell}{N},t)| \le 2\sqrt{\frac{q_2}{q_1}}\hat{\gamma}_6(\frac{\ell}{N}) \qquad j = 1,\dots,N-1$$

If the initial and boundary conditions are not "matched", this "unmatching" is propagated (propagation of singularities); for this reason only average approximation can be obtained.

We have to show now that system (7.4) can be obtained from (5.1)-(5.3) by a suitable discretization with respect to the space variable. First we give (5.1) a form which separates the characteristic directions

$$\frac{\partial v_2}{\partial t} + c\frac{\partial v_2}{\partial \lambda} + \sqrt{\frac{q_2}{q_1}}\left(\frac{\partial v_1}{\partial t} + c\frac{\partial v_1}{\partial \lambda}\right) = 0$$

$$\frac{\partial v_2}{\partial t} - c\frac{\partial v_2}{\partial \lambda} - \sqrt{\frac{q_2}{q_1}}\left(\frac{\partial v_1}{\partial t} - c\frac{\partial v_1}{\partial \lambda}\right) = 0$$

Afterwards we perform the usual Euler discretization according to the same rule as previously:

$$\frac{\partial v}{\partial \lambda}(\lambda,t) \simeq \frac{v(\lambda+h,t) - v(\lambda,t)}{h}$$

in the equations where $-c$ occurs, and

$$\frac{\partial v}{\partial \lambda}(\lambda,t) \approx \frac{v(\lambda,t) - v(\lambda-h,t)}{h}$$

in the equations where $+c$ occurs. Therefore we can write (formally):

$$\frac{d}{dt} v_2(j\tfrac{\ell}{N},t) + \frac{Nc}{\ell}\left[v_2(j\tfrac{\ell}{N},t) - v_2((j-1)\tfrac{\ell}{N},t)\right]$$

$$+\sqrt{\frac{q_2}{q_1}}\left\{\frac{d}{dt} v_1(j\tfrac{\ell}{N},t) + \frac{Nc}{\ell}\left[v_1(j\tfrac{\ell}{N},t) - v_1((j-1)\tfrac{\ell}{N},t)\right]\right\} = 0$$

$$j = 1,\ldots,N$$

$$\frac{d}{dt} v_2(j\tfrac{\ell}{N},t) - \frac{Nc}{\ell}\left[v_2((j+1)\tfrac{\ell}{N},t) - v_2(j\tfrac{\ell}{N},t)\right]$$

$$-\sqrt{\frac{q_2}{q_1}}\left\{\frac{d}{dt} v_1(j\tfrac{\ell}{N},t) - \frac{Nc}{\ell}\left[v_1((j+1)\tfrac{\ell}{N},t) - v_1(j\tfrac{\ell}{N},t)\right]\right\} = 0$$

$$j = 0,\ldots,N-1$$

Substituting in the above equations the expressions of $v_1(0,t)$ and $v_2(\ell,t)$ and of their derivatives, given by the boundary conditions (5.3), and using the following correspondence:

$$\xi_1^j(t) \sim v_1(j\tfrac{\ell}{N},t) \qquad j = 1,\ldots,N$$

$$\xi_2^j(t) \sim v_2(j\tfrac{\ell}{N},t) \qquad j = 0,\ldots,N-1$$

a certain system of ODE is obtained. When this system is given normal Cauchy form, one gets system (7.4).

We have in fact proved the convergence of the approximating functions obtained by applying the method of lines with the discretization rule mentioned above. It is obvious that the result is true at least for those boundary conditions which generate such functional equations that can be approximated by ODE (See Part I).

Now, the discretization rule, found here to correspond - via the normal hyperbolic form - to the approximation of time delays by ODE, is exactly the rule used by Courant, Isaacson and Rees [6] for solving only Cauchy problems, but for more general, nonlinear hyperbolic PDE.

A natural generalization of all these results is to apply the method of lines with the rule of discretization just mentioned to mixed problems for nonlinear hyperbolic PDE. Such generalization will require a suitable development of the theory of approximating functional equations by ODE if one likes to use the integration along the characteristics. Another way is the direct generalization of the results of [6].

REFERENCES

1. Abolinia, V. E., and Myshkis, A. D., A mixed problem for an almost linear hyperbolic system in the plane (in Russian), *Matem. Sbornik 50:92*, 423-442 (1960).
2. Banks, H. T., and Burns, J. A., Hereditary Control Problems: Numerical Methods Based on Averaging Approximations. January 1976 (Submitted to *SIAM J. Control and Optimization*)
3. Bremermann, H., Distributions, Complex Variables and Fourier Transforms. Addison Wesley, Reading, (1965).
4. Cooke, K. L., and Krumme, D. W., Differential-Difference Equations and Nonlinear Initial-Boundary Value Problems for Linear Hyperbolic Partial Differential Equations, *J. Math. Anal. Appl. 24*, 372-387 (1968).
5. Cooke, K. L., A linear mixed problem with derivative boundary conditions, *in* "Seminar on Differential Equations and Dynamical Systems", III, Lecture Series No 51, University of Maryland, College Park, pp. 11-17 (1970).
6. Courant, R., Isaacson, E., and Rees, M., On the Solutions of Nonlinear Hyperbolic Differential Equations by Finite Differences, *Comm. Pure Appl. Math. 5*, 243-255 (1952).
7. Fifer, St., Analogue Computation, Volume IV, McGraw Hill, New York, (1961).
8. Kappel, F., and Schappacher, W., Autonomous Nonlinear Functional Differential Equations and Averaging Approximations, *Nonlinear Analysis. Theory, Methods and Applications 2*, 391-422 (1978).
9. Krasovskii, N. N., The approximation of a problem of analytic design of controls in a system with time lag (in Russian), *Prikl. Mat. Mekh. 28*, 716-724 (1964). (English

version in *J. Appl. Math. Mech. 28*, 876-885 (1964).

10. Kunisch, K., Neutrale Funktional-Differentialgleichungen und Halbgruppentheorie, *Ber.math.-stat.Sekt.* Forschungszentrum Graz No 81 (1977).
11. Kunisch, K., Neutral Functional Differential Equations and Semi-Group Theory (1978, unpublished).
12. Manitius, A., Optimal Control of Hereditary Systems. Lecture notes for the "Autumn Mathematical Course, Intern. Centre for Theoretical Physics", Miramare, Trieste, Italy, December (1974).
13. Popov, Ye. P., Dynamics of automatic control systems (in Russian), Gostekhizdat, Moscow, (1954).
14. Răsvan, Vl., Absolute stability of time lag control systems, (in Romanian), Editura Academiei, Bucharest, (1975).
15. Repin, Yu. M., On the approximate replacement of systems with lag by ordinary differential equations (in Russian), *Prikl. Mat. Mekh. 29*, 226-235 (1965). (English version in *J. Appl. Math. Mech. 29*, 254-264 (1965).
16. Sobolev, S. L., Applications of functional analysis in mathematical physics (in Russian), Izd. Sib. Otdel. Akad. Nauk SSSR, Novosibirsk, 1962 (English version: AMS Translations of Mathem. Monographs, vol. 7, Providence, 1963).

LINEAR STIELTJES INTEGRO-DIFFERENTIAL EQUATIONS

Chaim Samuel Hönig

Instituto de Matematica e Estatistica
Universidade de São Paulo
São Paulo, Brasil

ABSTRACT

We consider linear Stieltjes integro-differential equations

$$y(t) - x + \int_s^t \cdot\, dA(\sigma)\cdot y(\sigma) = f(t) - f(s) \qquad t \in [a,b] \tag{L}$$

These equations generalize the usual linear differential equations. We take X a Banach space, $A : [a,b] \to L(X)$ a function of bounded variation (or, more generally, of bounded semivariation - see §2) and $y,f : [a,b] \to X$ regulated functions (i.e., functions that have only discontinuities of the first kind). We prove that (L) has always (i.e., for every s, x and f) one and only one solution if and only if for every $t \in [a,b[$ $(t \in]a,b])$ the operator $I_X + [A(t+) - A(t)]$ $(I_X - [A(t) - A(t-)])$ is invertible (in $L(X)$). We thus generalize and complete results of MacNerney [9], Hildebrandt [3] and Souza [10].

[1]Present address: Instituto de Matematica e Estatistica, Universidade de São Paulo, São Paulo, Brasil.

ISBN 0-12-186280-1

I. INTRODUCTION

The study of equation (L) started with Wall, [12], and MacNerney, [9]. In [9] the existence of a resolvent of (L) is proved in the case where $A : [a,b] \to L(X)$ is a continuous function of bounded variation (and the functions $y, f : [a,b] \to X$ are of bounded variation). In [3] Hildebrandt extended these results to the case where A allows discontinuities, i.e., A is a function of bounded variation. In [4] we proved the existence of a resolvent in the case where A is a continuous function of bounded semivariation (and y, f are regulated functions). In [10] and [1] these results were extended to certain cases where A allows discontinuities.

Here we present a summary of our results on equation (L); the complete proofs will appear in [7] where we prove also many results on the linear Volterra Stieltjes-integral equation

$$y(t) - x + \int_{t_0}^{t} d_\tau (K(t,\tau)\cdot y(\tau) = f(t) - f(t_0) \qquad (K)$$

For other articles on equation (K) see [2],[4],[11].

II. NOTATIONS

(See also [4],[5],[6]) - A *division* of $[a,b]$ is a finite sequence $d : t_0 = a < t_1 < \ldots < t_n = b$. We write $|d| = n$ and $\Delta d = \sup_{1 \le i \le n} |t_i - t_{i-1}|$. The set D of all divisions of $[a,b]$ is ordered by refinement and $\lim_{d \in D} x_d$ denotes the limit according to the associated net.

Let X and Y be Banach spaces and $\alpha : [a,b] \to L(X,Y)$. We define $SV[\alpha] = SV_{[a,b]}[\alpha] = \sup_{d \in D} SV_d[\alpha]$ where

$$SV_d[\alpha] = \sup \left\{ \left\| \sum_{i=1}^{|d|} [\alpha(t_i) - \alpha(t_{i-1})]\cdot x_i \right\| \; \middle| \; x_i \in X, \|x_i\| \le 1 \right\}.$$

If $SV[\alpha] < \infty$ we say that α is of *bounded semivariation* and we write $\alpha \in SV([a,b],L(X,Y))$; if we have furthermore $\alpha(a) = 0$ we write $\alpha \in SV_a([a,b],L(X,Y))$. If $Y = C$ we have

$$SV_d|\alpha| = V_d|\alpha| = \sum_{i=1}^{|d|} \|\alpha(t_i) - \alpha(t_{i-1})\|, \text{ hence } V[\alpha] = \sup_{d\in D} V_d[\alpha]$$

is the usual *variation* of α. If α is a function of bounded variation, i.e., if $V[\alpha] < \infty$, we write $\alpha \in BV([a,b],X')$.

If $f : [a,b] \to X$ is *regulated* (i.e., if for every $t \in [a,b[$ $\left(t\in]a,b]\right)$ there exists $f(t+)$ $[f(t-)]$ we write $f \in G([a,b],X)$. We obtain a Banach space when we endow $G([a,b],X)$ with the sup norm ($\|f\| = \sup_{a\leq t\leq b} \|f(t)\|$) and

$$G^-([a,b],X) = \{f \in G([a,b],X) \mid f(a+) = f(a),\ f(t-) = f(t) \text{ for } a < t \leq b\}$$

is a closed subspace.

Since in (L) the functions A and y may have common points of discontinuity, the usual Riemann-Stieltjes operator integral has to be replaced by a more general one. Hildebrandt in [3] uses the Young integral. We work with the interior integral (see the definition that follows) which is much simper and easier to deal with than the Young integral.

For $a \in SV([a,b],L(X,Y))$ and $f \in G([a,b],X)$ there exists the *interior integral*

$$\int_a^b \cdot d\alpha(t)\cdot f(t) = \lim_{d\in D} \sum_{i=1}^{|d|} [\alpha(t_i) - \alpha(t_{i-1})] \cdot f(\xi_i^\cdot) \quad \text{where } \xi_i^\cdot \in]t_{i-1},t_i[$$

see [4], Theorem I.4.12. If α and f have no common points of discontinuity (for instance, if one of them is continuous) the interior integral reduces to the usual Riemann-Stieltjes operator integral

$$\int_a^b d\alpha(t)\cdot f(t) = \lim_{\Delta d\to 0} \sum_{i=1}^{|d|} [\alpha(t_i) - \alpha(t_{i-1})]\cdot f(\xi_i)$$

where $\xi_i \in [t_{i-1},t_i]$. We recall that in this case we have the integration by parts formula

$$\int_a^b d\alpha(t)\cdot f(t) + \int_a^b \alpha(t)\cdot df(t) = \alpha(b)\cdot f(b) - \alpha(a)\cdot f(a)$$

which, in general, is not true for the interior integral; see [4], Theorem I.2.3.

The main justification of the notion of function of bounded semivariation and of the interior integral lies in the following generalization of the Riesz representation theorem for $F \in \mathcal{C}([a,b])'$:

The mapping $\alpha \in SV_a([a,b],L(X,Y)) \to F_\alpha \in L[G^-([a,b],X),Y]$ *is an isometry* (i.e., $\|F_\alpha\| = SV[\alpha]$) *of the first Banach space onto the second, where for every* $f \in G([a,b],X)$ *we define* $F_\alpha[f] = \int_a^b \cdot d\alpha(t)\cdot f(t)$. *For* $t \in]a,b]$ *and* $x \in X$ *whe have* $\alpha(t)\cdot x = F_\alpha[\chi_{[a,t]}x]$ - See [4], Theorem I.5.1.

If $I_f(t) = \int_a^t \cdot d\alpha(s)\cdot f(s)$, $t \in [a,b]$, and if $\alpha \in SV([a,b], L(X,Y))$ is regulated then $I_f \in G([a,b],Y)$ and for every $t \in [a,b[$ $(t \in]a,b])$ we have $I_f(t+) - I_f(t) = [\alpha(t+) - \alpha(t)]\cdot f(t+)$ $(I_f(t) - I_f(t-) = [\alpha(t) - \alpha(t-)]\cdot f(t-))$; see [4], Proposition I.4.14.

For $[c,d] \subset [a,b]$ we define $SV_{]c,d]}[\alpha] = \sup_{t \downarrow c} SV_{[t,d]}[\alpha]$ and analogously for $SV_{[c,d[}[\alpha]$.

For $a \in SV([a,b],L(X,Y))$ we consider the following property $(SV^\pm)$ - For every $t \in [a,b[$ $(t \in]a,b])$ we have

$$\lim_{\tau \downarrow t} SV_{]t,\tau]}[\alpha] = 0 \qquad \left(\lim_{\tau \uparrow t} SV_{[\tau,t[}[\alpha] = 0\right)$$

We ignore if every $\alpha \in SVG([a,b],L(X,Y)) = SV([a,b],L(X,Y) \cap G([a,b],L(X,Y))$ satisfies $(SV^\pm)$. This is true if Y reflexive or, more generally, weakly sequentially complete; see [8].

If $A \in SV([a,b],L(X))$ satisfies $(SV^\pm)$ we write $A \in SV^\pm([a,b],L(X))$ or simply $A \in SV^\pm$. For $A \in SV^\pm$ we consider the following property:

$(I^\pm)$ - For every $t \in [a,b[$ $(t \in]a,b])$ the operator $I_X + [A(t+) - A(t)]$ $(I_X - [A(t) - A(t-)])$ is invertible in $L(X)$.

Given a function $R : [a,b] \times [a,b] \to L(X)$ we write $R^t(s) = R_s(t) = R(t,s)$. We write $R \in G\cdot SV^u = G\cdot SV^u([a,b] \times [a,b],L(X))$ if we have

(G) - $R_s \in G([a,b],L(X))$ for every $s \in [a,b]$

$$(SV^u) - SV^u[R] = \sup_{a\leq t\leq b} SV_{[a,b]}[R^t] < \infty.$$

We say that R is a *quasi-resolvent* of (L) or of A if R satisfies (G) and

$$R(t,s) - I_X + \int_s^t \cdot dA(\sigma)\circ R(\sigma,s) = 0 \quad \text{for all} \quad s,t\in[a,b] \qquad (R^*)$$

If we have furthermore $R \in G\cdot SV^u$ then y given by

$$y(t) = f(t) + R(t,s)[x-f(s)] - \int_s^t \cdot d_\sigma R(t,\sigma)\cdot f(\sigma) \qquad (\rho)$$

is a regulated solution of (L) and we say that R is a *resolvent* of (L) or A. If R is unique so is the regulated solution of (L) and in this case we have

a) R satisfies

$$R(t,t) = I_X, \quad R(t,s)\circ R(s,r) = R(t,r), \quad R(s,t) = R(t,s)^{-1} \qquad (\circ)$$

b) A satisfies $(I^\pm)$

c) For every $u,v \in [a,b]$ we have

$$A(u) - A(v) = \int_u^v \cdot d_\tau R(\tau,s)\circ R(s,\tau) \quad \text{for any} \quad s \in [a,b]$$

If we define $r(t) = R(t,c)$ where the point $c \in [a,b]$ is fixed we have furthermore

d) For every $s \in [a,b]$ there exists $r(s)^{-1}$ and $R(t,s) = r(t)\circ r(s)^{-1}$.

e) $r,r^{-1} \in SVG([a,b],L(X))$

f) r satisfies

$$r(t) - r(s) + \int_s^t \cdot dA(\tau)\circ r(\tau) = 0 \quad (s,t \in [a,b]) \quad \text{and} \quad r(c) = I_X$$

Since, by hypothesis, A satisfies $(SV^\pm)$ if follows that r and r^{-1} satisfy $(SV^\pm)$ and that $R\in\mathcal{H}^\pm$, i.e. $R: [a,b] \times [a,b] \to L(X)$ satisfies $(\circ)$ and for all $s,t \in [a,b]$, R^t and R_s satisfy $(SV^\pm)$.

III. THE MAIN THEOREM

Reciprocally, if A satisfies $(I^{\pm})$ then (L) has a resolvent $R \in \mathcal{H}^{\pm}$ and R is unique. More precisely:

Theorem. *For* $A \in SV^{\pm}([a,b],L(X))$ *the following properties are equivalent:*

a) A *satisfies* $(I^{\pm})$

b) *There is only one quasi-resolvent* R *of the equation*

$$y(t) - x + \int_s^t \cdot dA(\tau)\cdot y(\tau) = f(t) - f(s), \quad a \le t \le b \tag{L}$$

and we have $R \in \mathcal{H}^{\pm}$ *(hence* R *is a resolvent of* (L)*)*

c) *For every* $s \in [a,b]$, $x \in X$ *and* $f \in G([a,b],X)$ *the equation* (L) *has one and only one regulated solution.*

Let us sketch the proof of a) $\Longrightarrow$ b): if A satisfies $(I^{\pm})$ we will prove the existence of a resolvent $R \in G\cdot SV^u$. It is immediate that there exists a division d of $[a,b]$ such that $SV_{]t_{i-1},t_i[}[A] \le c < \frac{1}{2}$; we define

$$A_i(t) = \begin{cases} A(t_{i-1}+) & \text{if} \quad t = t_{i-1} \\ A(t) & \text{if} \quad t_{i-1} < t < t_i \\ A(t_i-) & \text{if} \quad t = t_i \end{cases}$$

The proof procedes in five steps:

I - There exists one and only one

$$R_i \in G\cdot SV_i^u = G\cdot SV^u([t_{i-1},t_i] \times [t_{i-1},t_i],L(X))$$

that satisfies

$$R_i(t,s) - I_X + \int_s^t \cdot dA_i(\sigma)\circ R_i(\sigma,s) = 0 \tag{R_i^*}$$

for all $s,t \in [t_{i-1},t_i]$

Indeed, $G\cdot SV_i^u$ is a Banach space when endowed with the norm $|||U||| = \|U\|_\Delta + SV^u[U]$ where $\|U\|_\Delta = \sup\{\|U(t,t)\| \mid t_{i-1} \le t \le t_i\}$ and $SV^u[U] = \sup\{SV_{[t_{i-1},t_i]}[U^t] \mid t_{i-1} \le t \le t_i\}$. For $U \in G\cdot SV_i^u$ we define $\mathcal{T}U \in G\cdot SV_i^u$ by $(\mathcal{T}U)(t,s) = \int_s^t \cdot dA_i(\sigma)\circ U(\sigma,s)$. By

[4], (9') of II.1.7 (p. 73) we have

$$SV_{[t_{i-1},t_i]}[(\mathcal{T}U)^t] \leq 2SV_{[t_{i-1},t_i]}[A_i]\,|||U|||$$

and since $\|\mathcal{T}U\|_\Delta = 0$ it follows that

$$|||\mathcal{T}U||| \leq 2SV_{[t_{i-1},t_i]}[A_i]\,|||U||| \leq 2c|||U|||$$

hence $\mathcal{T}$ is contraction of $G\cdot SV_i^u$ since $2c < 1$. Then R_i is the fixed point of the mapping $U \to I_X - \mathcal{T}U$.

II - Since R_i is unique it follows easily that it satisfies

$$R_i(t,\sigma)\circ R_i(\sigma,s) = R_i(t,s) \quad \text{for all} \quad s,\sigma,t \in [t_{i-1},t_i]$$

If we fix a point $\bar{o} \in [t_{i-1},t_i]$ and define $r_i(t) = R_i(t,\bar{o})$ we have $R_i(t,s) = r_i(t)\circ r_i(s)^{-1}$ and r_i satisfies

$$r_i(t) - r_i(s) + \int_s^t \cdot dA_i(\sigma)\circ r_i(\sigma) = 0 \quad \text{for all} \quad s,t \in [t_{i-1},t_i];$$

r_i satisfies $(SV^{\pm})$ and we have $r_i \in SVG([t_{i-1},t_i],L(X))$.
r_i is not unique since r_iu_i satisfies the same conditions for every $u_i \in L(X)$ that is invertible.

III - For every $t \in [t_{i-1},t_i]$ the function R_i^t satisfies $(SV^{\pm})$.

Indeed, by [4], (8') of II.1.7 (p. 73), for $]c,d] \subset [t_{i-1},t_i]$ we have

$$SV_{]c,d]}[R_i^t] \leq SV_{]c,d]}[A_i]\,[1 + \sup_{c<\sigma\leq d} SV_{[\sigma,d]}[R_i^\sigma]]$$

$$+ SV_{]c,t]}[A_i] \sup_{c<\sigma\leq t} SV_{]c,d]}[R_i^\sigma]$$

and for $\varepsilon = SV_{]c,d]}[A_i] < \frac{1}{4}$ and $x = \sup_{t_{i-1}\leq t\leq t_i} SV_{]c,d]}[R_i^t]$

we get $x \leq \varepsilon(1+x) + SV_{[t_{i-1},t_i]}[A_i]x \leq \varepsilon(1+x) + \frac{1}{2}x$ hence

$x \leq \frac{2}{1-2} \leq 4\varepsilon$, and analogously for $SV_{[c,d[}[R_i^t]$ hence the result. If follows that r_i^{-1} satisfies $(SV^{\pm})$, therefore r_i^{-1} is regulated.

IV - We will now use the functions r_i and choose conveniently the invertible elements $u_i \in L(X)$ to obtain a function $r : [a,b] \to L(X)$ that satisfies

$$r(t) - r(s) + \int_s^t \cdot dA(\sigma) \circ r(\sigma) = 0 \quad \text{for all} \quad s,t \in [a,b] \qquad (r)$$

and such that there exists $r^{-1} \in SVG([a,b],L(X))$.

For this purpose we take as $r(a)$ any invertible element. Next we take r_1 such that $r_1(a)$ $(= r(a+))$ satisfies $[I_X + [A(a+)-A(a)]] \circ r_1(a) = r(a)$ (this is possible by $(I^{\pm})$) and we define $r(t) = r_1(t)$ for $a < t < t_1$. Then we take $r(t_1) = [I_X - [A(t_1)-A(t_1-)]]r_1(t_1-)$. We take r_2 such that $r_2(t_1)$ $(= r(t_1+))$ satisfies $I_X + [A(t_1+)-A(t_1)]] \circ r_2(t_1) = r(t_1)$ and we define $r(t) = r_2(t)$ for $t_1 < t < t_2$. If we proceed we obtain r. It is immediate that r satisfies (r) and that $r^{-1} \in SVG([a,b],L(X))$.

V - Finally we define $R(t,s) = r(t)\ r(s)^{-1}$. It is immediate that R satisfies (R^*) and that $R \in G \cdot SV^u$.

To complete the proof one can show that R is unique and that $R \in \mathcal{H}^{\pm}$.

Let is mention that *on* $\mathcal{H}^{\pm}$ *the seminorms* $SV^u[R] = \sup_{a \leq t \leq b} SV[R^t]$ *and* $SV_u[R] = \sup_{a \leq s \leq b} SV[R_s]$ *define the same separated topology* and that *the mapping that to every*

$$A \in \{\alpha \in SV_a([a,b],L(X)) \mid \alpha \in SV^{\pm}, \quad \alpha \text{ satisfies } (I^{\pm})\}$$

associates its resolvent $R \in \mathcal{H}^{\pm}$ *is bicontinuous from the first space onto the second.*

The solution y *of* (L) *depends continuously on* x, f *and* A.

All results are still true if we replace everywhere "bounded semivariation" by "bounded variation".

REFERENCES

1. Arbex, S. E., Equações integrais de Volterra-Stieltjes com núcleos descontínuos. Doctor Thesis, Instituto de Matemática e Estatística da Universidade de São Paulo, (1976).
2. Hinton, D. B., A Stieltjes-Volterra integral equation theory, *Canad. J. Math. 18*, 314-331 (1966).
3. Hildebrandt, T. H., On systems of linear differentio--Stieltjes integral equations, *Illinois J. Math. 3*, 352-373 (1959).
4. Hönig, C. S., Volterra Stieltjes-integral equations, Mathematics Studies 16, North-Holland Publ. Co., Amsterdam, (1975).
5. Hönig, C. S., A survey on Volterra-Stieltjes integral equations. Escola de Análise, Soc. Bras. Mat., São Paulo, (1977).
6. Hönig, C. S., Volterra-Stieltjes integral equations with linear constraints and discontinuous solutions, *Bull. Amer. Math. Soc. 81*, 593-598 (1975).
7. Hönig, C. S., The resolvent of a linear Stieltjes integro--differential equation. To appear.
8. Hönig, C. S., Semivariation and continuity. 3º Seminário Brasileiro de Análise, Sociedade Brasileira de Matemática. To appear.
9. Mac-Nerney, J. S., Stieltjes integrals in linear spaces, *Annals of Math. 61*, 354-367 (1955).
10. de Souza, M. I., Equações diferencio-integrais do tipo de Riemann-Stieltjes em espaços de Banach com soluções descontínuas. Master Thesis, IMEUSP, (1974).
11. Schwabik, St., On Volterra-Stieltjes integral equations, *Cas. pro pest. mat. 99*, 255-278 (1974).
12. Wall, H. S., Concerning harmonics matrices, *Arch. Math. 5*, 160-167 (1954).

A CRITICAL STUDY OF STABILITY OF NEUTRAL FUNCTIONAL DIFFERENTIAL EQUATIONS

A. F. Izé[1,2]

Universidade de São Paulo
São Carlos - S. P. Brasil

A. A. Freiria[1,3]
J. G. Dos Reis[1,3]

Universidade de São Paulo
Ribeirão Preto - S. P. Brasil

Suppose $r > 0$ is a given real number, $R = (-\infty,\infty)$, E^n is a real or complex n-dimensional linear vector space with norm $|.|$, $C([-r,0], E^n)$ is the Banach space of continuous functions mapping the interval $[-r,0]$ into E^n with the topology of uniform convergence given by the norm $\|\phi\| = \sup_{-r\leq\theta\leq 0} |\phi(\theta)|$. If $\zeta \in R$, $A > \theta$, and $x \in C([\zeta - r, \zeta + A], E^n)$ then for any

[1]Supported by CNPq, FAPESP and FINEP.

[2]Present address: Universidade de São Paulo, Instituto de Ciências Matematicas de São Carlos, Departamento de Matematica, 13.560 São Carlos, S. P. Brasil

[3]Present address: Universidade de São Paulo, Faculdade de Medicina de Ribeirão Preto, Departamento de Matematica, 14.1100 Ribeirão Preto, S. P. Brasil.

ISBN 0-12-186280-1

$t \in [\zeta, \zeta + A)$ we let $x_t \in C$ be defined by $x_t(\theta) = x_t(t+\theta)$, $-r < \theta < 0$. If Ω is an open subset of $R \times C$ and f, $D : \Omega \to E^n$ are given continuous functions we say that the relation

$$(d/dt)D(t,x_t) = f(t,x_t) \tag{1}$$

is a functional differential equation.

We say that (1) is a neutral functional differential equation if D is atomic at zero, that is, D has a continuous Fréchet derivative, $D'(t,\phi) = \int_{-r}^{0} d\mu(t,\phi,\theta)\Psi(\theta)$ and there is a scalar function $y(t,\phi,s,0)$ continuous for $(t,\phi) \in \Omega$, $s > 0$, $y(t,\phi,0,0) = 0$ such that

$$\left|\int_{s}^{0} d_\theta \mu(t,\phi,\theta)\Psi(\theta) - A(t,\phi,0)\Psi(0)\right| \leq y(t,\phi,s,0) \sup_{-s\leq\theta\leq 0} |\Psi(\theta)| \tag{2}$$

where μ is a matrix of bounded variation for $\theta \in [-r,0]$ and $A(t,\phi,0) = (t,\phi,0_+) - (t,\phi,0_-)$ with $\det A(t,\phi,0) \neq 0$.

If $D(t,\phi) = \phi(0)$ we have a retarded functional differential equation

$$\dot{x}(t) = f(t,x_t), \quad (.) = d/dt. \tag{3}$$

If we assume that f takes bounded sets of $R \times C$ into bounded sets then the solutions of (1) can be continued to the boundary of Ω. We assume then that a bounded solution of (1) is defined in the future, see [5].

I. SOME STABILITY IMPLICATIONS

Our aim in this section is to discuss the relationship between different concepts of stability in the sense of Lyapunov, for the solutions of (1), which is very important for the development of the theory and for applications in dissipative process. Since a functional differential equation can be stable at a point t_o and not stable at a point $t_1 > t_o$, as is shown by an example given by Zverkin [18], we assume here that if the solution $x \equiv 0$ is stable at a point t_o, it is stable for every $t > t_o$.

From the practical point of view there is no interest to consider equations that are stable at a point and not stable at other points.

Let $C_H = \{\phi \in C| \ ||\phi|| < H\}$ and $R^+ = [0,\infty)$ we assume that $D(t,0) = 0$ and $f(t,0) = 0$, $t \in R^+$.

DEFINITION I. We say that the solution $x \equiv 0$ of (1) is stable if given $\varepsilon > 0$ there is $\delta = \delta(\varepsilon, t_o) > 0$ such that $t_o > 0$, $\phi \in C_\delta$, implies $x_t(t_o,\phi) \in C_\varepsilon$, for $t > t_o$.

DEFINITION II. The solution $x \equiv 0$ of (1) is uniformly stable if the δ in Definition I is independent of t_o.

DEFINITION III. The solution $x \equiv 0$ is quasi-asymptotically stable if for each $t_o \geq 0$ there is $H_o = H_o(t_o)$ such that $\phi \in C_{H_o}$ implies $\lim_{t\to\infty} ||x_t(t_o,\phi)|| = 0$.

DEFINITION IV. The solution $x \equiv 0$ of (1) is quasi-uniform-asymptotically stable if the H_o in Definition III is independent of t_o.

DEFINITION V. The solution $X \equiv 0$ of (1) is asymptotically stable if it is stable and quasiasymptotically stable.

DEFINITION VI. The solution $x \equiv 0$ of (1) is equiasymptotically stable if given $t_o > 0$ there is a $\delta_o = \delta_o(t_o)$ and for each $\varepsilon > 0$ there is $T(\varepsilon,t_o)$ such that $\phi \in C_\delta$, implies $x_t(t_o,\phi) \in C_\varepsilon$ for $t \geq t_o + T(\varepsilon,t_o)$.

DEFINITION VII. The solution $x \equiv 0$ is uniform-equiasymptotically stable if the δ_o and T in Definition V are independent of t_o.

DEFINITION VIII. The solution $x \equiv 0$ is uniform-asymptotically stable if it is uniform stable and uniform-equiasymptotically stable.

DEFINITION IX. The solution $x \equiv 0$ of (1) is exponential asymptotically stable if there are constants, α, β, δ, such that $\phi \in C_\delta$ implies

$$\|x_t(t_o,\phi)\| \leq \beta e^{-\alpha(t-t_o)} \|\phi\|, \quad t \geq t_o.$$

Theorem 1. The following implications hold

(a) (IX) $\rightarrow$ (VII) $\rightarrow$ (II) $\rightarrow$ (I)

(b) (IX) $\rightarrow$ (VIII) $\rightarrow$ (VII) $\rightarrow$ (VI) $\rightarrow$ (III), (V) $\rightarrow$ (I)

(c) (IX) $\rightarrow$ (VIII) $\rightarrow$ (VII) $\rightarrow$ (IV) $\rightarrow$ (III).

The proof is simple and is similar to the proof given for ordinary differential equations [15,16]. In [14,15] we can find several examples which show that these concepts are not equivalent. In particular uniform equiasymptotic stability does not implies uniform asymptotic stability even for scalar linear ordinary differential equations and uniform stability and quasi-asymptotic stability does not imply uniform asymptotic stability as is seen by the example $\dot{x} = x/(t+1)$ whose general solution is $x = x_o(t_o+1)/(t+1)$ which is, however, equiasymptotically stable. In infinite-dimensional space we can not expect in general that asymptotic stability implies uniform asymptotic stability even for linear autonomous equations. The simple example $\dot{x}_n = -x_n/n$ in the space of sequences, given by Massera, shows that this is not true even for Hilbert spaces. Brumley [1] shows that the autonomous equation

$$\ddot{x}(t) - \ddot{x}(t-1) + 2x(t-1) + x(t-2) = 0 \tag{4}$$

which is of neutral type, has solutions that go to zero no faster than $t^{-\mu}$ where $\mu > 0$ is arbitrary, and then are not uniformly asymptotically stable, which is equivalent, for neutral linear autonomous equations, to exponential stability. We show here that for retarded equations uniform stability and quasi-asymptotic stability imply equiasymptotic stability. Furthermore, for a general class of periodic neutral functional differential equations, quasi-asymptotic stability and uniform stability imply uniform asymptotic stability.

Theorem 2. If the solution $x \equiv 0$ of (3) is uniformly stable and quasi-asymptotically stable it is equiasymptotically stable.

Proof. Since the solution $x \equiv 0$ is quasi-asymptotically stable, for each t_o there is $H_o(t_o) > 0$ such that, $\|\phi\| < H_o$ implies $\lim_{t\to\infty} \|x_t(t_o,\phi)\| = 0$. Let $0 < \varepsilon < \min\{H_o(t_o), H_o(t_o+t)\}$. By the uniform stability $x(t) \equiv 0$ there exists $\delta_1 = \delta(\varepsilon)$ such that $\|\phi\| < \delta_1$ implies $\|x_t(t_o,\phi)\| < \varepsilon$ for $t \geq t_o$.

We claim that any H for which $0 < H < \min\{\delta_1, H_o(t_o), H_o(t_o+r)\}$, satisfies the equiasymptotic stability condition.

Suppose that this is not true, that is, there is an $\varepsilon_1 > 0$ for which there is no $T(t_o,\varepsilon_1)$ satisfying the equiasymptotic stability condition. Then there are sequences $\{\phi_n\}$ and $\{t_n\}$, $\|\phi_n\| < H$ such that $t_n \to \infty$ for $n \to \infty$ and

$$\|x_{t_n}(t_o,\phi_n)\| \geq \varepsilon_1. \tag{5}$$

Consider now the sequence $\{x_n\} \in C([t_o,t_o+r], E^n)$, $x_n = x_{t_o+r}(t_o,\phi_n)$. Since $\|\phi_n\| < H \leq \delta_1$, the uniform stability hypotheses implies that $\|x_n\| < \varepsilon \leq H_o(t_o)$ and then the sequence $\{x_n\}$ is equicontinuous, because f takes bounded sets of $R \times C$ into bounded sets; hence there is a subsequence $\{x_{n_k}\}$ of $\{x_n\}$ converging to $\Psi \in C([t_o, t_o+r], E^n)$. Since $x_{n_k} \in C_H$, for every $n_k \in N$, $\Psi \in C_H$ and then $\lim_{t\to\infty} \|x_t(t_o,\Psi)\| = 0$.

By the uniform stability there is $0 < \delta(\varepsilon_1)$ such that $\|\phi\| < \delta(\varepsilon_1)$ implies $\|x_t(t_o,\phi)\| < \varepsilon_1$ for $t \geq t_o$.

Since $t_n \to \infty$ we can assume $t_n > t_o$ for all but a finite number of n's.

We claim that $\|x_t(t_o,r, x_{n_k})\| \geq \delta(\varepsilon_1)$ for $t \in [t_o,t_n]$. Assume that this is not true, that is, there exists $\bar{t}_n$, $t_o < \bar{t}_n < t_n$ with $\|x_{\bar{t}_n}(t_o+r, x_{n_k})\| < \delta(\varepsilon_1)$ and then $\|x_t(t_o+r, x_{n_k})\| < \varepsilon$ for every $t > \bar{t}_n$. At the point $t = t_n$ we would have $\|x_{n_t}(t_o+r, x_{n_k})\| = \|x_{t_n}(t_o, x_{n_k})\| < \varepsilon_1$ in contradiction with (5).

Choose an arbitrary $T > t_o+r$. We can assume $t_n > T$ for all but a finite number of n's. Consider the interval $[t_o+r, T]$. Since $x_{n_k} \to \Psi$, from the continuous dependence of solutions with respect to initial conditions, $x(t_o, x_{n_k})(t) \to x(t_o,\Psi)(t)$ and since

$$\|x_t(t_o,t_{n_k})(t)\| \geq \delta(\varepsilon_1), \|x_t(t_o,\Psi)\| \geq \delta(\varepsilon_1) \quad \text{on} \quad [t_o+r,T].$$

Since T is arbitrary and $\lim_{t\to\infty} \|x_t(t_o,\Psi)\| = 0$ we have a contradiction.

Theorem 3. If D and f are periodical in t the following implications hold.

(i) I $\to$ II.

(ii) VI $\to$ VII.

(iii) III $\to$ IV.

Proof. (ii) From the periodicity of D and f, if $\omega > 0$ is the period, $y_t = x_t(t_o - k\omega,\phi)$ is a solution of (1), $z_t = y_{t-k\omega}$ is also a solution of (1), and $z_{t_o} = y_{t_o-k\omega} = \phi$, for $t_o - k\omega > 0$, that is, $z_t = x_t$ and the two solutions have the same behavior.

By the equiasymptotic stability, for any $\varepsilon > 0$ there is a $\delta_o(\omega)$ and a $T(\omega,\varepsilon)$ such that $\|\phi\| < \delta_o(\omega)$ implies $\|x_t(\omega,\phi)\| < \varepsilon$ for $t \geq \omega + T(\omega,\varepsilon)$. By the theorem of continuity with respect to initial condition there is a $\delta = \delta(\omega,\delta_o)$ such that $\|\phi\| < \delta$ and $t_o \in [0,\omega]$ imply $\|x_t(t_o,\phi)\| < \varepsilon$ for every $t \geq \omega + T(\omega,\varepsilon)$. Then if $T(\varepsilon) = \omega + T(\omega,\varepsilon)$, $\|x_t(t_o,\phi)\| < \varepsilon$ for $t \geq t_o + T(\varepsilon)$. If $t_o \notin [0,\omega]$ we choose k such that $k\omega < t_o < (k+1)\omega$ and consider the solution $x_{t-k\omega}(t_o - k\omega,\phi)$. If $\|\phi\| < \delta_o$, $\|x_{t-k\omega}(t_o - k\omega,\phi)\| < \varepsilon$ for $t - k\omega > t_o - k\omega + T(\varepsilon)$. We have then uniform equiasymptotic stability with $\delta_o = \delta(\omega)$ and $T(\varepsilon) = \omega + T(\varepsilon,\omega)$ and (ii) is proved.

(i) and (iii) follow as in the proof above.

It should be remarked that for functional differential equations, stability at a point t_o does not imply stability at a point $t > t_o$ as is shown by an example given by Zverkin [18]. However, the argument in the above theorems implies that for autonomous neutral functional differential equations stability at a point t_o implies stability for any $t_1 > t_o$, since the behavior of the solutions is the same at all points.

Theorem 2 is not true in general even for linear autonomous neutral functional differential equation as is shown by Eq. (4) above which has bounded and therefore stable, uniformly stable solutions, that go to zero no faster then $t^{-\mu}$, $\mu > 0$ and then are not equiasymptotically stable. Hale and Cruz [3] introduced the following definition.

DEFINITION. We say that the operator D is uniformly stable if there exists a constant $R > 1$ such that for any $(\zeta,\phi) \in R^+ \times C$, $H \in (R^+,E^n)$ the solution $x(\zeta,\phi,H)$ of

$$D(t,x_t) = D(\zeta,\phi) + [H(t) - H(\zeta)], \quad t \geq \zeta, \quad x_\zeta = \phi$$

satisfies the condition

$$||x_t(\zeta,\phi,H)|| \leq R\,||\phi|| + \sup_{\zeta \leq u \leq t} |H(u) - H(\zeta)|, \qquad t \geq \zeta.$$

If the operator D is uniformly stable the solution $x \equiv 0$ of equation $(d/dt)D(t,x_t) = 0$, $x_\zeta = \phi$, is uniform asymptotically stable [3], and many stability properties that are true for retarded equations are also true for Eq. (1) with an uniformly stable D operator. In the proof of Theorem 2 we used strongly the fact that the operator $T(t)\phi = x_t(\phi)$ defined by the solution of Eq. (3) is compact and this is not true for neutral equation. However if the D operator is uniformly stable, $T(t)$ is an α-contraction in the sense of Darbo [4] and if system (1) is also periodic we will have the following stability implication.

<u>Theorem 4</u>. If D and f in (1) are periodic in t and D is linear and uniformly stable then if the solution $x \equiv 0$, of (1) is asymptotically stable it is uniform asymptotically stable.

<u>Proof</u>. Consider the operator T_ω defined by $T_\omega(\phi) = x_\omega(0,\phi)$. From Lemmas 6 and 8, Hale and Lopes [7] $T_\omega = S_\omega + U_\omega$, where S_ω is a linear operator with spectral ratio $r_\omega < 1$ and U_ω is a weakly compact operator. Since

by the asymptotic stability assumption we can choose a set A in such a way that $T_\omega(A)$ is always bounded, $T_\omega = S_\omega + U_\omega$ where U_ω is completely continuous. This implies that T_ω is an α-contraction, that is, if $\gamma(A)$ is the Kuratowski measure of noncompactness, $\gamma(T_\omega(A)) \leq k\gamma(A)$, $0 \leq k < 1$.

By the asymptotic stability there is $\delta_o > 0$ such that $||\phi|| < \delta_o$ implies $\lim_{t\to\infty} ||x_t(0,\phi)|| = 0$. By the uniform stability we can choose $0 < \delta_1 = \delta(\delta_o) < \delta_o$ such that $||\phi|| < \delta_1$ implies $||x_t(0,\phi)|| < \delta_o$ for $t \geq t_o \geq 0$.

We claim that any $\delta_2 > 0$, such that $\delta_2 < \delta_1$, satisfies the equiasymptotic stability condition. In fact, consider the set:

$$A_o = \overline{B(\delta_1) \cup T_\omega(B(\delta_1)) \cup T^2_\omega(B(\delta_1)) \cup \dots \cup T^n_\omega(B(\delta_1)) \cup \dots},$$

where $B(\delta_1)$ is a neighborhood of the origin of ratio δ_1. A_o is closed, bounded, and $A_o \subset B(\delta_o)$. Let us prove that A_1 defined by $A_1 = \overline{T_\omega(A_o)}$ is closed and $A_1 \subset A_o$. Assume $B_1 = T_\omega(A_o)$. If $\phi \in B_1$, $\phi = T_\omega\Psi$ where $\Psi \in \overline{T^k_\omega(B(\delta_1))}$ for some $k > 0$. Since T_ω is continuous $T_\omega(\overline{T^k_\omega(B(\delta_1))} \subset \overline{T_\omega(T^k_\omega(B(\delta_1))} = \overline{T^{k+1}_\omega(B(\delta_1))}$ and this implies that $\phi \in A_o$. Hence $A_1 = \bar{B}_1 \subset \bar{A}_o = A_o$.

Suppose that $A_{n-2} \supset A_{n-1}$, it follows then that $T(A_{n-2}) \supset T(A_{n-1})$ and therefore $\overline{T(A_{n-2})} \supset \overline{T(A_{n-1})}$. Thus $A_{n-1} \supset A_n$ and by the induction hypothesis $A_o \supset A_1 \supset A_2 \dots$.

The measure of noncompactness of A_o is $\gamma(A_o) < 2\delta_o$ and since $\gamma(A_1) = \gamma(T(A_o)) \leq k\gamma(A_o) \leq k2\delta_o$, by induction $\gamma(A_n) = \gamma(T(A_{n-1})) \leq k\gamma(A_{n-1}) \leq k^n 2\delta_o$, $0 \leq k < 1$ and then $0 \leq \gamma(A_n) \leq k^n 2\delta_o$. Thus $\gamma(A_n) \to 0$ as $n \to \infty$. Then, Kuratowski [12], there is a compact, nonempty set A, $A = \bigcap_{n>0} A_n$, such that A_n converges to A in the Hausdorff metric, that is, for any $\varepsilon > 0$ there is an $N(\varepsilon)$ such that $A_n \subset \{x \mid D_H(x,A) < \varepsilon\}$. The Hausdorff metric is defined by $D_H(A,B) = \max\{D_1(A,B), D_2(A,B)\}$ where $D_1(A,B) = \inf\{a_1 > 0 \quad B \subset N_{a_1}(A)\}$, $D_2(A,B) = \inf\{a_2 > 0 \mid A \subset N_{a_2}(A)\}$ and $N_r(C) = \{x \mid d(x,c) < r\}$.

From what we proved above we could complete the proof of the theorem, but, let us give a more interesting proof, that is, let us prove that A is invariant and $A = 0$. $A = \bigcap_{n>0} A_n$ implies that

$$T_\omega(A) = T_\omega(\bigcap_{n>0} A_n) \subset \bigcap_{n>0} T_\omega(A_n) \subset \overline{\bigcap_{n>0} T_\omega(A_n)} = \bigcap_{n>1} A_n = A,$$

since $A_o \supset A_1$. Let us show now that $A \subset T_\omega(A)$. If $a \in A$, since $A \subset T_\omega^n(A_o)$ for any $n \in N$, $a = T_\omega^n(a_o)$ where $a_o \in A_o$. But $a = T_\omega(T_\omega^{n-1}(a_o)) = T_\omega(b_n)$. Since $D_H(T_\omega^{n-1}(A_n), A) \to 0$ for $n \to \infty$, $d(b_n, A) \to 0$ for $n \to \infty$. Since A is compact, for each b_n there exists $C_n \in A$ such that $d(b_n, A) = d(b_n, C_n)$ and a subsequence $\{C_{n_k}\} \subset \{C_n\}$ such that $C_{n_k} \to C$.

Since $d(b_{n_k}, C) \leq d(b_{n_k}, C_{n_k}) + d(C_{n_k}, C)$ and $d(b_{n_k}, C_{n_k}) \to 0$, $C_{n_k} \to C$ as $k \to \infty$, $a = T_\omega(b_{n_k})$ for any k; hence $a = T_\omega(C)$ and $a \in T_\omega(A)$, that is, $A \subset T_\omega(A)$. Thus $T_\omega(A) = A$ and A is invariant by T_ω. To prove that $A = 0$ it is sufficient to prove that $A \subset B(\varepsilon)$ for any $\varepsilon > 0$. Given $\varepsilon > 0$ for any $x \in A$ there exists $N(x)$ such that $|T^{N(x)}(x)| < \varepsilon$ and there exists an open neighborhood $V(x)$ such that $T_\omega^{N(x)}(V(x)) \subset B_\varepsilon$. Since A is compact let $V(x_1), V(x_2), \ldots, V(x_k)$ an open finite covering of A, $\Omega = V(x_1) \cup V(x_2) \cup \ldots \cup V(x_k)$ and

$$N = \max \{N(x_1) \ \ldots \ N(x_k)\},$$

$A \subset \Omega$ implies that $A \equiv T_\omega^N(A) \subset T_\omega^N(\Omega) \subset B(k\varepsilon)$ for arbitrary ε, then $A \equiv 0$. Since $A_n \to 0$ for $n \to \infty$, given $\varepsilon > 0$ there exists $N(\varepsilon)$ such that $A_n \subset N_\varepsilon(0) = B(\varepsilon)$, that is, $\|x_t(0,\phi)\| < \varepsilon$ for $t \geq n\omega$.

II. DISSIPATIVE PROCESS

The method of proof of Theorem 4 can be applied to dissipative process as will be shown in the next theorem.

A continuous map $T : X \to X$ is said to be point dissipative if there is a bounded set $B \subset X$ with the property that for any $x \in X$ there is an integer $N(x)$ such that $T^n x \in B$ for $n \geq N(x)$.

If B satisfies the property that for any compact set $A \subset X$, there is an integer $N(A)$ such that $T^n(A) \subset B$ for $n \geq N(A)$ then T is said to be compact dissipative.

If B satisfies the property that for any $x \in X$ there is an open neighborhood 0_x of x and an integer $N(x)$ such that $T^n 0_x \subset B$ for $n \geq N(x)$ then T is said to be local dissipative. Obviously local dissipative implies compact dissipative implies point dissipative.

Theorem 5. Suppose $T : X \to X$ is point dissipative, a weak α-contraction, and satisfies the condition that for any $x \in X$, there is a neighborhood 0_x such that $\cap_{j=1} T^j 0_x$ is bounded. Then T is local dissipative.

Proof. Let $x \in X$ and $F_o = \overline{\cap_{j=1} T^j 0_x}$, F_o is closed and if we write $F_1 = \overline{T(F_o)}$ and $F_n = \overline{T(F_{n-1})}$ then $F_1 \supset F_2 \supset F_3 \supset \ldots \supset F_n \supset \ldots$. The boundedness of F_o implies that $\gamma(F_o) \leq k$, k finite, and $\gamma(F_n) \to 0$ as $n \to \infty$. By the Kuratowski theorem [12] there exists F compact such that $F_n \to F$ in the Hausdorff metric and as in the proof of Theorem 4 $a = T(b_{n_k}) = T(T^{n_k}(a_o))$ for every $k \in N$. Since there is a k such that $T(b_{n_k}) \in B$, $F \subset B$ and therefore $N_\varepsilon(F) \subset N_\varepsilon(B)$.

Since $F_n \to F$ there is an n_o such that $F_n \subset N_\varepsilon(F)$ for $n \geq n_o$. Thus $F_n \subset N_\varepsilon(F) \subset N_\varepsilon(B)$ is what implies that $T^n 0_x \subset N_\varepsilon(B)$. Since $N_\varepsilon(B)$ is bounded the proof is complete.

III. TOTAL AND INTEGRAL STABILITY

In discussing the existence of almost periodic solutions of retarded functional differential equations connecting with boundedness, there are two ways, The one is to assume a separation condition for bounded solutions, the other is to assume that an almost periodic system has a bounded solution with some kind of stability and so on. In particular the existence of a bounded totally stable solution implies the existence of an almost periodic solution but the existence of a uniformly asymptotically stable solution does not imply the existence of an almost periodic solutions [11]. Without uniqueness uniform asymptotic stability does not imply total stability. For neutral equations the above relationship between the existence of almost periodic solutions and some kind of stability is not well understood yet. It seems to be reasonable that if the operator D is stable in the sense defined in section I then the results obtained for retarded equations can be extended to neutral equations. We analyse in the following the relationship between uniform asymptotic stability, total stability and integral stability.

Consider the system of functional differential equations of neutral type

$$\frac{d}{dt} D(t,x_t) = f(t,x_t), \qquad x_\sigma = \phi \tag{6}$$

$$\frac{d}{dt} D(t,y_t) = f(t,y_t) + h(t,y_t), \qquad y_\sigma = \phi \tag{7}$$

where $D : [\sigma,\infty) \times C \to R^n$ is a linear continuous operator, $f, h : \Omega \to R^n$, $\Omega \subset R \times C$, are continuous and take bounded sets of $R \times C$ into bounded sets. We assume also that $f(t,\phi)$ satisfy a local Lipschitz condition with respect to ϕ uniformly for t in bounded sets of $[\sigma,\infty)$ and $f(t,0) = h(t,0) = 0$. Under the above condition systems (7) and (8) have a unique solution $x_t(\sigma,\phi)$, $x_\sigma(\sigma,\phi) = \phi$ through (σ,ϕ) and these solutions can be continued to a maximal interval of existence as long as they exists and a bounded solution of (7) or (8) exists in $[\sigma,\infty)$.

In what follows we assume that D is a uniformly stable operator. The following Lemma [13a] and Theorem 6 that is a corollary of Theorem 1 of [9] is used to prove the following results of this paper on total and integral stability.

<u>Lemma 1</u>. Assume that $f(t,\phi)$ satisfy a Lipschitz condition with respect to ϕ in a neighborhood of the origin uniformly for t in bounded sets and $f(t,0) = 0$. If $x_t(\sigma,\phi)$ and $x_t(\sigma,\psi)$ are solutions of (1) then there exist constants $K_o > 0$, $L_o > 0$ such that

$$|x_t(\sigma,\phi) - x_t(\sigma,\psi)| \le K_o e^{L_1(t-t_o)} |\phi-\psi| \qquad t \ge \sigma.$$

If $V : [\tau,\infty) \to R$ is a continuous function we define the "derivative" $\dot{V}(t,\phi)$ along the solutions of (1) by

$$\dot{V}(t,\phi) = \dot{V}_{(1)}(t,\phi) = \overline{\lim_{h\to 0_+}} \frac{1}{h}[V(t+h,x_{t+h}(t,\phi) - V(t,\phi)]$$

Theorem 6. If the zero solution of (1) is uniformly asymptotically stable then there are constants $\delta_o > 0$, $k = K(\delta_o) > 0$, $M > 0$ and continuous non decreasing positive functions $u(s)$, $c(s)$, $b(s)$, $w(s)$, $v(s)$ for $0 \leq s \leq \delta_o$, $u(0) = c(0) = b(0) = w(0) = v(0)$ such that:

i) $u(|D(t,\phi)|) \leq V(t,\phi)$

ii) $c(|\phi|) \leq V(t,\phi) \leq b(|\phi|)$

iii) $\dot{V}_{(7)}(t,\phi) \leq \dot{V}_{(6)}(t,\phi) + M|h(t,\phi)|$

iv) $\dot{V}(t,\phi) \leq -w(|D(t,\phi)|)$; $\dot{V}(t,\phi) \leq -V(|\phi|)$

v) $|V(t,\phi) - V(t,\psi)| \leq K|\phi - \psi|$ for every $t \geq \tau$, ϕ, $\psi \in C$, $|\phi|$, $|\psi| \leq \delta_o$.

Proof. The derivative of V along the solutions of (7) is

$$\dot{V}_{(7)}(t,\phi) \leq \dot{V}_{(6)}(t,\phi) + \frac{k}{1-\rho(s_o)}|h(t,\phi)|$$
$$= \dot{V}_{(6)}(t,\phi) + M|h(t,\phi)|, \qquad t \geq \sigma, \quad |\phi| \leq \delta_o.$$

Since in the condition of atomicity (2), because D is linear, we can take $\gamma(t,\phi, s,0) = \rho(s)$, $\rho(0) = 0$ and small enough in such a way that $0 < \rho(s_o) < 1$, $K > 0$ the constant in condition d) of Theorem 1 of (4). Condition IV follows easily since $|D(t,\phi)| \leq K(|\phi|)$.

DEFINITION X. The solution $x \equiv 0$, of (1) is totally stable if for every $\varepsilon > 0$ there exists $\eta_1(\varepsilon) > 0$, $\eta_2(\varepsilon) > 0$ such that $|\phi| < \eta$ and $|h(t,\phi)| < \eta_2$ implies that $|y_t(\sigma,\phi)| < \varepsilon$ where $y_t(\sigma,\phi)$ is a solution of (2).

It is well known Coppel [2], Yoshizawa [17] that if a retarded functional differential equation $\dot{x} = f(t,x_t)$ has a bounded solution which is totally stable then it is almost periodic. It has been conjectured that if the bounded solution is uniformly asymptotically stable then it is almost periodic. A counterexample given by Kato [11] shows that this is no longer true in general. The following theorem proved for neutral equations shows that if f satisfies a Lipschitz condition then uniform asymptotic stability implies total stability so in particular for a general class of retarded equations the existence of a bounded solution which is uniform asymptotically stable implies the existence of an almost periodic solution. It is an interesting problem to extend the results of Miller, Coppel and Yoshizawa to neutral equations however there are some serious difficulties involved because the operator associated with the neutral equation is not compact.

Theorem 7. If the solution $x \equiv 0$ of (1) is uniformly asymptotically stable it is totally stable.

Proof. Let $x_t(\sigma,\phi)$ and $y_t(\sigma,\phi)$ be solutions (7) and (8), from iii) and iv) of Theorem 6, we have

$$\dot{V}_{(7)}(t,\phi) \leq \dot{V}_{(6)}(t,\phi) + M|h(t,\phi)|, \qquad t \geq \sigma, \quad |\phi| \leq \delta_o$$

$$\dot{V}_{(7)}(t,\phi) \leq -w(|\phi|) + M|h(t,\phi)|, \qquad t \geq \sigma, \quad |\phi| \leq \delta$$

and $|h(t,\phi)| \leq \eta_2$ then

$$\dot{V}_{(7)}(t,\phi) \leq -w(|\phi|) + M\eta_2. \tag{*}$$

Choose $\varepsilon > 0$ small enough in such a way that $\delta_o \leq \frac{\varepsilon}{2}$ and $c(\frac{\varepsilon}{2}) > \ell$, $\ell > 0$ we will show that there exist constants $\eta_1 > 0$, $\eta_2 > 0$ such that $|\phi| < \eta_1$, $|h(t,\phi)| < \eta_2$ implies that $|y_t(\sigma,\phi)| < \varepsilon$ for $t \geq \sigma$ where $y_t(\sigma,\phi)$ is a solution of (7).

Choose $\eta_1 > 0$ such that $b(\eta_1) \leq \ell$ and assume that there is $t_1 > \sigma$ such that $|y_{t_1}(\sigma,\phi)| \geq \varepsilon$. From Theorem 6 we have

$$v(t_1,y_{t_1}(\sigma,\phi)) \geq c(|y_{t_1}(\sigma,\phi)|) \geq c(\varepsilon) \geq c(\tfrac{\varepsilon}{2}) > \ell$$

$$v(\sigma,y_\sigma(\sigma,\phi)) \leq b(|\phi|) \leq b(\eta_1) \leq \ell$$

Since $V(t,\phi)$ is continuous there exists t_2, $\sigma < t_2 < t_1$ such that $V(t_2, y_{t_2}(\sigma,\phi)) = \ell$ and $V(t, y_t(\sigma,\phi)) > \ell$ for $t > t_2$. Hence $b(|y_{t_2}(\sigma,\phi)|) \geq V(t_2, y_{t_2}(\sigma,\phi)) \geq c(|y_{t_2}(\sigma,\phi)|)$.

If $\psi = y_{t_2}(\sigma,\phi)$ we have, $b(|\psi|) \geq \ell \geq c(|\psi|)$ and $b(\eta_1) \leq \ell < c(\varepsilon/2)$ then since $b(s)$, $c(s)$ are continuous positive nondecreasing functions with $b(0) = c(0) = 0$ it follows that

$$\eta_1 \leq |\psi| < \varepsilon$$

and from inequality (*) and choosing $\eta_2 < w(\eta_1)/2$ we have

$$\dot{V}_{(7)}(t_2,\psi) \leq -w(|\psi|) + M_{\eta_2} < 0$$

We also have

$$\dot{V}_{(7)}(t_2,\psi) = \overline{\lim_{h\to 0_+}} \frac{1}{h} [V(t_2+h, y_{t_2+h}(t_2,\psi)) - V(t_2,\psi)] \geq 0$$

since $V(t_2, y_{t_2}(\sigma,\phi)) = V(t_2,\psi) = \ell$ and $V(t, y_t(\sigma,\phi)) > \ell$ for $t > t_2$. This is a contradiction, then $|y_t(\sigma,\phi)| < \varepsilon$ for $t \geq \sigma$ and the proof is complete.

IV. INTEGRAL STABILITY

DEFINITION XI. The solution $x = 0$ of (6) is integrally stable if given $\varepsilon > 0$, there exist $\delta_1 > 0$ and $\delta_2 > 0$ such that

$$|\phi| < \delta_1, \quad \int_\sigma^\infty \sup_{|\psi|<\varepsilon} |h(t,\psi)|dt < \delta_2$$

implies $|y_t(\sigma,\phi)| < \varepsilon$ for $t \geq \sigma$ where $y_t(\sigma,\phi)$ is a solution of (7).

DEFINITION XII. The solution $x = 0$ of (6) is asymptotically integrally stable if it is integrally stable and given

$\varepsilon > 0$ there exists $\eta > 0$ and functions $T = T(\eta,\varepsilon) > 0$ and $\gamma = \gamma(\eta,\varepsilon) > 0$ such that

$$|\phi| < \eta \quad \text{and} \quad \int_\sigma^\infty \sup_{|\psi|<\varepsilon} |h(t,\psi)|dt < \gamma(\eta,\varepsilon)$$

implies $|y_t(\sigma,\phi)| < \varepsilon$ for $t \geq \sigma + T(\eta,\varepsilon)$ where $y_t(\sigma,\phi)$ is a solution of (7).

Lemma 2. The solution $x \equiv 0$ of (6) is asymptotically integrally stable if only if it is integrally stable and there exists functions $T(\eta,\varepsilon) > 0$, $\gamma(\eta,\varepsilon) > 0$, $0 < \eta < \eta_o$, $\varepsilon > 0$ such that for every continuous function $\alpha(t)$ with $\int_\sigma^\infty |\alpha(t)|dt < \gamma(\eta,\varepsilon)$, $|\phi| < \eta$ and $t \geq \sigma + T(\eta,\varepsilon)$ the solution $y_t(\sigma,\phi)$ of the system

$$\frac{d}{dt} D(t,y_t) = f(t,y_t) + \alpha(t) \tag{8}$$

satisfies the inequality $|y_t(\sigma,\phi)| < \varepsilon$.

Proof. If the zero solution of (6) is asymptotically integrally stable. The result follows immediately. Assume that the converse is not true, that is, there exist $\delta' < \delta_o$ and $\varepsilon' > 0$ such that for any $T(\delta',\varepsilon') > 0$ and $\gamma(\delta',\varepsilon') > 0$ there exist ϕ with $|\phi| < \delta'$ and $h(t,\psi)$ satisfying

$$\int_\sigma^\infty \sup_{|\psi|<\varepsilon'} |h(t,\psi)|dt < \gamma(\delta',\varepsilon')$$

and t_1, $t_1 > \sigma + T(\delta',\varepsilon')$ for which $|y_{t_1}(\sigma,\phi)| \geq \varepsilon'$ where $y_t(\sigma,\phi)$ is a solution of (7).

If we take $h(t,y_t(\sigma,\phi)) = \alpha(t)$ for $t \in [\sigma,t_1]$:

$$\begin{aligned}\int_\sigma^{t_1} |\alpha(t)|dt &= \int_\sigma^{t_1} |h(t)y_t(\sigma,\phi)|dt \\ &\leq \int_\sigma^{t_1} \sup_{|\psi|<\varepsilon} |h(t,\psi)|dt \\ &\leq \int_\sigma^{\infty} \sup_{|\psi|\leq\varepsilon} |h(t,\psi)|dt < \gamma(\delta',\varepsilon') .\end{aligned}$$

Let $z_t(\sigma,\phi)$ a solution of (8) with $\alpha(t)$ chosen as above. From the hypotheses, $\int_\sigma^\infty |\alpha(t)|dt < \gamma(\delta',\varepsilon')$, $|\phi| < \delta'$ and $t_1 > \sigma + T(\delta',\varepsilon')$ imply that $|z_{t_1}(\sigma,\phi)| < \varepsilon'$. But for $t \in [\sigma,t_1]$, $y_t(\sigma,\phi) \equiv z_t(\sigma,\phi)$ and then $|y_{t_1}(\sigma,\phi)| < \varepsilon'$, a contradiction.

Corollary 1. If the solution $x \equiv 0$ of (6) is integrally stable and if for every δ, $0 < \delta < \delta_o$, $\varepsilon > 0$, there exist $T(\delta,\varepsilon) > 0$, $\gamma(\delta,\varepsilon) > 0$ such that for every continuous function $\alpha(t)$ on $[\sigma,t_1]$, $t_1 > \sigma + T(\delta,\varepsilon)$ and such that

$$|\phi| < \delta \quad \int_\sigma^{t_1} |\alpha(t)|dt < \gamma(\delta,\varepsilon) \quad \text{implies} \quad |y_t(\sigma,\phi)| < \varepsilon \quad \text{on}$$

$[\sigma + T, t_1]$ where $y_t(\sigma,\phi)$ is a solution of (8), then the solution $x \equiv 0$ of (6) is asymptotically integrally stable.

Proof. Since the solution $x \equiv 0$ of (6) is integrally stable let $T(\sigma,\varepsilon) > 0$, $\gamma(\delta,\varepsilon) > 0$ and $\alpha(t)$ a continuous function satisfying $\int_\sigma^\infty |\alpha(t)|dt < \gamma(\delta,\varepsilon)$. We will show that if $|\phi| < \delta$ the solution $y_t(\sigma,\phi)$ of (8) satisfies $|y_t(\sigma,\phi)| < \varepsilon$ for $t \geq \sigma + T$.

Assume that this is not true, that is, there exists t_1, $t_1 \geq \sigma + T$ such that $|y_{t_1}(\sigma,\phi)| \geq \varepsilon$.

Since $t_1 \geq \sigma + T$, from the hypotheses above we have

$$\int_\sigma^{t_1} |\alpha(t)|dt \leq \int_\sigma^\infty |a(t)|dt < \gamma(\delta,\varepsilon)$$

and this implies that $|y_{t_1}(\sigma,\phi)| < \varepsilon$ on $[\sigma+T,t_1]$ what is a contradiction.

Theorem 8. Let $V(t,\phi)$ a Lyapunov functional defined for $t \geq 0$, $|\phi| \leq \delta_o$ satisfying:

1) $V(t,\phi) \geq a(|\phi|)$ where a is a continuous function, nonnegative, nondecreasing, $a(0) = 0$, $V(t,0) = 0$.

2) $|V(t,\phi) - V(t,\psi)| \leq M_1|\phi - \psi|$, $M_1 > 0$, $|\phi|$, $|\psi| \leq \delta_o$

3) $\overline{\lim\limits_{h\to 0_+}} \frac{1}{h}[V(t+h,x_{t+h}(t,\phi) - V(t,\phi)] \leq g(t)V(t,\phi)$ where

$\int_0^\infty g(t)dt < \infty$, $g(t) \geq 0$, $x_t(\sigma,\phi)$ is a solution of (6) then the solution $x \equiv 0$ of (6) is integrally stable.

Proof. Let $x_t(\sigma,\phi)$ and $y_t(\sigma,\phi)$ be solutions of (6) and (7) respectively. From Theorem 6 we have

$$\dot{V}_{(7)}(t,\phi) \leq \dot{V}_{(6)}(t,\phi) + M|h(t,\phi)|$$

for $t \geq \sigma$, $|\phi| \leq \delta_0$, $M > 0$.
Hence $\dot{V}_{(7)}(t,\phi) \leq g(t)V(t,\phi) + M|h(t,\phi)|$. Let $y_t(\sigma,\phi)$ a solution of (7) and let us solve the inequality

$$\dot{V}(t,y_t(\sigma,\phi)) - g(t)V(t,y_t(\sigma,\phi)) \leq M|h(t,y_t(\sigma,\phi))|$$

$$\cdot\; e^{-\int_\sigma^t g(s)ds} \cdot \dot{V}(t,y_t(\sigma,\phi))$$

$$+\; e^{-\int_\sigma^t g(s)ds} (-g(t))V(t,y_t(\sigma,\phi))$$

$$\leq M\, e^{-\int_\sigma^t g(s)ds} |h(t,y_t(\sigma,\phi)|.$$

The left side of the inequality above is less than or equal to the derivative of

$$V(t,y_t(\sigma,\phi))\; e^{-\int_\sigma^t g(s)ds}$$

Then the solution of the differential inequality is

$$V(t,y_t(\sigma,\phi)) \leq V(\sigma,\phi)\; e^{\int_\sigma^t g(s)ds}$$

$$+\; M\, e^{\int_\sigma^t g(s)ds} \int_\sigma^t e^{-\int_\sigma^t g(s)ds} |h(t,y_t(\sigma,\phi))|dt\,.$$

If $|\phi| < \eta_1$, $\int_\sigma^\infty g(s)ds = k_1$

$$\int_\sigma^t |h(s,y_s(\sigma,\phi)\, ds < \eta_2, \quad \text{and}$$

if we take $\eta = \max\{\eta_1,\eta_2\}$ and $\bar{M} = \max\{M_1,M\}$ we will have

$$a(|y_t(\sigma,\phi)|) \le V(t,y_t(\sigma,\phi)) \le M_1|\phi|e^{k_1+M_1}e^{k_1}\eta_2$$

$$a(|y_t(\sigma,\phi)|) \le V(t,y_t(\sigma,\phi)) \le M_1\eta_1 e^{k_1} + Me^{k_1}\eta_2$$

or

$$a(|y_t(\sigma,\phi)|) \le V(t,y_t(\sigma,\phi)) \le \bar{M}\eta e^{k_1} + \bar{M}e^{k_1}\eta = 2\bar{M}e^{k_1}\eta.$$

Since $a(s)$ is a continuous function, positive, nondecreasing for $s > 0$, $a(0) = 0$, it follows that there exists a continuous positive monotone function B such that

$$|y_t(\sigma,\phi)| \le B(\eta) \quad \text{for} \quad t \ge \sigma.$$

Then the solution $x \equiv 0$ of (6) is integrally stable.

<u>Theorem 9</u>. Let $V(t,\phi)$ be a Lyapunov functional defined for $t \ge 0$, $|\phi| \le \delta_o$, satisfying:

1) $V(t,\phi) \ge a(|\phi|)$ where a is a continuous nonnegative nondecreasing function, $a(0) = 0$, $V(t,0) = 0$.

2) $|V(t,\phi) - V(t,\psi)| \le M_2|\phi - \psi|$, $M_2 > 0$, $|\phi|$, $|\psi| \le \delta_o$

3) $\overline{\lim_{h\to 0_+}} \frac{1}{h}[V(t+h,x_{t+h}(t,\phi)) - V(t,\phi)] \le -W(|\phi|)$

where $W(s)$ is a continuous positive, nondecreasing, $W(0) = 0$, and $x_t(\sigma,\phi)$ is a solution of (6). Then the solution $x = 0$ of (6) is asymptotically integrally stable.

<u>Proof</u>. From Theorem 8, the solution $x \equiv 0$ of (6) is integrally stable, then given $\varepsilon > 0$ we choose $\eta > 0$ in such a way that

$$|y_{t_2}(\sigma,\phi)| < \eta, \qquad \int_{t_2}^{t_3} |h(t,y_t(\sigma,\phi)|dt < \eta$$

implies $|y_t(\sigma,\phi)| < \varepsilon$ for t_1, $t_2 \le t \le t_3$. Let $\eta > 0$ such that $B(\eta) = \varepsilon$, $\bar{\eta} = \min\{\delta,\eta\}$, $\ell = W(\bar{\eta})$, $T = T(\delta,\varepsilon) = \dfrac{M_2(\delta+\eta)}{\ell}$ and $\alpha(t) = h(t,y_t(\sigma,\phi)$ a continuous function on $[\sigma,t_1]$, $t_1 > \sigma + T$, $|\phi| < \delta$ and $\int_\sigma^{t_1} |h(t,y_t(\sigma,\phi))|dt < \bar{\eta}$.

If $|y_t(\sigma,\phi)| \ge \bar{\eta}$ on $[\sigma,\sigma+T]$ we would have

$$\begin{aligned}
&\overline{\lim_{h\to 0_+}} \frac{1}{h}\left[V(t+h,y_{t+h}(t,y_t(\sigma,\phi)) - V(t,y_t(\sigma,\phi))\right] \\
&\le \overline{\lim_{h\to 0_+}} \frac{1}{h}\left[V(t+h,x_{t+h}(t,y_t(\sigma,\phi)) - V(t,y_t(\sigma,\phi))\right] \\
&+ \overline{\lim_{h\to 0_+}} \frac{1}{h}\left[V(t+h,y_{t+h}(t,y_t(\sigma,\phi)) - V(t+h,x_{t+h}(t,y_t(\sigma,\phi))\right] \\
&\le -W(|y_t(\sigma,\phi)|) + M_2|h(t,y_t(\sigma,\phi))|.
\end{aligned}$$

By integration of the inequality above on $[\sigma,\sigma+T]$ we have

$$\begin{aligned}
V(\sigma+T,y_{\sigma+T}(\sigma,\phi)) &\le V(\sigma,\phi) + M_2\int_\sigma^{\sigma+T} |h(t,y_t(\sigma,\phi))|dt \\
&- \int_\sigma^{\sigma+T} W(|y_t(\sigma,\phi)|)dt \le M_2|\phi| + M_2\bar{\eta} - \int^{\sigma+T} W(\bar{\eta})dt \\
&\le M_2(|\phi| + \bar{\eta}) - T\ell \le 0
\end{aligned}$$

what is a contradiction, then there exists $\bar{t}$, $\bar{t} \in [\sigma,\sigma+T]$ such that

$$|y_{\bar{t}}(\sigma,\phi)| < \eta \le \bar{\eta}.$$

Since

$$\int_t^{t_1} |h(t,y_t(\sigma,\phi))|dt < \int_t^{t_1} |h(t,y_t(\sigma,\phi))| < \bar{\eta} \le \eta$$

from the choice of η it follows that $|y_t(\sigma,\phi)| < \beta(\eta) = \varepsilon$ for t, $\bar{t} \le t \le t_1$, but this is true also for $t \in [\sigma+T,t_1]$. Then from Corollary 1 the solution $x \equiv 0$ of (6) is asymptotically integrally stable.

Theorem 10. Assume that $f(t,\phi)$ satisfies a Lipschitz Condition with respect to ϕ in a neighborhood of the origin uniformly with respect to t in bounded sets, $f(t,0) = 0$, then if the solution $x = 0$ of (6) is uniform asymptotically stable it is asymptotically integrally stable.

Proof. The hypotheses of the Theorem imply the hypotheses of Theorem 6 and Corollay 1 and then of Theorem 9 and the solution $x \equiv 0$ of (6) is asymptotically integrally stable.

Examples:

1) The solution $x = 0$ of

$$\frac{d}{dt}\left[x(t) + cx(t-r)\right] = -ax(t) \tag{9}$$

where a, c are constants, $a > 0$, $|c| < 1$ is asymptotically integrally stable.

2) The solution $x = 0$ of

$$\frac{d}{dt}\left[x(t) + cx(t-r)\right] = -a(t)x(t) \tag{10}$$

where c is a constant $|c| < 1$, $a(t)$ is a continuous function for $t > 0$, $a(t) > \delta > 0$ and $a(t) > c^2a(t+r)$, $t > 0$, is asymptotically integrally stable.

3) A transmission line without loss with two differential elements in the terminals [14b] can be described by a system of two equations

$$C_1 \frac{d}{dt} D(t,x_t) = -\frac{1}{z}\,x(t) - \frac{q}{z}\,x(t-r) - g(D(t,x_t))$$

$$L_1 \frac{d}{dt} i(t) = R_1 i(t) + D(t,x_t)$$

where C_1, L_1, R_1, z and q are constants, $|q| < 1$, and $D(t,\phi) = \phi(0) - q\phi(-r)$, and there exists H such that

$$\inf_{|x|>H} \frac{q(x)}{x} = M > -\frac{1}{z}\frac{1-|q|}{1+|q|} .$$

Then $x = 0$ is asymptotically integrally stable.

REFERENCES

1. Brumley, W. F., On the asymptotic behavior of solutions of differential-difference equations of neutral type. *J. Differential Equations 7*, 175-188 (1970).
2. Coppel, W. A., Almost periodic properties of ordinary differential equations. *Ann. Mat. Pura e Appl. 76*, 27-49 (1967).
3. Cruz, M. A. and Hale, J. K., Stability of functional differential equations of neutral type. *J. Differential Equations 7*, 334-351 (1970).
4. Darbo, G., Punti uniti in trasformazioni a codominio non compatto. *Rend. Sem. Mat. Univ. Padova 24*, 84-92 (1955).
5. Hale, J. K., Forward and backward continuation for neutral functional differential equations. *J. Differential Equations 9*, 168-181 (1971).
6. Hale, J. K., Ordinary Differential Equations. Interscience, New York, 391-402.
7. Hale, J. K. and Lopes, O. F., Fixed point theorems and dissipative process. *J. Differential Equations 13*, 391-402 (1973).
8. Izé, A. F. and Freiria, A. A., Integral stability for neutral functional differential equations. (To appear).
9. Izé, A. F. and Dos Reis, J. G., Stability of perturbed neutral functional differential equations. *J. of Nonlinear Analysis 2*, 563-571 (1978).
10. Izé A. F. and Dos Reis, J. G., Contributions to stability of neutral functional differential equations. *J. Differential Equations 28*, (1978).
11. Kato, J. and Sibuya, Y., Catastrophic deformation of a flow and non existence of almost periodic solution. *J. Faculty of Sc. Univ. of Tokyo, Ser. I A, 24*, 267-280 (1977).
12. Kuratowski, K., Topology, Vol. I. New York, (1966).

13. Lopes, O. F., a) Existência e estabilidade de oscilações forçadas de equações diferenciais funcionais. Tese de Livre-Docência, ICMSC-USP (1975).

b) Periodic solutions of perturbed neutral differential equations. *J. Differential Equations 15*, n. 1 (1974).

14. Massera, J. L., On Liapunoff's conditions of stability. *Ann. of Math. 5*, 705-721 (1949).

15. Massera, J. L., Contributions to stability theory. *Ann. of Math. 64*, 182-206 (1956).

16. Yoshizawa, T., Stability Theory by Liapunov's a Second Method. Math. Soc. of Japan, (1966).

17. Yoshizawa, T., Asymptotically almost periodic solutions of an almost periodic system, *Funkcial. Ekvac.*, 23-40 (1969).

18. Zverkin, A. M., Dependence of the stability of the solutions of differential equations with a delay on the choice of the initial instant. *Vestnik Moscov. Univ. Ser. I: Mat. Mech. 5*, 15-20 (1959).

NONLINEAR PERTURBATIONS OF LINEAR PROBLEMS WITH INFINITE DIMENSIONAL KERNEL

R. Kannan[1]

Department of Mathematics
University of Texas at Arlington
Arlington

I. INTRODUCTION

In this paper we consider the problem of existence of solutions of nonlinear problems of the type

$$Ex = Nx \ , \tag{1}$$

$x \in X$, X a real Hilbert space, E is a linear operator with domain $D(E) \subset X$ and N a nonlinear operator over X. We consider here the case when the kernel of E, denoted by X_0, is such that $\dim X_0 = \infty$.

In the recent years there has been an extensive literature on the question of existence of solutions to problems of type (1) when X_0 is finite dimensional and the partial inverse of E restricted to its range is compact. It is the purpose of this paper to show that analogous abstract existence theorems can be obtained in the case when X_0 is infinite dimensional.

[1] Present address: Department of Mathematics, University of Texas at Arlington, Arlington, Texas 76019

RECENT ADVANCES IN DIFFERENTIAL EQUATIONS

ISBN 0-12-186280-1

Essentially our idea is as follows. We approximate (1) by a sequence of problems

$$E_n x = N_n x \tag{2}$$

where $x \in X_n$, X_n being finite dimensional subspaces of X, E_n being a linear operator with finite dimensional kernel and N_n is a suitably modified operator obtained from N. We can now apply the abstract existence results mentioned above and obtain solutions $x_n \in X_n$ of (2). Applying a passage to the limit argument we obtain a solution x of (1).

For the details of the results indicated here we refer to [4,5]. In these papers one can also see references to some of the related literature on problems of type (1).

II. AN ABSTRACT EXISTENCE THEOREM FOR THE CASE $\dim X_0 < \infty$

Let X, Y be a real Hilbert spaces, $E : D(E) \subset X \to Y$ be a linear operator and $N : X \to Y$ be a nonlinear operator. Let X_0 be the kernel of E and Y_0 the range of E. We assume that $P : X \to X_0$ and $Q : Y \to Y_0$ be projection operators such that

$$PX = X_0, \quad QY = Y_0 \quad \text{and} \quad Y = Y_0 \oplus Y_1, \quad X = X_0 \oplus X_1 .$$

We assume that Y is a space of linear operators on the elements x of X. Let $\langle y,x\rangle$ denote the operation y applied to x such that

a) $|\langle y,x\rangle| \leq k\|y\| \; \|x\|$ for some constant k and all $x \in X$, $y \in Y$,

b) for $y \in Y$ we have $y \in Y_1$ i.e., $QY = 0$ if and only if $\langle Qy,x^*\rangle = 0$ for all $x^* \in X_0$.

The linear operator $X : Y_1 \to D(E) \cap X_1$ can now be defined. We assume that E, H, P and Q satisfy the relations: (i) $H(I - Q)Ex = (I - P)x$ (ii) $QEx = EPx$ and (iii) $EH(I - Q)x = (I - Q)x$. Since $X_0 = \text{kernel } E$, $QEx = EPx = 0$

and hence writing $x = x_0 + x_1$, $x_0 = Px \in X_0$, $x_1 \in X_1$, equation (1) can be seen to be equivalent to the system of equations given by

$$x_1 = H(I - Q)N(x_0+x_1) , \tag{3}$$

$$0 = QN(x_0+x_1) . \tag{4}$$

If $S : Y_0 \to X_0$ is a continuous linear operator such that $S^{-1}(0) = 0$, then (4) is equivalent to

$$0 = SQN(x_0+x_1) . \tag{5}$$

We now assume that X_0 is finite dimensional and H is compact. Further let $L = ||H||$. We then have:

Theorem 2.1 [3]. Let $N : X \to Y$ be continuous and let R, r be positive numbers such that (a) for all $x_0 \in X_0$, $x_1 \in X_1$, $||x_0|| \leq R$, $||x_1|| = r$ we have $||N(x_0+x_1)|| \leq L^{-1}r$ and (b) for all $x_0 \in X_0$, $x_1 \in X_1$, $||x_0|| = R$, $||x_1|| \leq r$ we have $\langle SQN(x_0+x_1), x_0 \rangle \geq 0$ (or ≤ 0).

Then the equation $Ex = Nx$ has at least one solution $x \in X$, $x = x_0+x_1$, $||x_0|| \leq R$, $||x_1|| \leq r$ and $||x|| \leq (R^2+r^2)^{1/2}$.

The details of the proof may be seen in [3]. In his survey paper, Cesari [1] discusses how this theorem includes as particular cases several results of the recent extensive literature on problems of type (1) where X_0 is finite dimensional.

The proof follows by an application of the Schauder fixed point theorem to the system of equations (3) and (5). Thus we consider the map T of the space $X_0 \oplus X_1$ into itself defined by

$$[x_1, x_0] \xrightarrow{T} [H(I - Q)N(x_1+x_0), \quad x_0 + SQN(x_1+x_0)] .$$

By assumption H is compact and S, Q are continuous operators with finite dimensional range. Hence T is compact and we are in a setting to apply the Schauder fixed point theorem.

III. PRELIMINARIES FOR THE CASE $\dim X_0 = \infty$

Let E, N be operators from their domains $D(E)$, $D(N)$ in $\mathcal{X}$ to $\mathcal{Y}$, both $\mathcal{X}$ and $\mathcal{Y}$ real Banach or Hilbert spaces and let us consider the equation $Ex = Nx$. According to the choice of $\mathcal{X}$ we may be obtaining usual solutions or generalized solutions. Let X, Y be real Hilbert spaces such that $X \subset \mathcal{X}$, $Y \subset \mathcal{Y}$ and let the inclusion map $j : X \to \mathcal{X}$ be compact.

We note first that if $\{x_k\}$, $x_k \in X$ is a sequence of elements such that $\|x_k\| \leq M$ then there exists a subsequence which we denote by $\{x_k\}$ again for sake of simplicity such that $\{jx_k\}$ converges strongly in X toward some element ζ. On the other hand, X is Hilbert and thus we can obtain a subsequence still denoted by $\{x_k\}$ such that x_k converges weakly to x in X. Then $\zeta = jx$, because $x_k \to x$ weakly in X implies $jx_k \to jx$ weakly in $\mathcal{X}$.

Let us assume that X_1 and X_0, as introduced in Section 2, contain finite dimensional subspaces X_{1n}, X_{0n} such that $X_{1n} \subset X_{1,n+1} \subset X_1$, $X_{0n} \subset X_{0,n+1} \subset X_0$, $n = 1,2,\ldots$ with $\cup X_{1n} = X_1$, $\cup X_{0n} = X_0$. Further let $R_n : X_1 \to X_{1n}$, $R_n X_1 = X_{1n}$ and $S_n : X_0 \to X_{0n}$, $S_n X_0 = X_{0n}$ be orthogonal projection operators such that $\|R_n x_1\|_X \leq \|x_1\|_X$, $\|S_n x_0\|_X \leq \|x_0\|_X$ for all $x_1 \in X_1$, $x_0 \in X_0$.

Let $[v_1, v_2, \ldots, v_n, \ldots]$ be a complete orthonormal system in X_0 and $x_{0n} = \mathrm{span}(v_1, v_2, \ldots, v_n)$, $n = 1,2,\ldots$. Further let $[w_1, w_2, \ldots, w_n, \ldots]$ be a complete orthonormal system in Y such that $\langle w_i, v_j \rangle = 0$ for all $i \neq j$. Let $Y_n = \mathrm{span}(w_1, \ldots, w_n)$ and S_n' be the orthogonal projection of Y onto Y_{0n}. Then, $S_n' QNx = 0$ if and only if $\langle QNx, v_j \rangle = 0$, $j = 1,2,\ldots,n$ and this holds for all $n = 1,2,\ldots$.

Analogous to the system of equations (3) and (5) we consider the coupled system of operator equations

$$x = S_n Px + R_n H(I - Q)Nx , \tag{6}$$

$$0 = S_n' QNx . \tag{7}$$

Let $\alpha_n : Y_{0n} \to X_{0n}$ be defined by $\alpha_n y = \sum_1^n \langle y, v_i\rangle v_i$. Then we have $0 = S_n'QNx$ if and only if $0 = \alpha_n S_n' QNx$. We conclude that system (6,7) is equivalent to the system

$$x = S_n Px + R_n H(I-Q)Nx ,$$

$$0 = \alpha_n S_n' QNx .$$

Analogous to Theorem 2.1, we now obtain:

Theorem 3.1. Let R, r be positive constants such that

a) for all $x_0 \in X_0$, $x_1 \in X_1$, $\|x_0\| \leq R$, $\|x_1\| \leq r$ we have $\|N(x_0+x_1)\| \leq L^{-1} r$

b) for all $\|x_0\| = R_0$, $\|x_1\| \leq r$ we have $\langle \alpha_n S_n' QN(x_0+x_1), x_0\rangle \geq 0$ (or ≤ 0).

Then, for every n, the system (6,7) has at least one solution $x_n \in D(E) \cap (X_{0n} \oplus X_{1n})$, $x_n = x_{0n} + x_{1n}$, $S_n Px_n = x_{0n}$ with $\|x_n\| \leq M = (R^2 + r^2)^{1/2}$, M independent of n.

The proof follows exactly the lines of Theorem 2.1, by replacing X_0 and X_1 by X_{0n} and X_{1n}, respectively.

IV. AN ABSTRACT THEOREM FOR THE CASE $\dim X_0 = \infty$

We now obtain a sufficient condition for the existence of a solution of equation (1) when $\dim X_0 = \infty$ by adopting a "passage to the limit" argument. We assume that both the Hilbert spaces X and Y are contained in real Banach or Hilbert spaces $\mathcal{X}$ and $\mathcal{Y}$ with compact injections $j : X \to \mathcal{X}$, $j' : Y \to \mathcal{Y}$. Actually we can limit ourselves to the consideration of the spaces $\bar{\mathcal{X}}$ and $\bar{\mathcal{Y}}$ made up of limit elements from sequences in X and Y as discussed in Section 3. Hence $\bar{\mathcal{X}}$ is identical to X and $\bar{\mathcal{Y}}$ is identical to Y, though they may have different topologies. We shall write $\bar{\mathcal{X}} = jX$, $\bar{\mathcal{Y}} = jY$.

Analogously we have $\mathcal{X}_0 = jX_0$, $\mathcal{Y}_0 = j'Y_0$, $\mathcal{X}_1 = jX_1$, $\mathcal{Y}_1 = j'Y$, and the linear operators $\rho : \bar{\mathcal{X}} \to \mathcal{X}_0$, $\beta : \bar{\mathcal{Y}} \to \mathcal{Y}_0$ are then defined by $\rho x = x_0$ in $\bar{\mathcal{X}}$ if $Px = x_0$ in X and $\beta y = y_0$ in $\mathcal{Y}$ if $Qy = y_0$ in Y.

We now assume the following:

(*) $x_n \to x$ weakly in X and $jx_n \to jx$ strongly in $\mathcal{X}$ implies that $Nx_n \to Nx$ in $\mathcal{Y}$, $S_nPx_n \to Px$ strongly in $\mathcal{X}$ and $R_nx_n \to x$ strongly in $\mathcal{X}$.

By the results of Section 3, there are elements $x_n \in X_n$ such that

$$x_n = S_nP_nx_n + R_nH(I-Q)Nx_n , \tag{8}$$

$$0 = \alpha_nS_n'QNx_n \tag{9}$$

where $\|x_n\| \leq M$ for all n. Hence there exists a subsequence, denoted by $\{x_n\}$, such that $x_n \to x$ weakly in X and $jx_n \to jx$ strongly in $\mathcal{X}$. Then by (8) and (9), proceeding to the limit, we have

$$x = Px + H(I-Q)Nx , \qquad 0 = QNx , \qquad x \in \tilde{\mathcal{X}} .$$

We note that in $\tilde{\mathcal{X}}$ the concept of solution is to be appropriately understood. Note that $x \in \tilde{\mathcal{X}}$ and thus is still an element of X. By virtue of the hypotheses on P and H, we have

$$Ex = EPx + EH(I-Q)Nx + QNx = Nx .$$

Thus we have:

<u>Theorem 4.1</u>. Under the hypotheses above, if there are positive constants R, r such that (a) for all $x_0 \in X_0$, $x_1 \in X_1$, $\|x_0\| \leq R$, $\|x_1\| \leq r$ we have $\|N(x_0+x_1)\| \leq L^{-1}r$ and (b) for all $\|x_0\| = R$, $\|x_1\| \leq r$ we have $\langle QN(x_0+x_1), x_0\rangle \geq 0$ (or ≤ 0), then the equation $Ex = Nx$ has at least one solution x satisfying $\|x\| \leq (R^2+r^2)^{1/2}$.

It must be noted that we can replace the bifurcation equation $\alpha_nS_n'QNx = 0$ by the equation

$$J_n\alpha_nS_n'QNx = 0 \tag{10}$$

where $J_n : X_{0n} \to X_{0n}$ is an invertible operator. In this case the sufficient condition is replaced by the inequality

$\langle J_n \alpha_n S_n' QNx, x_0 \rangle \geq 0$ (or ≤ 0)

Proceeding similarly as in the above theorem we can also obtain the following:

Theorem 4.2. Let $N : x \to y$ be continuous and let $\alpha(R)$, $\beta(R)$, $R \geq 0$ be monotone nondecreasing functions such that (a) $x \in X$, $\|x\| \leq R$ implies $\|Nx\| \leq \alpha(R)$ and (b) $x_1, x_2 \in X$, $\|x_1\|, \|x_2\| \leq R$ implies $\|Nx_1 - Nx_2\| \leq \beta(R)\|x_1 - x_2\|$. Let us assume further that (c) there are positive numbers R_0, r such that $L\alpha((R_0^2+r^2)^{1/2}) \leq 1$ and $R_0 + L(\alpha(R_0^2+r^2)^{1/2}) \leq r$ and (d) $\langle QN(x_0+x_1), x_0 \rangle \geq 0$ (or ≤ 0) for all $\|x_0\| \leq R_0$ and $\|x_1\| \leq r$. Then the equation $Ex = Nx$ has at least one solution $x = x_0 + x_1$, $\|x_0\| \leq R_0$, $\|x_1\| \leq r$.

We refer to [5] for details of the proof.

V. TWO APPLICATIONS

We first consider the existence of solutions $u(t,x)$, periodic in t of period 2π, of the hyperbolic nonlinear problem

$$u_{tt} + u_{xxxx} = f(t,x,u,\ldots), \quad 0 < x < \pi, \quad -\infty < t < +\infty \tag{11}$$

$$u(t,0) = u(t,\pi) = u_{xx}(t,0) = u_{xx}(t,\pi) = 0 \tag{12}$$

$$u(t,2\pi,x) = u(t,x), \quad 0 < x < \pi, \quad -\infty < t < +\infty . \tag{13}$$

Let $I = [0,\pi] \times [0,2\pi]$, $G = [0,\pi] \times R$. Let D denote the set of all real valued functions $u(t,x)$, 2π periodic in t, of class C^∞ in G and such that $D_x^{2k}\psi(t,0) = D_x^{2k}\psi(t,\pi) = 0$, $k = 0,1,2,\ldots$. Let A_m denote the completion of D under the norm

$$\|u\|_m = (\textstyle\iint_I [(D_t^m u)^2 + (D_x^{2m} u)^2] dt dx)^{1/2} .$$

Thus, A_m is a real Hilbert space with inner product

$$(u,v)_m = (D_t^m u, D_t^m v) + (D_x^{2m} u, D_x^{2m} v), \quad u,v \in A_m ,$$

where $(,)$ denotes the inner product in $L_2(I)$, and we denote $L_2(I)$ by A_0. Let E denote the operator defined by $Eu = u_{tt} + u_{xxxx}$.

For $g \in A_m$, we consider the linear problem $Eu = g$. We say that u is a weak solution of this problem with boundary conditions (12), (13) if $u \in A_m$ and $(u,EQ)_m = (g,Q)_m$ for all $Q \in D$. Thus both the equation $Eu = g$ and the boundary conditions (12) and (13) are understood in the weak sense. A complete orthonormal system in $A = L_2(I)$

$$\{e_{k\ell}^{(0)}\} = 2^{1/2}\pi^{-1}\sin kt \sin \ell x,\ 2^{1/2}\pi^{-1}\cos kt \sin \ell x, \pi^{-1}\sin \ell x$$

whose elements can be indexed by $\ell = 1,2,\ldots$, $k = 0, \pm 1, \pm 2, \ldots$ as usual. A complete orthonormal system in A_m is

$$e_{k\ell}^{(m)} = 2^{1/2}\pi^{-1}\ (k^{2m} + \ell^{4m})^{-1/2}\sin kt \sin \ell x ,$$

$$2^{1/2}\pi^{-1}\ (k^{2m}+\ell^{4m})^{-1/2}\ \cos kt \sin \ell x , \pi^{-1}\ell^{-2m}\sin \ell x .$$

In order to apply the ideas of the previous sections we introduce the spaces X, Y etc. Let $X = A_2$ and let $X_0 = \ker E$ i.e., X_0 is spanned by $e_{k\ell}^{(2)}$, $k^2 = \ell^4$. Let X_{0n} be the finite dimensional subspaces of X_0 generated by $e_{k\ell}$, $k^2 = \ell^4$, $\ell = 1,2,\ldots,n$. Then let X_{1n} be the subspace of X_1 generated by $e_{k\ell}$, $k^2 \neq \ell^4$, $\ell = 1,2,\ldots,n$, $k = 0,1,\ldots,n$. Let R_n, S_n be the appropriate orthogonal projections given by $R_n : X_0 \to X_{0n}$ and $S_n : X_0 \to X_{0n}$. Finally let $J : X_0 \to X_0$ be the linear operator defined as follows: for any $u \in X_0$ $u = \Sigma\, u_{k\ell}e_{k\ell}$ where Σ ranges over all k and ℓ with $k^2 = \ell^4$ we set $Ju = \Sigma\, k^{-1}u_{k\ell}e'_{k\ell}$ where $e'_{k\ell}$ is obtained by replacing $\cos kt$ by $\sin kt$ and $\sin kt$ by $-\cos kt$. Then J is an isomorphism and $(Ju)_t = u$. We can now apply Theorem 4.1 and the remarks following it to obtain the following:

Theorem 5.1. Let $\gamma(R)$, $R \geq 0$, be a monotone nondecreasing positive function such that, $u \in X$, $\|u\|_X \leq R$ implies $\|f(t,x,u,u_t,u_x)\| \leq \gamma(R)$. Also let $L = \|H\|$. Further let $f(t,x,u,u_t,u_x) \in C^2(G\times R^3)$ and let R_0, r be positive constants such that (a) for all $u_0 \in X_0$, $u_1 \in X_1$, $\|u_1\| \le r$, $\|u^*\| = R_0$ we have

$$\int_I f(t,x,u,u_t,u_x)_t (u_0)_{tt} dtdx \le 0 \quad (\text{or} \ge 0)$$

and (b) $L\gamma(R) \le r$ where $R^2 = R_0^2 + r^2$.

Then the quasilinear hyperbolic problem (11-13) has at least one solution $u(t,x) \in A_2$.

For the details of the proof and the choice of the spaces $\mathcal{X}$ and $\mathcal{Y}$ we refer to [5]. If the nonlinearity involves a small parameter i.e., $f = \varepsilon\, g(t,x,u,u_t,u_x)$ then we can establish that the results of Petzeltova |9| follow by an application of Theorem 4.2 and 5.1.

We now consider the nonlinear hyperbolic problem

$$u_{tt} - u_{xx} + \beta u_t + \psi(u) = h(t,x) \ , \quad \beta \neq 0 \tag{14}$$

$$u(t+2\pi,x) = u(t,x) = u(t,x+2\pi) \tag{15}$$

where, in comparison with the preceding example, the nonlinearity $f(t,x,u,u_t,u_x)$ is of the form $-\beta u_t - \psi(u) + h(t,x)$. The function $\psi : R \to R$ is continuous and $h : R^2 \to R$ is a given 2π-periodic function, periodic in t and x. Let E be the linear operator defined by $Eu = u_{tt} - u_{xx}$, and N be the nonlinear operator defined by $Nu = -\beta u_t - \psi(u) + h(t,x)$. Let $X = L_2(I)$ where $I = [0,2\pi] \times [0,2\pi]$. As before let $e_{k\ell}$ stand for the functions $e^{ikt}e^{i\ell t}$, $k, \ell = 0, \pm 1, \pm 2, \dots$. Any $u \in X$ can be written as $\Sigma\, u_{k\ell} e_{k\ell}$. The kernel X_0 of E in X is the subspace generated by $e_{k\ell}$ with $|k| = |\ell|$. Finally let $J : X_0 \to X_0$ be defined by

$$Ju = \Sigma\, (ik)^{-1} u_{k\ell} e_{k\ell} \ .$$

Then if $u = \Sigma\, u_{k\ell} e_{k\ell}$, $|k| = |\ell|$, we can write $u = u_0 + \Sigma' u_{k\ell} e_{k\ell}$ Σ' being extended over all $|k| = |\ell| \neq 0$. In the sense of distributions we can then note that $Ju_t = u - u_0$. We can then show, by applying Theorem 4.1, the following:

<u>Theorem 5.2</u>. If $\psi : R \to R$ is a continuous function with limits $\psi(-\infty) < \psi(\infty)$ and

$$\psi(-\infty) < (2\pi)^{-2} \int_0^{2\pi}\int_0^{2\pi} h(t,x)\,dt\,dx < \psi(\infty)$$

then (14-15) has at least one weak solution.

Once again we refer to [5] for the details. The above result was also obtained in [6].

VI. REMARKS

1) It is evident from Section 3 and 4 that the existence of a solution of the equation $Ex = Nx$ when $\dim X_0 = \infty$ may also be stated as follows: there exists a bounded open ball B in X with the origin as centre such that the system of equations (8) and (9) has a solution in the restriction B_n of B to the space $X_{0n} \oplus X_{1n} = X_n$ and the operator N has the appropriate continuity properties.

2) From Section 5 it follows that the study of the linear hyperbolic problems are important to obtain analytical conditions implying the sufficient abstract conditions. In [8] Naparstek studies several linear hyperbolic problems for the existence of weak solutions. In particular extensive studies are made of the operator $u_{tt} - \sigma^2 u_{xx}$. He then applies these ideas to obtain the existence of solutions of nonlinear hyperbolic problems involving nonlinearities of the type $\varepsilon\, g(t,x,u)$ by following the lines of [2,12].

2) We will consider in a later paper the autonomous hyperbolic problems of the type considered in Section 5 and use will be made of the ideas in Naparstek [8].

REFERENCES

1. Cesari, L., Functional analysis, nonlinear differential equations, *in* "Nonlinear Functional Analysis and Differential Equations" (L. Cesari, R. Kannan, J. D. Schuur, eds.), pp. 1-197. M. Dekker, New York, (1976).
2. Cesari, L., Existence in the large of periodic solutions

of hyperbolic partial differential equations, *Archive Rat. Mech. Anal. 20*, 170-190 (1965).

3. Cesari, L. and Kannan, R., An abstract existence theorem at resonance, *Proc. Amer. Math. Soc. 63*, 221-225 (1977).
4. Cesari, L. and Kannan, R., Existence of solutions of nonlinear hyperbolic problems, *Ann. Scuola Norm. Sup. Pisa*, to appear.
5. Cesari, L. and Kannan, R., Solutions of nonlinear hyperbolic equations at resonance, *J. Nonl. Anal.*, to appear.
6. Fucik, S. and Mawhin, J., Generalized periodic solutions of nonlinear telegraph equations, *J. Nonl. Anal. Meth. Appl. 2*, 609-618 (1978).
7. Hale, J. K., Periodic solutions of a class of hyperbolic equations, *Archive Rat. Mech. Anal. 23*, 380-398 (1967).
8. Naparstek, A., Periodic solutions of certain weakly nonlinear hyperbolic partial differential equations, Univ. of Michigan, Ph.D. thesis, (1968).
9. Petzeltova, H., Periodic solutions of the equation $u_{tt} + u_{xxxx} = f(.,.,u,u_t)$, *Czechoslovak. Math. Journ. 23 (98)*, 269-285 (1973).

COMPARISON RESULTS FOR REACTION-DIFFUSION EQUATIONS

V. Lakshmikantham[1,2]

Department of Mathematics
University of Texas at Arlington
Arlington, Texas

I. INTRODUCTION

Let u, v denote certain measures of total population such as number of individuals, mass, area of shade cast by plants, etc. A simple model describing populations which are diffusing through Ω as well as interacting with each other is

$$\begin{cases} u_t = a_1 \Delta u + uM(u,v) , \\ v_t = a_2 \Delta v + vN(u,v) , \end{cases} \tag{1.1}$$

where $a_1, a_2 \geq 0$, Δ is the Laplace operator in x and M, N are suitable functions. Equations (1.1) are considered with the boundary conditions

$$\frac{\partial u}{\partial \tau} = 0, \quad \frac{\partial v}{\partial \tau} = 0 \quad \text{on} \quad R_+ \times \partial\Omega \tag{1.2}$$

---

[1]Research partially supported by U.S. Army Research Grant DAAG29-77-G0062.

[2]Present address: Department of Mathematics, University of Texas at Arglington, Arlington, Texas 76019, U.S.A.

ISBN 0-12-186280-1

and the initial conditions

$$u(0,x) = u_0(x), \quad v(0,x) = v_0(x) \quad \text{on} \quad \Omega . \tag{1.3}$$

Models of the type (1.1) to (1.3) occur in studies of population genetics [2,28], conduction of nerve impulses [2,10], chemical reactions [1,11] and several other biological questions [1,27]. See also [5,6,7,9,19,20,21,25,26,27,30].

If we confine our attention to equations which are independent of space variable x, the equations (1.1) become the Kolmogorov form of ordinary differential equations describing interactive growth of two populations. This approach has a long history dating from the pioneering work of Lotka and Volterra and still occupies a central portion in mathematical ecology [20,21].

It is therefore clear that many models in chemical, biological and ecological processes involve a system of parabolic differential equations of the form

$$u_t^i = a^i \Delta u^i + f_i(t,x,u_1,\ldots,u_N,u_{x_i}^i,\ldots,u_{x_n}^i) \tag{1.4}$$

subject to appropriate initial boundary conditions. Such systems are called weakly coupled systems.

No model of the form (1.4) can be relied on to be in quantitative agreement with real world systems. Volterra himself recognized this fact. He emphasized consistently that differential equations are, at best, only rough approximations of actual ecological systems. They would apply only to animals without age or memory, which eat all the food they encounter and immediately convert it into offspring. Nevertheless, system (1.4) is worth studying since it might offer better qualitative understanding of real-world problems and since such study might help in constructing and analyzing more complex models.

One of the important and effective techniques in the qualitative analysis of dynamical systems is the comparison method or theory of differential inequalities [16,29]. This method is also a useful tool for the qualitative analysis of reaction-diffusion systems [3,4,8,12,13,15,18]. Unfortunately, comparison theorems for systems impose monotonicity requirements on the reaction terms which are physically unreasonable and thus not suited to the problems at hand. To circumvent this monotonic restrictions, one has two alternatives:

(i) since this difficulty is essentially due to the choice of the cone relative to the system, namely, the cone of nonnegative elements of R^n, a possible answer lies in choosing an appropriate cone other than R^n to work in a given situation [14,17].

(ii) extend suitably the important classical result of Müller [2] which leads to the new notion of quasi-solutions that have computational advantage.

In this lecture, we shall develop these ideas. We consider reaction-diffusion equations in a Banach space and work with arbitrary cones. This general approach unifies several results and offers a flexible mechanism. We demonstrate the results by means of simple examples.

II. MAIN RESULTS

In order to develop the theory of comparison theorems for reaction-diffusion equations in a Banach space E, we need to introduce the concept of a cone which induces a partial ordering in E.

Let E be a real Banach space with $\|\cdot\|$ and let E^* denote the set of continuous linear functionals. A proper subset K of E is said to be a cone if

$$\lambda K \subset K, \quad \lambda \geq 0, \quad K + K \subset K, \quad K = \bar{K} \quad \text{and} \quad K \cap (-K) = \{0\}.$$

Here $\bar{K}$ denotes the closure of K. Let us denote by K^0 the interior of K and assume that K^0 is nonempty. The cone K induces a partial ordering on E defined by

$$u \leq v \quad \text{iff} \quad v - u \in K \quad \text{and} \quad u < v \quad \text{iff} \quad v - u \in K^0 .$$

A linear functional $\phi \in E^*$ is said to be a positive linear functional if $\phi(x) \geq 0$ whenever $x \in K$. Let K^* denote the set of positive linear functionals. It then follows that K is contained in the closed half space

$$C_\phi = [x \in E: \phi(x) \geq 0, \quad \phi \text{ a positive linear functional}].$$

Thus the positive linear functionals are support functionals and since K is a cone in E, K is the intersection of all the closed half spaces which support it. If $S \subset K^*$ and $K = \cap [C_\phi : \phi \in S]$ then S is said to generate the cone K. Let $S_0 = [\phi \in S: \|\phi\| \leq \gamma_0]$ and $\bar{S}_0$ be the closure of S_0 in the weak star topology.

Let Ω be a bounded domain in R^n and let $H = (t_0, \infty) \times \Omega$, $t_0 \geq 0$. Suppose that the boundary ∂H of H is split into two parts ∂H_0, ∂H_1 such that $\partial H = \partial H_0 \cup \partial H_1$ and $\partial H_0 \cap \partial H_1$ is empty.

A vector τ is said to be an outer normal at $(t,x) \in \partial H_1$, if $(t, x-h\tau) \in \Omega$ for small $h > 0$. The outer normal derivative is then given by $\frac{\partial u}{\partial \tau}(t,x) = \lim_{h \to 0} \frac{u(t,x) - u(t,x-h\tau)}{h}$. We shall always assume that an outer normal exists on ∂H_1 and the functions in question have outer normal derivatives on ∂H_1.

If $u \in C[\bar{H}, E]$ and if the partial derivatives u_t, u_x, u_{xx} exist and are continuous in H, then we shall say that $u \in C$. As before, E is a real Banach space and we shall assume K is a cone in E generated by S_0.

The following lemmas are basic to our discussions.

Lemma 2.1. Assume that

(i) $m, n \in C$, and $m(t,x) < 0 < n(t,x)$ on ∂H_0;

(ii) $\frac{\partial m}{\partial \tau}(t,x) < 0 < \frac{\partial n}{\partial \tau}(t,x)$ on ∂H_1, where τ is the outer normal on ∂H_1;

(iii) if $(t_1, x_1) \in H$ and $\phi \in \bar{S}_0$ such that $m(t_1,x_1) \leq 0 \leq n(t_1,x_1)$ and either $\phi(m(t_1,x_1)) = 0$, $\phi(m_{x_i}(t_1,x_1)) = 0$, $i = 1,2,\ldots,n$ and the quadratic form $\sum_{i,j=1}^{n} \lambda_i \lambda_j \phi(m_{x_i x_j}(t_1,x_1)) \leq 0$, $\lambda \in R^n$, or $\phi(n(t_1,x_1)) = 0$ $\phi(n_{x_i}(t_1,x_1)) = 0$, $i = 1,2,\ldots,n$, and the quadratic form $\sum_{i,j=1}^{n} \lambda_i \lambda_j \phi(n_{x_i x_j}(t_1,x_1)) \geq 0$, $\lambda \in R^n$, then either $\phi(m_t(t_1,x_1)) < 0$ or $\phi(n_t(t_1,x_1)) > 0$ respectively.

Under these conditions we have

$$m(t,x) < 0 < n(t,x) \quad \text{on} \quad \bar{H}.$$

Lemma 2.2. Assume that

(i) $m \in C$ and $m(t,x) < 0$ on ∂H_0 ($n \in C$ and $n(t,x) > 0$ on ∂H_0);

(ii) $\frac{\partial m}{\partial \tau}(t,x) < 0$ on ∂H_1 ($\frac{\partial n}{\partial \tau}(t,x) > 0$ on ∂H_1);

(iii) if $(t_1,x_1) \in H$ and $\phi \in \bar{S}_0$ such that $m(t_1,x_1) \leq 0$, $\phi(m(t_1,x_1)) = 0$, $\phi(m_{x_i}(t_1,x_1)) = 0$, $i = 1,2,\ldots,n$ and $\sum_{i,j=1}^{n} \lambda_i\lambda_j\phi(m_{x_ix_j}(t_1,x_1)) \leq 0$, $\lambda \in R^n$, then $\phi(m_t(t_1,x_1)) < 0$ ($n(t_1,x_1) \geq 0$, $\phi(n(t_1,x_1)) = 0$, $\phi(n_{x_i}(t_1,x_1)) = 0$, $i = 1,2,\ldots,n$ and $\sum_{i,j=1}^{n} \lambda_i\lambda_j\phi(n_{x_ix_j}(t_1,x_1)) \geq 0$, then $\phi(n_t(t_1,x_1)) > 0$).

Under these conditions, we have

$$m(t,x) < 0 \quad \text{on} \quad \bar{H} \quad (\, n(t,x) > 0 \quad \text{on} \quad \bar{H}).$$

Let $f \in C[\bar{H} \times E \times E^n \times E^{n^2}, E]$, where E^i stands for $E \times E \times E \ldots$, i times. A function f is said to be quasimonotone nondecreasing in u relative to K for each $(t,x) \in \bar{H}$, if for any $u,v \in E$, $u_x, v_x \in E^n$, $u_{xx}, v_{xx} \in E^{n^2}$ and $\phi \in \bar{S}_0$ such that $u \leq v$, $\phi(u) = \phi(v)$, $\phi(u_{x_i}) = \phi(v_{x_i})$ $i = 1,2,\ldots,n$ and $\sum_{i,j=1}^{n} \lambda_i\lambda_j\phi(u_{x_ix_j} - v_{x_ix_j}) \leq 0$, $\lambda \in R^n$, we have

$$\phi(f(t,x,u,u_x,u_{xx})) \leq \phi(f(t,x,v,v_x,v_{xx})).$$

For the case $E = R^N$ and $K = R^N_+$, the quasimonotone condition on f implies that $f_i(t,x,u,u_x,u_{xx}) = f_i(t,x,u,u^i_x,u^i_{xx})$, for each i, $1 \leq i \leq N$.

The following comparison results can be proved by means of Lemma 2.2.

Theorem 2.1. Suppose that

(i) $v,w \in C$, $f \in C[\bar{H} \times E \times E^n \times E^{n^2}, E]$, f is quasimonotone nondecreasing in u relative to K and

$$v_t \le f(t,x,v,v_x,v_{xx}), \quad w_t \ge f(t,x,w,w_x,w_{xx}) \quad \text{on} \quad H,$$

where $v = v(t,x)$, $w = w(t,x)$;

(ii) (a) $v(t,x) < w(t,x)$ on ∂H_0

(b) $\frac{\partial v(t,x)}{\partial \tau} < \frac{\partial w}{\partial \tau}(t,x)$ on ∂H_1.

Then $v(t,x) < w(t,x)$ on $\bar{H}$, if one of the inequalities in (i) is strict.

We can dispense with the strict inequality needed in Theorem 2.1 if f satisfies the following condition:

(C_0) $z \in C$, $z > 0$ on $\bar{H}$, $\frac{\partial z}{\partial \tau}(t,x) \ge \gamma > 0$ on ∂H_1 and for sufficiently small $\varepsilon > 0$, either

(a) $\varepsilon z_t > f(t,x,v,v_x,v_{xx}) - f(t,x, v-\varepsilon z, v_x-\varepsilon z_x, v_{xx}-\varepsilon z_{xx})$,

or

(b) $\varepsilon z_t > f(t,x, w+\varepsilon z, w_x+\varepsilon z_x, w_{xx}+\varepsilon z_{xx}) - f(t,x,w,w_x,w_{xx})$

on H, where $v,w, \in C$.

Theorem 2.2. Let the assumption (i) of Theorem 2.1 hold. Suppose further that the condition (C_0) is satisfied. Then the relations

(ii) (a) $v(t,x) \le w(t,x)$ on ∂H_0,

(b) $\frac{\partial v}{\partial \tau}(t,x) \le \frac{\partial w}{\partial \tau}(t,x)$ on ∂H_1,

imply $v(t,x) \le w(t,x)$ on $\bar{H}$.

As an example, consider the case when $E = R$,

$f(t,x,u,u_x,u_{xx}) = au_{xx} + bu_x + F(t,x,u)$ where $\sum_{i,j=1}^{n} a_{ij}u_{x_i x_j}$, $bu_x = \sum_{j=1}^{n} b_j u_{x_j}$, and suppose that F is Lipschitzian in u for a constant $L > 0$. Suppose that the boundary ∂H_1 is regular, that is, there exists a function $h \in C$ such that $h(x) \ge 0$, $\frac{\partial h(x)}{\partial \tau} \ge 1$ on ∂H_1 and h_x, h_{xx} are bounded. Let $M > 1$, $H(x) = e^{LMh(x)} \ge 1$, $z(t,x) = e^{Nt}H(x)$, where $N = LM + A$, $|aH_{xx} + bH_x| \le A$. Then

$$\frac{\partial z}{\partial \tau}(t,x) = LM\frac{\partial h(x)}{\partial \tau} \cdot z(t,x) \ge LM = \gamma > 0 \quad \text{on} \quad \partial H_1 \quad \text{and}$$

$$z_t - az_{xx} - bz_x \geq [N - A]z = LMz > Lz.$$

Consequently, using Lipschitz condition of F, we arrive at

$$\varepsilon z_t > \varepsilon[az_{xx} + bz_x] + F(t,x, w+\varepsilon z) - F(t,x,w),$$

which is exactly the condition (b) of (C_0).

Remark. If ∂H_1 is empty so that $\partial H_0 = \partial H$, the assumption (C_0) in Theorem 2.2 can be replaced by a weaker hypothesis, namely a one sided Lipschitz condition of the form

$$(C_1) \quad f(t,x,u,P,Q) - f(t,x,v,P,Q) \leq L(u-v), \quad u > v.$$

In this case, it is enough to set $\tilde{w} = w + e^{2Lt}$ so that

$$v(t,x) < \tilde{w}(t,x) \quad \text{on} \quad \partial H$$

and

$$\begin{aligned}
\tilde{w}_t &\geq f(t,x,w,w_x,w_{xx}) + 2\varepsilon Le^{2Lt} \\
&\geq f(t,x,\tilde{w},\tilde{w}_x,\tilde{w}_{xx}) + \varepsilon Le^{2Lt} \\
&> f(t,x,\tilde{w},\tilde{w}_x,\tilde{w}_{xx}) \quad \text{on} \quad H.
\end{aligned}$$

Even when ∂H_1 is not empty, the condition (C_1) is enough provided (ii)(b) is strengthened to $\frac{\partial v}{\partial \tau} + Q(t,x,v) \leq \frac{\partial w}{\partial \tau} + Q(t,x,w)$ on ∂H_1, where $Q \in C[\bar{H} \times E, E]$ and $Q(t,x,u)$ is strictly increasing in u. To see this observe that $\tilde{w} > w$ and hence $Q(t,x,w) < Q(t,x,\tilde{w})$ which gives the desired strict inequality needed in the proof. Of course, if Q is not strictly increasing or $Q \equiv 0$, then the condition (C_0) becomes essential.

Let us next consider the mixed problem

$$u_t = f(t,x,u,u_x,u_{xx}), \tag{2.1}$$

$$u(t,x) = u_0(t,x) \quad \text{on} \quad \partial H_0 \quad \text{and} \quad \frac{\partial u}{\partial \tau}(t,x) = 0 \quad \text{on} \quad \partial H_1, \tag{2.2}$$

and assume that the solutions of (2.1) and (2.2) exist on $\bar{H}$.

A closed set F E is said to be flow-invariant relative

to the system (2.1), (2.2) if for every solution $u(t,x)$ of (2.1), (2.2), we have

$$u_0(t,x) \in F \text{ on } \partial H_0 \text{ implies } u(t,x) \in F \text{ on } \bar{H}.$$

The function $f(t,x,u,u_x,u_{xx})$ is said to be quasi-nonpositive (quasi-nonnegative) if $u \le 0$ $(u \ge 0)$, $\phi(u) = 0$, $\phi(u_{x_i}) = 0$, $i = 1,2,\ldots,n$ and $\sum_{i,j=1}^{n} \lambda_i\lambda_j\phi(u_{x_ix_j}) \le 0$ $\left(\sum_{i,j=1}^{n} \lambda_i\lambda_j\phi(u_{x_ix_j}) \ge 0\right)$, $\lambda \in R^n$ for some $\phi \in \bar{S}_0$, then $\phi(f(t,x,u,u_x,u_{xx})) \le 0$ $(\phi(f(t,x,u,u_x,u_{xx})) \ge 0)$.

A result on flow-invariance which is useful in obtaining bounds on solutions of (2.1), (2.2) can be proved in a similar way.

Theorem 2.3. Assume that f is quasi-nonpositive and that the condition (C_0)(a) holds with $v = u$, where $u = u(t,x)$ is any solution of (2.1), (2.2). Then the closed set $\bar{Q}$ is flow-invariant relative to the system (2.1), (2.2), where $Q = [u \in E;\ u < 0]$.

The following corollaries are useful in some situations, whose proofs also we omit.

Corollary 2.1. Assume that f is quasi-nonnegative and that the condition (C_0)(b) holds with $w = u(t,x)$. Then the closed set $\bar{Q}$ is flow invariant relative to (2.1), (2.2) where $Q = [u \in E:\ u > 0]$.

Corollary 2.2. Suppose that the condition (C_0) holds with $v = w = u$. Assume also that the following condition holds:

if $u \le b$, $\phi(u) = \phi(b)$, $\phi(u_{x_i}) = 0$, $i = 1,2,\ldots,n$ and $\sum_{i,j=1}^{n} \lambda_i\lambda_j\phi(u_{x_ix_j}) \le 0$, $\lambda \in R^n$ for some $\phi \in \bar{S}_0$, then $\phi(f(t,x,u-b,u_x,u_{xx})) \le 0$, and

if $a \le u$, $\phi(u) = \phi(a)$, $\phi(u_{xi}) = 0$, $i = 1,2,\ldots,n$ and $\sum_{i,j=1}^{n} \lambda_i\lambda_j\phi(u_{x_ix_j})) \ge 0$, $\lambda \in R^n$ for some $\phi \in \bar{S}_0$, then $\phi(f(t,x,u-a,u_x,u_{xx})) \ge 0$.

Then the closed set $\bar{w}$, where $w = [u \in E: a < u < b,\ a,b \in E]$ is flow-invariant relative to (2.1), (2.2).

We shall next consider a comparison result which yields upper and lower bounds for solutions of (2.1), (2.2) in terms of solutions of ordinary differential equations.

Theorem 2.4. Assume that

(i) $u = u(t,x)$ is any solution of (2.1), (2.2) and the condition (C_0) holds with $v = w = u$;

(ii) $g_1, g_2 \in C[R_+ \times E, E]$, $g_1(t,r)$, $g_2(t,r)$ are quasimonotone nondecreasing in r relative to K and for $(t,x,u) \in \bar{H} \times E$, $\phi \in \bar{S}_0$ if $\phi(u_{x_i}) = 0$, $i = 1,2,\ldots,n$, $\sum_{i,j=1}^{n} \lambda_i \lambda_j \phi(u_{x_i x_j}) \le 0$, $\lambda \in R^n$,

$$\phi(f(t,x,u,u_x,u_{xx})) \le \phi(g_1(t,u)),$$

and if $\phi(u_{x_i}) = 0$, $i = 1,2,\ldots,n$, $\sum_{i,j=1}^{n} \lambda_i \lambda_j \phi(u_{x_i x_j}) \ge 0$, $\lambda \in R^n$,

$$\phi(g_2(t,u)) \le \phi(f(t,x,u,u_x,u_{xx})).$$

(iii) $r(t)$, $\rho(t)$ are solutions of

$$r' = g_1(t,r),\quad r(t_0) = r_0,\quad \rho' = g_2(t,\rho),\quad \rho(t_0) = \rho_0,$$

respectively existing on $[t_0,\infty)$ such that

$$\rho_0 \le u_0(t,x) \le r_0, \quad \text{on } \partial H_0.$$

Then $\rho(t) \le u(t,x) \le r(t)$ on $\bar{H}$.

Corollary 2.3. If $\bar{w}$ is flow-invariant relative to the system (2.1), (2.2), there exist functions g_1, g_2 satisfying the assumptions of Theorem 2.4.

Proof. We construct g_1, g_2 as follows: for $\phi \in \bar{S}_0$,

$$\phi(g_1(t,u)) = \sup[\phi(f(t,x,v,v_x,v_{xx})): \ x \in \bar{\Omega}, \ a \le v \le u,$$
$$\phi(v) = \phi(u), \ \phi(v_{x_i}) = 0, \ i = 1,2,\ldots,n$$
$$\text{and} \ \sum_{i,j=1}^{n} \lambda_i \lambda_j \phi(v_{x_i x_j}) \le 0, \ \lambda \in R^n],$$

and

$$\phi(g_2(t,u)) = \inf[\phi(f(t,x,v,v_x,v_{xx})): \ x \in \bar{\Omega}, \ u \le v \le b,$$
$$\phi(v) = \phi(u), \ \phi(v_{x_i}) = 0, \ i = 1,2,\ldots,n$$
$$\text{and} \ \sum_{i,j=1}^{n} \lambda_i \lambda_j \phi(v_{x_i x_j}) \ge 0, \ \lambda \in R^n].$$

We shall next discuss comparison results which offer better bounds under much weaker assumptions. These results are extensions of the classical result of Müller [22]. The proof employs Lemma 2.1.

Theorem 2.5. Assume that

(i) for $\phi \in \bar{S}_0$, $v,w \in C$, $\phi(v_t) \le \phi(f(t,x,\sigma,\sigma_x,\sigma_{xx}))$ for all $\sigma \in E$, such that $v \le \sigma \le w$, $\phi(v) = \phi(\sigma)$, $\phi(v_{x_i}) = \phi(\sigma x_i)$, $i = 1,2,\ldots,n$ and $\sum_{i,j=1}^{n} \lambda_i \lambda_j \phi(\sigma_{x_i x_j} - v_{x_i x_j}) \ge 0$, $\lambda \in R^n$, and

$$\phi(w_t) \ge \phi(f(t,x,\sigma,\sigma_x,\sigma_{xx})$$

for all $\sigma \in E$ such that $v \le \sigma \le w$, $\phi(w) = \phi(\sigma)$, $\phi(w_{x_i}) = \phi(\sigma_{x_i})$, $i = 1,2,\ldots,n$ and $\sum_{i,j=1}^{n} \lambda_i \lambda_j \phi(\sigma_{x_i x_j}) \le 0$;

(ii) the condition (C_0) holds with (a), (b), of (C_0) replaced by the weaker conditions

(a*) $\phi(\varepsilon z_t) > \phi(f(t,x,\sigma,\sigma_x,\sigma_{xx}) - f(t,x, \sigma-\varepsilon z, \sigma_x-\varepsilon z_x), \sigma_{xx}-\varepsilon z_{xx}))$ whenever $v \le \sigma \le w$, $\phi(v) = \phi(\sigma)$, $\phi(v_{x_i}) = \phi(\sigma_{x_i})$, $i = 1,2,\ldots,n$ and $\sum_{i,j=1}^{n} \lambda_i \lambda_j \phi(\sigma_{x_i x_j} - v_{x_i x_j}) \ge 0$, $\lambda \in R^n$ and (b*) $\phi(\varepsilon z_t) > \phi(f(t,x,\sigma+\varepsilon z,\sigma_x+\varepsilon z_x,\sigma_{xx}+\varepsilon z_{xx}) - f(t,x,\sigma,\sigma_x,\sigma_{xx}))$ whenever $v \leq \sigma \leq w$, $\phi(w) = \phi(\sigma)$, $\phi(w_{x_i}) = \phi(\sigma_{x_i})$, $i = 1,2,\ldots,n$ and $\sum_{i,j=1}^{n} \lambda_i\lambda_j\phi(\sigma_{x_ix_j} - w_{x_ix_j}) \leq 0$.

(iii) $u(t,x)$ is any solution of (2.1), (2.2) such that $v \leq u_0 \leq w$ on ∂H_0 and $\frac{\partial v}{\partial \tau} \leq \frac{\partial u}{\partial \tau} \leq \frac{\partial w}{\partial \tau}$ on ∂H_1.

Then $v(t,x) \leq u(t,x) \leq w(t,x)$ on $\bar{H}$.

<u>Corollary 2.4</u>. Let the assumptions (i), (ii) of Theorem 2.4 hold without g_1, g_2 being quasimonotone nondecreasing. Suppose that the conditions (ii), (iii) of Theorem 2.5 are satisfied. Assume further that

$$\phi(w'(t)) \geq \phi(g_1(t,\sigma)) \quad \text{for all} \quad v \leq \sigma \leq w \quad \text{and} \quad \phi(\sigma) = \phi(w),$$

and

$$\phi(v'(t)) \leq \phi(g_2(t,\sigma)) \quad \text{for all} \quad v \leq \sigma \leq w \quad \text{and} \quad \Phi(\sigma) = \Phi(v).$$

Then $v(t) \leq u(t,x) \leq w(t)$ on $\bar{H}$.

It is important to note that the functions v, w do not depend on the space variable x in the foregoing corollary.

As an example of Corollary 2.4, consider $E = R^2$ and $K = R_+^2$. Suppose that $g_1(t,u_1,u_2)$ is nondecreasing in u_2 and $g_2(t_1,u_1,u_2)$ is nonincreasing in u_1. Then the functions v, w have to satisfy

$$w_1' \geq g_1(t,w_1,w_2), \qquad v_1' \leq g_1(t,v_1,v_2)$$

$$w_2' \geq g_2(t,v_1,v_2), \qquad v_2' \leq g_2(t,w_1,v_2).$$

III. APPLICATIONS

A significant amount of attention has been given to systems of nonlinear reaction-diffusion equations of the form

$$u_t = A\Delta u + F(t,u) \tag{3.1}$$

in $R_+ \times \Omega$ with the initial condition

$$u(0,x) = u_0(x) \quad \text{on} \quad \bar{\Omega} \tag{3.2}$$

and the Neumann boundary condition

$$\frac{\partial u}{\partial \tau}(t,x) = 0 \quad \text{on} \quad (0,\infty) \times \partial\Omega. \tag{3.3}$$

See for example [2,7,18,23,24,30]. In (3.1) Δ denotes the Laplace operator in $x \varepsilon R^n$, $u, F, u_0 \varepsilon R^N$ and A is a diagonal matrix. Equations of the type (3.1) have been very important in the modeling and study of population dynamics, nuclear and chemical reactions, and several models of nerve conduction [1,2,10,11,27]. For example, $u(t,x)$ may represent the concentration of the chemical at time t and location x. The boundary condition (3.3) implies that no chemical flows in or out of the boundary $\partial\Omega$. In population growth, this boundary condition means that there is no migration across the boundary.

Consider the standard cone in R^N, namely, $K = R^N_+ = [u \varepsilon R^n : u_i \geq 0, \; i = 1,2,\ldots,N]$. Clearly the set $S = [\phi \varepsilon K : \phi(u) = u_i, \; i = 1,2,\ldots,N] \subset K^*$ generates the cone K. Let us note that the weak coupling of the system (3.1) suggests the choice of this special cone. Thus the inequality $u \leq v$ implies the componentwise inequalities $u_i \leq v_i$, $i = 1,2,\ldots,N$.

One of the simplest results that can be deduced from the results of Section 4, concerning the problem (3.1) to (3.3), is the following:

Theorem 3.1. Assume that

(i) $A \geq 0$ and $u(t,x)$ is any solution of (3.1) to (3.3) existing on $R_+ \times \bar{\Omega}$;

(ii) $F(t,x) - F(t,y) \leq B(x-y)$, $x \geq y$, where B is an $n \times n$, nonnegative matrix;

(iii) the boundary $\partial\Omega$ is regular, i.e., there exists a $h \varepsilon C$ such that $h(x) \geq 0$ on $\bar{\Omega}$, $\frac{\partial h(x)}{\partial \tau} \geq \gamma > 0$ on $\partial\Omega$ and h_x, h_{xx} are bounded.

Then the following conclusions are valid:

(a) If $u_i = 0$, $u_j \geq 0$, $j \neq i$, $j = 1,2,\ldots,N$, implies $F_i(t,u) \geq 0$, then $u(t,x) \geq 0$ on $R_+ \times \bar{\Omega}$ provided $u_0(x) \geq 0$ on $\bar{\Omega}$;

(b) If $F(t,u)$ is quasimonotone nondecreasing in u relative to R_+^N, that is, for each i, $1 \leq i \leq N$, $F_i(t,u)$ is nondecreasing in u_j, $j \neq i$ and if the solutions $r(t)$, $\rho(t)$ of $y' = F(t,y)$ with $r(0) = \bar{y}_0$, $\rho(0) = \underline{y}_0$ exist on R_+, then

$$\rho(t) \leq u(t,x) \leq r(t) \quad \text{on} \quad R_+ \times \bar{\Omega}, \tag{3.4}$$

provided that $\underline{y}_0 \leq u_0(x) \leq \bar{y}_0$ on $\bar{\Omega}$;

(c) If $F(t,u)$ is quasimonotone nondecreasing in u, $F(t,0) \equiv 0$, then

$0 \leq u_0(x)$ on $\bar{\Omega}$ implies that $u(t,x) \geq 0$ on $R_+ \times \bar{\Omega}$,

$0 < u_0(x)$ on $\bar{\Omega}$ implies that $u(t,x) > 0$ on $R_+ \times \bar{\Omega}$, and

$0 \leq u_0(x) \leq y_0$ on $\bar{\Omega}$ implies that $0 \leq u(t,x) \leq r(t)$ on $R_+ \times \bar{\Omega}$, where $r(t)$ is the same function assumed in (b);

(d) If $F(t,u)$ is not quasimonotone and if the closed set $\bar{w} = [u \in R^N: a \leq u \leq b]$ is flow invariant relative to (3.1) to (3.3), then the estimate (3.4) holds $r(t)$, $\rho(t)$ are now being the solutions of

$$r' = g_1(t,r), \quad r(0) = \bar{y}_0, \quad \rho' = g_2(t,\rho), \quad \rho(0) = \underline{y}_0,$$

where $g_{1i}(t,u) = \max[F_i(t,v): a \leq v \leq u, \ v_i = u_i]$, $g_{2i}(t,u) = \min[F_i(t,v): u \leq v \leq b, \ v_i = u_i]$, $1 \leq i \leq N$,

(e) If $F(t,u)$ is not quasimonotone and $\bar{w}$ is not known to be flow invariant, then (3.4) holds if $r(t)$, $\rho(t)$ satisfy the relations

$$r_i' \geq F_i(t,\sigma) \quad \text{for all } \sigma \text{ such that } \rho \leq \sigma \leq r \text{ and } \sigma_i = r_i,$$

$$\rho_i' \leq F_i(t,\sigma) \quad \text{for all } \sigma \text{ such that } \rho \leq \sigma \leq r \text{ and } \sigma_i = \rho_i,$$

$$1 \leq i \leq N.$$

Proof. The conclusion (a) follows from Corollary 2.1. Theorem 2.4 yields (b) with che choice $F = g_1 = g_2$. Uniqueness of solutions of $y' = F(t,y)$ together with the fact that $F(t,0) \equiv 0$, implies (c). Corollary 2.3 gives the conclusion (d) whereas (e) follows from Corollary 2.4.

Let us next demonstrate the advantage of employing a suitable cone other than R_+^N in the study of reaction-diffusion equations. Consider the system

$$\begin{cases} \dfrac{\partial u_1}{\partial t} = a_{11}\Delta u_1 + a_{12}\Delta u_2 + F_1(t,u_1,u_2), \\ \dfrac{\partial u_2}{\partial t} = a_{21}\Delta u_1 + a_{22}\Delta u_2 + F_2(t,u_1,u_2), \end{cases} \tag{3.5}$$

with (3.2) and (3.3) where $u = (u_1,u_2)$. Clearly this system is not weakly coupled in the sense of system (3.1). Consequently, if we choose to work relative to the cone R_+^2, we can not draw any conclusions concerning (3.5) as in Theorem 3.1. However, suppose we notice that for some $\alpha,\beta > 0$ with $\beta > \alpha$, we have the relations

$$b_1 \equiv \alpha a_{22} - \alpha^2 a_{12} = \alpha a_{11} - a_{21} \geq 0,$$

$$b_2 \equiv \beta a_{22} - \beta^2 a_{12} = \beta a_{11} - a_{21} \geq 0.$$

Then, considering the cone $K = [u \in R^2: u_2 \leq \beta u_1,$ and $u_2 \geq \alpha u_2]$ and noting that $S = [\phi: \phi_1(u) = u_2 - \alpha u_1, \phi_2(u) = \beta u_1 - u_2]$ generates the cone K, we can write (3.5) as

$$\begin{cases} \dfrac{\partial}{\partial t}(u_2 - \alpha u_1) = b_1\Delta(u_2 - \alpha u_2) + \tilde{F}_1(t,u_1,u_2) \\ \dfrac{\partial u}{\partial t}(\beta u_1 - u_2) = b_2\Delta(\beta u_1 - u_2) + \tilde{F}_2(t,u_1,u_2) \end{cases} \tag{3.6}$$

where $\tilde{F}_1 = F_2 - \alpha F_1$ and $\tilde{F}_2 = \beta F_1 - F_2$. It is easy to see that (3.6) is weakly coupled relative to K. We therefore have the following result observing that $K \quad R_+^2$.

Theorem 3.2. Assume that $\tilde{F} = (\tilde{F}_1,\tilde{F}_2)$ is quasimonotone nondecreasing relative to K and $\tilde{F}$ satisfies a uniqueness condition as in (ii) of Theorem 3.1. Suppose that (iii) of Theorem 3.1 holds. Then

$$\underline{y}_{20} - \alpha\underline{y}_{10} \le u_{20}(x) - \alpha u_{10}(x) \le \bar{y}_{20} - \alpha\bar{y}_{10},$$

$$\beta\underline{y}_{10} - \underline{y}_{20} \le \beta u_{10}(x) - u_{20}(x) \le \beta\bar{y}_{10} - \bar{y}_{20}, \quad \text{on} \quad \bar{\Omega}$$

implies

$$\rho_1(t) \le u_1(t,x) \le r_1(t),$$

$$\rho_2(t) \le u_2(t,x) \le r_2(t) \quad \text{on} \quad R_+ \times \bar{\Omega}.$$

REFERENCES

1. Aris, R., The mathematical theory of diffusion and reaction in permeable catalysts. Clarendon Press, Oxford, (1975).
2. Aronson, D. G. and Weinberger, H. F., Nonlinear diffusion in population genetics, combustion and nerve propagation. Lecture Notes, Vol. 446, Springer Verlag, New York, (1975).
3. Chandra, J., and Davis, P. W., Comparison theorems for systems of reaction diffusion equations. To appear in Proc. Int. Conf. on Applied Nonlinear Analysis.
4. Chandra, J., Some comparison theorems for reaction diffusion equations. To appear in Proc. Int. Conf. on "Recent trends in differential equation", Trieste, Aug., 1978.
5. Cohen, D. S., Multiple stable solutions of nonlinear boundary value problems arising in chemical reactor theory. *SIAM J. Appl. Math. 20*, 1-13 (1971).
6. Cohen, D. S., and Laetsch, T. W., Nonlinear boundary value problems suggested in chemical reactor theory. *J. Diff. Eq. 7*, 217-226 (1970).
7. Conway, E. D., and Smoller, J. A., Diffusion and the predator-prey interaction. *SIAM J. Appl. Math. 33*, 673-686 (1977).
8. Conway, E. D., and Smoller, J. A., A comparison technique for systems of reaction diffusion equations. *Comm. Part. Diff. Eq. 2*, 679-697 (1977).
9. Crank, J., The Mathematics of Diffusion, Chap. VIII. Clarendon Press, Oxford, (1955).
10. Fitzhugh, R., Mathematical models of excitation and pro-

pagation in nerves, *in* "Biological Engineering" (H. P. Schwann, ed.). McGraw-Hill, New York, (1969).

11. Fife, P. C., Pattern formation in reacting and diffusing systems. *J. Chem. Phys. 64*, 554-564 (1976).
12. Lakshmikantham, V., and Vaughn, R. L., Reaction-diffusion inequalities in cones. To appear in *Appl. Anal.*
13. Lakshmikantham, V., and Vaughn, R. L., Parabolic differential inequalities in cones. To appear in *J. Math. Anal. Appl.*
14. Lakshmikantham, V., and Leela, S., Cone valued Lyapunov functions. *Nonlinear Anal. 1*, 215-222 (1977).
15. Lakshmikantham, V., Leela, S., and Vatsala, A.S., Reaction diffusion equations in abstract cones. To appear in Proc. of Int. Conf. on Applied Nonlinear Analysis.
16. Lakshmikantham, V., and Leela, S., Differential and Integral Inequalities, Vols. 1 and II. Academic Press, New York, (1979).
17. Lakshmikantham, V., On the method of vector Lyapunov functions. Proc. Twelfth Allerton Conf. on Circuit and System Theory, 71-76, (1974).
18. McNabb, A., Comparison and existence theorems for multicomponent diffusion systems. *J. Math. Anal. Appl. 3*, 133-144 (1961).
19. Maynard-Smith, J., Mathematical Ideas in Biology. Cambridge Univ. Press, London, (1968).
20. Maynard-Smith, J., Models in Ecology. Cambridge Univ. Press, Cambridge, (1974).
21. May, R., Stability and complexity in model ecosystems. Princeton Univ. Press, Princeton,(1973).
22. Müller, M., Über das Fundamentaltheorem in der Theorie der gewöhnlichen Differentialgleichungen. *Math. Z. 26*, 619-645 (1926).
23. Pao, C. V., Positive solution of a nonlinear diffusion system arising in chemical reactors. *J. Math. Anal. Appl. 46*, 820-835 (1974).
24. Rosen, G., Solutions to systems of nonlinear reaction diffusion equations. *Bull. Math. Biol. 37*, 277-289 (1975).
25. Rauch, J., and Smoller, J. A., Qualitative theory of the Fitzhugh-Nagumo equation. *Advances in Math. 27*, 12-14 (1978).

26. Samuelson, P., Generalized predator-prey oscillations in ecological and economic equilibrium. *Proc. Nat. Acad. Sci. U.S.A. 68*, 980-983 (1971).

27. Turing, A. M., On the chemical basis of morphogenesis. *Phil. Trans. Roy. Soc. London Ser. B, 237*, 37-52 (1952).

28. Waltman, P. E., The equations of growth. *Bull. Math. Biophys. 26*, 39-43 (1964).

29. Walter, W., Differential and Integral Inequalities. Springer Verlag, Berlin, (1970).

30. Williams, S. A., and Chow, P. Z., Nonlinear reaction diffusion models for interacting populations. *J. Math. Anal. Appl. 62*, 157-169 (1978).

ON THE SYNTHESIS OF SOLUTIONS OF INTEGRAL EQUATIONS

J. J. Levin[1,2]

University of Wisconsin
Madison, Wisconsin

Consider

$$y'(t) + \int_{-\infty}^{\infty} g(y(t-\xi))dA(\xi) = \mu \qquad \left(' = \frac{d}{dt}, \ -\infty < t < \infty \right), \qquad (1^*\mu)$$

where $\mu \in \mathbb{C}^N$, under the assumption

$$g \in C(\mathbb{C}^N, \mathbb{C}^N) \qquad (2)$$

$$A = (A_{ij}), \quad A_{ij} \in BV(\mathbb{R}^1, \mathbb{C}^1) \qquad (i,j = 1,\ldots,N) \qquad (3)$$

If

$$f \in L^\infty(\mathbb{R}^1, \mathbb{C}^N), \quad \lim_{t\to\infty} f(t) = f(\infty) \text{ exists}, \qquad (4)$$

then $(1^*f(\infty))$ (i.e., $(1^*\mu)$ with $\mu = f(\infty)$) is called the limit equation associated with

---

[1]This research was supported by the U. S. Army Research Office.

[2]Present address: University of Wisconsin, Madison, Wisconsin 53706, USA

ISBN 0-12-186280-1

$$x'(t) + \int_{-\infty}^{\infty} g(x(t-\xi))\, dA(\xi) = f(t) \qquad (-\infty < t < \infty). \tag{1f}$$

It is well known that $(1^* f(\infty))$ often plays an important role in the analysis of the asymptotic behavior as $t \to \infty$ of the bounded solutions of (1f). Similarly, if

$$y(t) + \int_{-\infty}^{\infty} g(y(t-\xi))\, dA(\xi) = \mu \qquad (-\infty < t < \infty), \tag{$5^*\mu$}$$

then $(5^* f(\infty))$ is the limit equation for

$$x(t) + \int_{-\infty}^{\infty} g(x(t-\xi))\, dA(\xi) = f(t) \qquad (-\infty < t < \infty) \tag{5f}$$

under assumptions (2)-(4) and

$$f \in \mathcal{B}(\mathbb{R}^1, \mathbb{C}^N). \tag{6}$$

Elsewhere, [2], we have reported on the problem of analyzing bounded solutions of (1f) and (5f) in terms of solutions of their respective limit equations. Problems of an inverse type are considered here. That is, from a sequence of solutions, $\{y_m\}$, of $(1^*\mu)$, which satisfy a certain "closeness condition" ((12) below), functions $x, f : \mathbb{R}^1 \to \mathbb{C}^N$ are constructed (synthesized) so that the pair x, f satisfy (1f),

$$f(-\infty) = f(\infty) = \mu \tag{7}$$

and, sometimes,

$$\int_{-\infty}^{\infty} |f(t) - f(\infty)|\, dt < \infty. \tag{8}$$

Similar problems, but with (8) replaced by

$$f \in BV(\mathbb{R}^1, \mathbb{C}^N),$$

are considered for $(5^*\mu)$, (5f). It turns out, as in [2], that the key aspects of this study are not tied to a specific equation but are meaningful in a more general setting. As a by-product of working in the general setting we improve some synthesis results for $(1^*\mu)$, (1f) obtained in [1] as well as obtaining analogous results for $(5^{*}\mu)$, (5f).

Although we shall not discuss them here, the present results together with those of [2] have applications to the study of the asymptotic behavior of the bounded solutions of (1f) and (5f). In particular, they sometimes allow for the dropping of hypothesis (8) for (1f) and (9) for (5f), keeping only (7), in such asymptotic problems. In [1] an application of this type was given for a Volterra equation which could be studied by means of an equation of the form (1f). We refer to [1],[2], [3] for references to studies involving limit equations.

We begin by summarizing the notation and definitions of [2] which are employed here. Let $\mathcal{Y} = \{y : \mathbb{R}^1 \to \mathbb{C}^N\}$, where $\mathbb{R}^1 = (-\infty,\infty)$, $\mathbb{C}^1$ = complex-plane, $\mathbb{C}^N = \{(z_1,\ldots,z_N) \mid z_k \in \mathbb{C}^1\}$, $|z| = \Sigma_{k=1}^N |z_k|$ $(z \in \mathbb{C}^N)$, $x_n \to x$ $(x_n, x \in \mathbb{C}^N)$ if $|x_n - x| \to 0$, $\bar{A}$ = closure of A $(A \subset \mathbb{C}^N)$ with respect to the metric associated with $|\cdot|$, $B_r = \{x \in \mathbb{C}^N \mid |x| < r\}$. In limits such as $x_n \to x$ it will be understood that $n \to \infty$. Elements of $\mathcal{Y}$ will be called curves whether or not they are continuous. The subsets $C(\mathbb{R}^1, \mathbb{C}^N)$, $C_u(\mathbb{R}^1, \mathbb{C}^N)$, $C_u^{(k)}(\mathbb{R}^1, \mathbb{C}^N)$, $L^\infty(\mathbb{R}^1, \mathbb{C}^N)$, $\mathcal{B}(\mathbb{R}^1, \mathbb{C}^N)$, ... of $\mathcal{Y}$ of continuous, uniformly continuous, uniformly continuous k^{th} derivative, essentially bounded, Borel measurable, ... curves will for brevity be denoted by C, C_u, $C_u^{(k)}$, L^∞, $\mathcal{B}$, ... The tauberian curves, $T = T(\mathbb{R}^1, \mathbb{C}^N)$, and an important subset, $\tilde{T} \subset T$, are defined by

$$x \in T \quad \text{if} \quad \lim_{t\to\infty,\ \tau\to 0} |x(t+\tau) - x(t)| = 0$$

$$x \in \tilde{T} \quad \text{if} \quad x \in T \quad \text{and} \quad \limsup_{t\to\infty} |x(t)| < \infty .$$

For each $x \in \mathcal{Y}$ and $s \in \mathbb{R}^1$ let $x_s \in \mathcal{Y}$ be defined by

$$x_s : t \to x(t+s) .$$

For each $x \in \mathcal{Y}$ and $Y \subset \mathcal{Y}$ let

$$R(x) = \text{range of } x, \qquad R(Y) = \bigcup_{y\in Y} R(y) ,$$

$$\Omega(x) = \{\omega \in \mathbb{C}^N \mid x(t_n) \to \omega \text{ for some } t_n \to \infty\},$$

$$\Gamma(x) = \{y \in \mathcal{Y} \mid x_{t_n} \to y \text{ for some } t_n \to \infty \text{ uniformly on every compact subset of } \mathbb{R}^1\} .$$

$Y \subset \mathcal{Y}$ is said to be translation invariant if $y_t \in Y$ for every $y \in Y$ and $t \in \mathbb{R}^1$. $\Gamma(x)$ is translation invariant for every $x \in \mathcal{Y}$.

A bounded solution of (1f) is, by definition, in $L^\infty \cap AC_{loc}$ and satisfies (1f) a.e. on $\mathbb{R}^1$. A bounded solution of (5f) is in $\mathcal{B} \cap L^\infty$ and satisfies (5f) on $\mathbb{R}^1$. For each $Q \subset \mathbb{C}^N$ let

$$S_Q(1^*\mu) = \{y \in L^\infty \cap C_u^{(1)} \mid y \text{ satisfies } (1^*\mu),\ R(y) \subset Q\}$$

$$S_Q(5^*\mu) = \{y \in L^\infty \cap C_u \mid y \text{ satisfies } (5^*\mu),\ R(y) \subset Q\}$$

denote certain solution subsets of $\mathcal{Y}$. They are translation invariant.

<u>Definition</u>. $\{t_m\} \subset \mathbb{R}^1$ is an expanding t-sequence if $t_1 < t_2 < \ldots$ and $t_m - t_{m-1} \to \infty$.

<u>Definition</u>. $\psi_m \in C^\infty(\mathbb{R}^1, [0,1])$ $(m = 1,2,\ldots)$ is a ψ-sequence associated with an expanding t-sequence $\{t_m\}$ if

$$\sum_{m=1} \psi_m(t) \equiv 1 \qquad (t \in \mathbb{R}^1)$$

$$\lim_{m\to\infty} \|\psi_m^{(j)}\|_\infty = 0, \quad \lim_{t\to\infty} \sum_{m=1}^{\infty} |\psi_m^{(j)}(t)| = 0 \qquad (j = 1,2,\ldots),$$

$$\psi_1(t) \equiv 1, \quad (t \le t_1), \quad \psi_1(t) \equiv 0 \quad (t_2 \le t), \quad \psi_1'(t) \le 0 \quad (t_1 \le t \le t_2),$$

$$\psi_m(t) \equiv 0 \quad (t \le t_{m-1} \text{ and } t \ge t_{m+1}), \quad \psi_m(t_m) = 1,$$

$$\psi_m'(t) \ge 0 \quad (t_{m-1} \le t \le t_m), \quad \psi_m'(t) \le 0 \ (t_m \le t \le t_{m+1}) \ (m = 2,3,\ldots).$$

Some motivation for the present theorems is provided by the following abbreviated form of results from [2]. Theorem 1 analyzes the behavior of a curve $x \in \tilde{T}$ in terms of its limit set $\Gamma(x)$. Theorems 1a and 1b are consequences of Theorem 1 for bounded tauberian solutions of (1f) and (5f).

<u>Theorem 1</u>. *If* $x \in \tilde{T}$ *then there exist a sequence* $\{y_m\} \subset \Gamma(x)$ *and an expanding* t-*sequence* $\{t_m\}$ *such that*

$$\lim_{m\to\infty} \{ \sup_{t_{m-2}\le t\le t_{m+2}} |x(t) - y_m(t)| \} = 0 . \tag{10}$$

Moreover, for any ψ*-sequence,* $\{\psi_m\}$*, associated with* $\{t_m\}$

$$x = \sum_{m=1}^{\infty} \psi_m y_m + \eta \tag{11}$$

where $\eta(t) \to 0 \quad (t \to \infty)$.

Theorem 1a. *If* (2)-(4) *hold and* x *is a bounded solution of* (1f), *then* $x \in \tilde{T}$, *the conclusions of Theorem 1, and*

$$\Gamma(x) \subset S_{\Omega(x)}(1^*f(\infty)), \qquad \Omega(x) = R(S_{\Omega(x)}(1^*f(\infty))),$$

$$\lim_{m\to\infty} \{ \operatorname{ess\,sup}_{t_{m-1}\le t\le t_{m+1}} |x'(t) - y_m'(t)| \} = 0,$$

$$\lim_{t\to\infty} \{ \operatorname{ess\,sup}_{t\le\tau<\infty} |\eta'(\tau)| \} = 0$$

are satisfied.

Theorem 1b. *If* (2)-(4), (6) *hold and* $x \in T$ *is a bounded solution of* (5f), *then* $x \in \tilde{T}$, *the conclusions of Theorem 1, and*

$$\Gamma(x) \subset S_{\Omega(x)}(5^*f(\infty)), \qquad \Omega(x) = R(S_{\Omega(x)}(5^*f(\infty)))$$

are satisfied.

If follows immediately from (10), the monotonicity of the t_m, and the triangle inequality that

$$\lim_{m\to\infty} \{ \sup_{t_{m-1}\le t\le t_{m+2}} |y_m(t) - y_{m+1}(t)| \} = 0 . \tag{12}$$

Thus, as $m \to \infty$ successive y_m are closer and closer on longer and longer t-intervals. This closeness property, (12), will be important for sequences $\{y_m\} \subset \mathcal{Y}$ which do not arise in the framework of Theorem 1.

In Theorem 2 a curve, x, is synthesized from a sequence of curves, $\{y_m\}$, which satisfies (12). Note that the definition of x is formula (11) of Theorem 1 whith $\eta = 0$. The remaining hypotheses on $\{y_m\}$ and the definition of $q \in \mathcal{Y}$

reflect the applications of Theorem 2 to integral equations which are made in Theorems 2a and 2b. As well as Theorem 1, various geometric situations, one of which is discussed later, motivate the hypotheses of Theorem 2.

Theorem 2. *Let* (2), (3), $K \in \mathbb{R}^+$, $\{t_m\}$ *is an expanding t-sequence,* $\{\psi_m\}$ *is an associated* ψ*-sequence,* $\{y_m\} \subset \mathcal{Y}$ *satisfies* (12) *and*

$$y_m \in \mathcal{B} \quad (m = 1,2,\dots), \qquad \sup_m \{ \sup_{-\infty<t<\infty} |y_m(t)| \} \le K$$

all hold. Define $x, q \in \mathcal{Y}$ *by*

$$x = \sum_{m=1}^{\infty} \psi_m y_m \tag{13}$$

$$q(t) = \int_{-\infty}^{\infty} g(x(t-\xi))\, dA(\xi) - \sum_{m=1}^{\infty} \psi_m(t) \int_{-\infty}^{\infty} g(y_m(t-\xi))\, dA(\xi) . \tag{14}$$

Then

$$x \in \mathcal{B} , \qquad \sup_{-\infty<t<\infty} |x(t)| \le K ,$$

$$q \in \mathcal{B} , \qquad \sup_{-\infty<t<\infty} |q(t)| < \infty , \qquad q(-\infty) = q(\infty) = 0$$

are satisfied.

Theorem 2a. *Let the hypotheses of Theorem 2, including definitions* (13) *and* (14), *as well as:* $\mu \in \mathbb{C}^N$,

$$y_m \in S_{\bar{B}_K}(1^*\mu) \qquad (m = 1,2,\dots)$$

and the definitions of $f, p \in \mathcal{Y}$ *by* (1f) *and*

$$p = \sum_{m=1}^{\infty} \psi_m' y_m ,$$

respectively, hold. Then the conclusions of Theorem 2 and

$$x \in C_u^{(1)}, \qquad x' \in L^{\infty}, \qquad q \in C_u,$$

$$p \in L^{\infty} \cap C_u^{(1)}, \qquad p' \in L^{\infty}, \qquad p(-\infty) = p(\infty) = 0,$$

$$f = \mu + q + p, \qquad f \in L^{\infty} \cap C_u, \qquad f(-\infty) = f(\infty) = \mu$$

are satisfied.

Theorem 2b. *Let the hypotheses of Theorem 2, including definitions* (13) *and* (14), *as well as:* $\mu \in \mathbb{C}^N$,

$$y_m \in S_{\bar{B}_K}(5^*\mu) \qquad (m = 1,2,\ldots)$$

and the definition of $f \in \mathcal{Y}$ *by* (5f) *hold. Then the conclusions of Theorem 2 and*

$$x \in C, \qquad q \in C_u,$$

$$f = \mu + q, \qquad f \in L^{\infty} \cap C_u, \qquad f(-\infty) = f(\infty) = \mu$$

are satisfied.

The key conclusion of Theorems 2a and 2b is that (7) is satisfied; it rests on $q(\pm\infty) = 0$, established in Theorem 2. Thus, the f's constructed in Theorems 2a and 2b may be regarded as perturbations of the assigned μ. With some additional assumptions, f can be constructed to also satisfy (8) and (9). Here we will only pursue this point in the more concrete framework of Theorem 3.

The following definition formalizes the notion of when a set of curves $Y \subset \mathcal{Y}$, having the property that both $y(\infty)$ and $y(-\infty)$ exist for each $y \in Y$, contains a loop. It is useful in studying those integral equations for which the solutions of the limit equation possess limits at $\pm\infty$. An application of this type is made in [1].

Definition. $Y \subset \mathcal{Y}$ is said to be of type 2 if either

there exists a $y \in Y$ such that

$$y(-\infty) = y(\infty) = c, \quad y(t) \not\equiv c \tag{15}$$

or

$$\text{there exist } \{y^{(1)},\dots,y^{(n)}\} \subset Y \quad (n \geq 2) \text{ such that } y^{(j)}(-\infty) = c_j \quad (1 \leq j \leq n),$$
$$y^{(j)}(\infty) = c_{j+1} \quad (1 \leq j \leq n-1), \quad y^{(n)}(\infty) = c_1, \tag{16}$$
$$\text{with } c_i \neq c_j \text{ for } i \neq j .$$

<u>Theorem 3</u>. (i) *Let* (2), (3), $K \in \mathbb{R}^+$, *and* Y *is of type* 2 *hold. Further, let*

$$y \in \mathcal{B}, \quad \sup_{-\infty<t<\infty} |y(t)| \leq K \qquad (\textit{if } (15) \textit{ holds})$$

$$y^{(j)} \in \mathcal{B} \quad (1 \leq j \leq n), \quad \max_{1\leq j\leq n} \{ \sup_{-\infty<t<\infty} |y^{(j)}(t)| \} \leq K \qquad (\textit{if } (16) \textit{ holds}).$$

Then there exist $\{t_m\}$, $\{\psi_m\}$, $\{y_m\}$ *satisfying the hypotheses of Theorem 2 such that, with definitions* (13) *and* (14), *the conclusions of Theorem 2 and*

$$y_m = \begin{cases} y_{\lambda_m} & (m = 1,2,\dots) \quad (\textit{if } (15) \textit{ holds}) \\ y_{\lambda_m}^{(\theta_m)} & (m = 1,2,\dots) \quad (\textit{if } (16) \textit{ holds}) \end{cases}$$

$$\Omega(x) = \begin{cases} \overline{R(y)} & (\textit{if } (15) \textit{ holds}) \\ \bigcup_{j=1}^{n} \overline{R(y^{(j)})} & (\textit{if } (16) \textit{ holds}) \end{cases}$$

where $1 \leq \theta_m \leq n$ *and* $\lambda_m \in \mathbb{R}^1$ $(m = 1,2,\dots)$, *are satisfied. In particular,* $x(\infty)$ *does not exist.*

(ii) *Let the assumptions of* (i) *and*

$$\int_{-\infty}^{0} |t|V(t)\,dt < \infty, \quad \int_{0}^{\infty} t|V(\infty) - V(t)|dt < \infty,$$

where

$$V(t) = \sum_{i,j=1}^{N} \{\text{total variation of } A_{ij} \text{ on } (-\infty,t]\}$$

hold. Then the conclusions of (i) *and*

$$q \in L^1 \tag{17}$$

are satisfied.

(iii) *Let the assumptions of* (i) *and either*

$$A' \in AC(\mathbb{R}^1, \mathbb{C}^{N^2}), \quad \int_{-\infty}^{\infty} |t|^k |A^{(j)}(t)| \, dt < \infty$$
$$(k = 0,1,2; \; j = 1,2)$$

or

$$A(t) = \int_0^t a(\xi) d\xi \quad (0 \le t < \infty), \quad A(t) = 0 \quad (-\infty < t \le 0)$$
$$a \in AC(\mathbb{R}^+, \mathbb{C}^{N^2}), \quad \int_0^{\infty} t^k |a^{(j)}(t)| dt < \infty \tag{18}$$
$$(k = 0,1,2; \; j = 0,1)$$

hold. Then the conclusions of (i) *and*

$$q \in L^1 \cap BV \cap AC, \qquad q' \in L^\infty \cap L^1 \tag{19}$$

are satisfied.

Hypothesis (18) is particularly suited for applications to Volterra equations.

Theorema 3a. (i) *Let the hypotheses of Theorem 3* (i) *as well as:* $\mu \in \mathbb{C}^N$ *and*

$$Y \subset S_{\bar{B}_K}(1^* \mu)$$

hold. Then, with f *and* p *defined as in Theorem 2a, the conclusions of Theorems 3* (i) *and 2a are satisfied.*

(ii) *Let the assumptions of* (i) *and Theorem 3* (ii) *hold. Then the conclusions of* (i), *Theorem 3* (ii), $p \in L^1$, *and*

$$\int_{-\infty}^{\infty} |f(t) - \mu| \, dt < \infty$$

are satisfied.

Theorem 3b. (i) *Let the hypotheses of Theorem 3* (i) *as well as:* $\mu \in \mathbb{C}^N$ *and*

$$Y \subset S_{\bar{B}_K}(5^* \mu)$$

hold. Then, with f *defined as in Theorem 2b, the conclusions of Theorem 3* (i) *and 2b are satisfied.*

Let the assumptions of (i) *and Theorem 3* (ii) *hold. Then the conclusions of* (i), *Theorem 3* (ii), *and* (20) *are satisfied.*

Let the assumptions of (i) *and Theorem 3* (iii) *hold. Then the conclusions of* (i), *Theorem 3* (iii), (20), *and* $f \in BV$ *are satisfied.*

Assertion (20) of Theorems 3a and 3b rests on (17); $f \in BV$ of Theorem 3b rests on (19).

REFERENCES

1. Levin, J. J., On some geometric structures for integrodifferential equations, *Advances in Math. 22*, 146-186 (1976).
2. Levin, J. J., On the asymptotic behavior of solutions of integral equations, Proceedings of the Helsinki Symposium on Integral Equations (to appear).
3. Levin, J. J. and Shea, D. F., On the asymptotic behavior of the bounded solutions of some integral equations, I, II, III, *J. Math. Anal. Appl. 37*, 42-82, 288-326, 537-575 (1972).

LOCAL EXACT CONTROLLABILITY OF NONLINEAR EVOLUTION EQUATIONS

K. Magnusson[1]
A. J. Pritchard[1]

Control Theory Centre
University of Warwick

I. INTRODUCTION

In this paper we consider abstract nonlinear evolution equations of the form

$$\dot{z} = Az + Nz + Bu, \quad z(0) = z_o \tag{1}$$

where A is a closed linear operator on a Banach space Z and is the generator of a strongly continuous semigroup $S(t)$. N is a nonlinear operator and the control $u \in L^m|0,T;U|$, $1 \leq m < \infty$ where U is also a Banach space. We interpret the solution of (1) in the mild form

$$z(t) = S(t)z_o + \int_o^t S(t-s)\,Nz(s)ds + \int_o^t S(t-s)\,Bu(s)ds \tag{2}$$

The problem is to find conditions under which the origin

[1] Present address: Control Theory Centre, University of Warwick, Warwick, G.B.

RECENT ADVANCES IN DIFFERENTIAL EQUATIONS

ISBN 0-12-186280-1

$z_o = 0$ can be steered via equation (2) to any point in a neighbourhood of the origin in a time T. We would not expect this neighbourhood to be in Z, since it is well known that for the linearized system

$$z(t) = S(t)z_o + \int_0^t S(t-s)\, Bu(s)ds \tag{3}$$

it is rarely the case that for any z_o in Z we can steer via (3) to any other point $z_1 \varepsilon Z$, such that $z(T) = z_1$. Moreover since the main tool in our analysis will be the inverse function theorem which requires that the linearized system is onto, we will need to characterize these linear subspaces of Z which can be reached through equation (3). At the same time we will require a good local existence theorem for equation (2) which enables us to consider a large class of nonlinearities N and control operators B. For this we use the following slight generalization of a theorem proved by Ichikawa and Pritchard |1|.

Theorem 1. Let $Z, V, \bar{Z}$ be Banach spaces with $V \subset Z$, $V \subset \bar{Z}$ $p, \bar{p}, r, s, q, a, b, \varepsilon$ be positive real numbers, such that

$$p \geq r \geq 1, \quad \bar{p} \geq q \geq 1 \quad \frac{1}{r} = \frac{1}{q} + \frac{1}{s} - 1 .$$

Assume the following

(i) $S(t) \ \varepsilon \ L(Z,V) \cap L(\bar{Z},V), \quad t > o$

$$\|S(t)z\|_V \leq \bar{g}(t) \ \|z\|_Z \qquad g \ \varepsilon \ L^p|0,T|$$

$$\|S(t)z\|_V \leq \bar{g}(t) \ \|z\|_{\bar{Z}} \qquad \bar{g} \ \varepsilon \ L^{\bar{p}}|0,T|$$

(ii) $N : V \to \bar{Z}$ and if

$$z \ \varepsilon \ L^r|0,T;V|, \quad \|z\|_{L^r|0,T;V|} \leq a$$

then $Nz \ \varepsilon \ L^s|0,T;\bar{Z}|$, $\|Nz\|_{L^s|0,T;\bar{Z}|} \leq b$

(iii) $$\|Nz-N\hat{z}\|_{L^s|0,T;\bar{Z}|} \leq k(\|z\|_{L^r|0,T;V|}, \|\hat{z}\|_{L^r|0,T;V|}) \ \|z-\hat{z}\|_{L^r|0,T;V|}$$

$k : R^+ \times R^+ \to R$ is continuous and

$k(x,y) = k(y,x)$ with $k(x,o) \to 0$ as $x \to 0$

(iv) If $\|z\|_{L^r|0,T;V|}$, $\|\hat{z}\|_{L^r|0,T;V|} \leq a$

$$\|\bar{g}\|_{L^q|0,T|} \; k(\|z\|_{L^r|0,T;V|}, \|\hat{z}\|_{L^r|0,T;V|}) < 1$$

(v) $$\|g\|_{L^r|0,T|} \|z_o\| + \|\bar{g}\|_{L^q|0,T|} \; b \leq a - \varepsilon$$

(vi) $$\left\| \int_o^t S(t-s)\; Bu(s)ds \right\|_{L^r|0,T;V|} \leq R \|u\|_{L^m|0,T;U|}$$

If conditions (i)-(vi) hold and $\|u\|_{L^m|0,T:U|} \leq {}^{\varepsilon}/_{R}$, then there exists a unique solution, z of (2) in $L^r|0.T;V|$ with $\|z\|_{L^r|0.T:V|} \leq a$.

<u>Corollary 1</u>. Let Zo be a Banach space such that $S(t) \in L(\bar{Z}, Zo)$, $t > o$ with

$$\|S(t)z\|_{Z_o} \leq g_o(t) \|z\|_{\bar{Z}} , \quad g_o \in L^{\beta}|0,T|$$

$\frac{1}{\beta} + \frac{1}{s} = 1$ and

$$\left\| \int_o^T S(T-s)\; Bu(s)ds \right\|_{Z_o} \leq \bar{R} \|u\|_{L^m|0,T;U|}$$

Then if $zo = 0$ under the conditions of theorem 1, we have $z(t) \in Z_o$.

The above results allow us to consider systems modelled by differential equations, delay equations and partial differential equations. Also the conditions imposed on N and B are such that unbounded operators can be included, for example nonlinearities and controls acting on the boundary of a region or a system described by a partial differential equations (see |1|,|2|).

Example

$$\dot{z} = z_{xx} + zz_x + u, \quad u \in L^2|0.T:L^2|0.1||$$

$$z(0,t) = z(1,t) = 0 \quad z(x,o) = 0$$

By taking $V = H_o^{'}(0,1)$, $\bar{Z} = L^2|0.1|$ it is easy to show that

$$||Nz - N\tilde{z}||_{L^\infty|0.T,\bar{Z}|} \leq \bar{K}(||z||_{L^\infty|0.T:V|} + ||\tilde{z}||_{L^\infty|0.T:V|})\ ||z - z||_{L^\infty|0,T;V|}$$

and

$$||\int_o^t S(t-s)\ u(s)ds||_{L^\infty|0,T:V|} \leq K\ ||u||_{L^2|0.T:U|}$$

where $U = L^2|0.1|$.
Furthermore

$$||\int_o^T S(T-s)\ u(s)ds||_{H_o^{'}(0.1)} \leq K\ ||u||_{L^2|0.T:U|}$$

and hence all the conditions of theorem 1 and corollary 1 are satisfied with $r = s = \infty$, $\beta = q = 1$ and $Zo = H_o^{'}(0.1)$

Thus we have a solution of the equation in $L^\infty|0.T;H_o^{'}(0,1)|$ $z(T) \in H_o^{'}(0,1)$.

II. THE LINEARIZED SYSTEM

In this section we characterize those subspaces of Z which can be reached from the origin through the linearized system (3). For this we require the following theorem relating the ranges of various operators with conditions on their adjoints.

Theorem 2. Let W, V, Z be reflexive Banach spaces and F, G bounded linear operators on W, V both with range in Z.

Then

$$\text{Range }(F) \subset \text{Range }(G)$$

if and only if $\exists\ \gamma > 0$, such that

$$\|G^* z^*\|_{V^*} \geq \gamma \|F^* z^*\|_{W^*}$$

for all $z^* \in Z^*$.

Now let $Gu = \int_0^T S(T-s)\, Bu(s)ds$ and W be continuously injected in Z, then we have

<u>Theorem 3</u>. Suppose $u \in L^m|0,T;U|$ with $1 < m < \infty$ and U a reflexive Banach space, then for any $w \in W$, there exists a control u, such that $w = Gu$, if and only if there exists $\gamma > 0$, such that

$$\|B^* S^*(.) z^*\|_{L^{m'}|0,T;U^*|} \geq \gamma \|z^*\|_{W^*} \tag{4}$$

where $\frac{1}{m} + \frac{1}{m'} = 1$.

<u>Proof</u>. Since $Gu = \int_0^T S(T-s)Bu(s)ds$ for any $z^* \in Z^*$, we have

$$<z^*, Gu>_{Z^*,Z} = <z^*, \int_0^T S(T-s)Bu(s)ds>_{Z^*,Z} .$$

Then for a class of operators B (see |2|), we have

$$<z^*, Gu>_{Z^*,Z} = \int_0^T <B^* S^*(T-s)z^*, u(s)_{U,U^*} ds$$

where $< , >_{Z,Z^*}$, $< , >_{U,U^*}$ denote the duality pairings between Z^*, Z and U^*, U respectively.

Thus

$$(G^* z^*)\ (t) = B^* S^* (T-t) z^* .$$

Now set $Fw = w$ and $V = L^m|0.T:U|$ and the result follows from theorem 2.

In order to apply the above Theorem we calculate the left hand side of (4) and then choose W^* so that (4) holds.

This tells us that

$$W \subset \text{Range } (G) \ .$$

In some cases we can conlude more than this, i.e. if in addition

$$\|B^* S^*(.) z^*\|_{L^{m'}|0.T:U^*|} \leq \alpha \ \|z^*\|_{W^*} \qquad (5)$$

for some $\alpha > 0$ and all $z^* \varepsilon Z^*$ then

$$\text{Range } (G) = W \ .$$

In the case when we cannot get a simple characterisation of W we can always put a Banach space topology on Range (G) via the following (see |3|).

Define the normed space

$$X_{G^*} = Z^* / \ker G^*$$

with norm $\|\tilde{z}\|_{X_{G^*}} = \|G^* z^*\|_{L^{m'}|0.T:U^*|}$

where $\tilde{z}$ is the equivalence class of X_{G^*} containing z^*.
We now define X^{G^*} as the space of linear functionals on Z^* continuous with respect to the pseudo-norm

$$\|z^*\|_{G^*} = \|G^* z^*\|_{L^{m'}|0.T:U^*|}$$

on Z^*. There exists an isometry between $(X_{G^*})^*$ and X^{G^*} and furthermore

$$X^{G^*} = \text{Range } (G^{**}) = \text{Range } (G) \ .$$

Thus we have a Banach space topology on Range (G) and it is easy to show that

$$\|Gu\|_{\text{Range }(G)} \leq K \|u\|_{L^m|0.T:U|} .$$

Thus with this topology the condition of corollary 1 is always satisfied with Z_o = Range G.

Example. Let $\{\lambda_n\}$ be a sequence of negative real numbers and $\{\Phi_n\}$ be an orthonormal sequence on a separable Hilbert space H. Define the operator A by

$$Az = \sum_{n=1}^{\infty} \lambda_n \Phi_n < \Phi_n, z>_H$$

with $D(A) = \{z : \sum_{n=1}^{\infty} |\lambda_n < \Phi_n, z>|^2 < \infty\}$.

The A generates a strongly continuous semigroup

$$S(t)z = \sum_{n=1}^{\infty} e^{\lambda_n t} \Phi_n < \Phi_n, z>_H$$

If $B = I$, the identity operator on H and $u \varepsilon L^2|0.T:U| = V$ we have

$$\|G^* z^*\|^2_{V^*} = \sum_{n=1}^{\infty} \frac{1 - e^{2\lambda_n T}}{2|\lambda_n|} | < \Phi_n, z^* >_H |^2$$

so the norm on Range (G) is

$$\|z\|^2_{\text{Range }(G)} = \sum_{n=1}^{\infty} \frac{2|\lambda_n|}{1-e^{2\lambda_n T}} | < \Phi_n, z >_H |^2$$

but this norm is equivalent to

$$\|z\|^2_W = \sum_{n=1}^{\infty} |\lambda_n| \, |< \Phi_n, z >_H|^2 .$$

Now

$$\alpha \sum_{n=1}^{\infty} \frac{1}{|\lambda_n|} | < \Phi_n, z^* >_H |^2 \leq \|G^* z^*\|^2_{V^*}$$

$$\geq \beta \sum_{n=1}^{\infty} \frac{1}{|\lambda_n|} | < \Phi_n, z^* >|^2$$

i.e.

$$\alpha \; \| z^* \|_{W^*} \geq \| G^* z^* \|_{V^*} \geq \beta \; \| G^* z^* \|_{W^*}$$

and hence we conclude that

$$\text{Range } (G) = W = D \; ((-A)^{1/2})$$

and the norm on Range (G) is equivalent to norm on W. That it we can steer to the space $D((-A)^{1/2})$ via the control system

$$z(t) = \int_0^t S(t-s) \; u(s)ds \; .$$

For example if $\lambda_n = -n^2\pi^2$, $\phi_n(x) = \sqrt{2} \sin n\pi x$ and $H = L^2|0.1|$ the controlled system is the mild solution of

$$\dot{z} = z_{xx} + u$$

$z(o,t) = z(1,t) = 0 \qquad z(x,o) = 0$

and we can control exactly to $H_o^{'}(0,1)$, i.e. Range (G) = $= H_o^{'}(0,1)$.

III. EXACT CONTROLLABILITY

The two previous sections have set up the necessary machinery for us to apply the inverse function theorem which states

Theorem 4. Let V, W be two Banach spaces, Φ a continuously differentiable map from a neighbourhood of the origin in V into W such that $\Phi(o) = 0$. Assume $d\Phi(o)$ is onto, then the range of Φ contains a neighbourhood of the origin in W.

Now let us assume that the conditions of Theorem 1, corollary 1, hold, with $z_o = 0$ and the range of G being exactly Z_o with the same topology. We also assume that Z is a reflexive Banach space.

Finally assume the following

$$\|Nz - N\tilde{z} - \quad (\tilde{z})(z-\tilde{z})\|_{L^s|0.T:\bar{Z}|}$$

$$= o(\|z - \tilde{z}\|_{L^r|0.T:V|}) \tag{6}$$

and $z \to dN(z)$

$$L^r|0.T:V| \to L(L^r|0,T:V| \; ; L^s|0.T:\bar{Z}|) \tag{7}$$

is continuous

$$N(0) = 0 \qquad dN(0) = 0 \; .$$

We then have the following theorem

Theorem 5. The nonlinear system (2) can be steered from the origin to any point in a neighbourhood of the origin in $Zo = W$ by controls $u \in L^m|0.T:U|$. (We mean by this statement that there exists an ε such that for all $z \in Zo$, $\|z\|_{Z_o} < \varepsilon$, there exists a control $u \in L^m|0.T:U|$ such that the solution of (2) $z(T) = z$).

Proof. Let us define a map Φ_T by

$$\Phi_T(u) = z(T)$$

Range $(G) = W = Z_o$, $G \in L(L^m|0.T:U|:W)$ which follows from the remarks in §2, and so applying theorem 1 and corollary 1 it is easy to show that Φ_T maps a neighbourhood of zero in $L^m|0.T:U|$ into $W = Z_o$ and $\Phi_T(o) = 0$.

By using (6) and (7) we can show that Φ_T is continuously differentiable in a neighbourhood of the origin in $L^m|0.T:U|$ and that

$$d\Phi_T(o)u = y(T)$$

where $y(T) = \int_0^T S(T-s)\, Bu(s)ds = Gu$.

But our assumption from §2 is that $d\Phi_T(0)$ is a continuous linear map from $L^m|0.T:U|$ *onto* Z_o and hence the inverse function theorem (theorem 4) is applicable and this completes the proof.

Example

$$\dot{z} = z_{xx} + zz_x + u$$

$$z(0,t) = z(1,t) = 0 \qquad z(x,o) = 0$$

It is easy to show that

$Nz = zz_x$ satisfies (6) and (7)

with $r = s = \infty$, $V = H_o'(0.1)$, $\bar{Z} = L^2|0.1|$.

We have seen that the linearised system is controllable to $H_o'(0.1)$

i.e. Range (G) $= H_o'(0.1) = Zo$

and thus the system is exactly controllable to a neighbourhood of the origin in $H_o'(0.1)$.

REFERENCES

1. Ichikawa, A. and Pritchard, A. J., Existence, uniqueness and stability of nonlinear evolution equations. Report No. 65, Control Theory Centre, University of Warwick. To appear in *J. Math. Anal. Appl.*
2. Curtain, R. F. and Pritchard, A. J., Infinite dimensional linear systems theory. Vol. 8, Lecture Notes in Control and Information Sciences, Springer Verlag, Berlin, (1978).
3. Dolecki, S. and Russell, D. L., A general theory of observation and control. *SIAM J. Control 15*, 185-221.
4. Magnusson, K. and Pritchard, A. J., Controllability of nonlinear evolution equations. Report No. 75, Control Theory Centre, University of Warwick.

TOPOLOGICAL DEGREE AND THE STABILITY OF A CLASS OF VOLTERRA INTEGRAL EQUATIONS

Patrizia Marocco[1]

Istituto di Matematica
Università degli Studi, Trieste

In this paper I will report some results of my doctor thesis. The starting point is the following problem raised by Prof. Vidossich during a course on integral equations.

Let us consider a Volterra equation of the form

$$x(t) = y(t) + \int_0^t k(t,s)f(s,x(s))ds \tag{1}$$

where $f : R^+ \times R^n \to R^n$ is a continuous function, $k(t,s)$ is a continuous matrix function, $y : R^+ \to R^n$ is a continuous function such that there exists the $\lim_{t \uparrow \infty} y(t) = y(\infty)$, f satisfies the following inequality:

$$\|f(t,x)\| \leq A\|x\|g(t) + Bg(t)$$

for certain constants A and B and a continuous function $g : R^+ \to]0,+\infty[$, and the pair (C_g, C_∞) is admissible with respect to the linear integral operator

$$Kx(t) = \int_0^t k(t,s)x(s)ds$$

---

[1]Present address: Istituto di Matematica, Università degli Studi, Piazzale Europa 1, Trieste, Italy.

ISBN 0-12-186280-1

generated by the kernel $k(t,s)$. Does it exist a solution x of (1) such that $\lim_{t \uparrow \infty} x(t)$ exists?

The interest in this question lies on the fact that it would unify some earlier results. In [1] there are two theorems of existence of convergent solutions of the equation (1): a theorem of Bantas in which it is assumed that f is Lipschitz with a suitably small constant, and a theorem of Corduneanu in which it is assumed that f is bounded. Therefore an affirmative answer to the above question would contain the two theorems I mentioned, generalizing the first one to arbitrary Lipschitz constants.

In [2] it is presented an affirmative answer to the question. The proof uses the topological degree as the fundamental tool and this subject suggested the idea of looking for a stability result using the continuous dependence of the topological degree.

The idea is the following: let T be the operator defined by

$$Tx(t) = \int_0^t k(t,s) f(s,x(s))\,ds$$

with $f(t,0) \equiv 0$. If it can be shown that

$$\deg(I - T, B(0,\varepsilon), 0)$$

is different from zero for every $\varepsilon > 0$, then the equation

$$x - Tx - y = 0 \tag{2}$$

will have a solution in the ball $B(0,\varepsilon)$ for $\|y\|_\infty$ sufficiently small in view of the continuous dependence of the topological degree.

Since (2) and (1) are equivalent, we may conclude that for each $\varepsilon > 0$ there exists $\varepsilon > 0$ such that $\|y\|_\infty < \delta$ implies $\|x\|_\infty < \delta$ for at least one solution of (1). Therefore we have the required stability if (1) has a unique solution for every forcing term y. But since the more interesting case is when no uniqueness assumption is made, there are some technical difficulties to handle.

By this technique the following result can be proved:

Theorem. *Consider the equation*

$$x(t) = \int_0^t k(t,s)f(s,x(s))ds$$

with $f : R^+ \times U \to R^n$, $U \subseteq R^n$ *neighbourhood of the origin, such that the Nemytskii operator* $F(x)(t) = f(t,x(t))$ *is a continuous map of a neighbourhood of the origin in* C_∞ *with values in* C_g, $k(t,s)$ *a continuous matrix function such that the pair* (C_g, C_∞) *be admissible with respect to the operator*

$$Kx(t) = \int_0^t k(t,s)x(s)ds .$$

If the identically zero function is the unique solution of the given equation, then it is stable with respect to the space C_∞, *that is for every* $\varepsilon > 0$ *there exists* $\delta > 0$ *such that every* $y \in C_\infty$ *and* $\|y\|_\infty < \delta$ *every solution of*

$$x(t) = y(t) + \int_0^t k(t,s)f(s,x))ds$$

has norm $< \varepsilon$.

Outline of the proof. Fix $\varepsilon > 0$ arbitrarily. We want to determine $\delta > 0$, for which the condition of stability is satisfied.

Since $F : V \to C_g$ is continuous, V being a suitable neighbourhood of the origin in C_∞, there exists $\varepsilon_0 > 0$ such that the ball $B(0,\varepsilon_0)$ in C_∞ is contained in V and $F(B(0,\varepsilon_0))$ is a bounded subset of C_g. To prove the theorem it is sufficient to assume $\varepsilon < \varepsilon_0$, and so we do.
Since the equation

$$x - K \circ F(x) = 0$$

has only the identically zero solution by hypothesis, we have

$$\|x - K \circ F(x)\|_\infty \neq 0 \qquad \text{for} \quad \|x\|_\infty = \varepsilon .$$

Since $K \circ F$ is a compact operator on bounded sets, $I - K \circ F$

is a closed operator on bounded sets. It follows that

$$\eta_o = \inf_{\|x\|_\infty = \varepsilon} \|x - K F(x)\|_\infty > 0 \ .$$

Let $\eta = \min(\eta_o, \varepsilon)$. We choose $\delta = \frac{\eta}{4}$.
Let $y \varepsilon C_\infty$ such that $\|y\|_\infty < \delta$. We want to prove that all the possible solutions of

$$x(t) = y(t) + \int_o^t k(t,s)f(s,x))ds$$

have norm less than ε (and therefore, in particular, they exist if ε_o is so small that the ball of center the origin and radius ε_o in R^n is contained in U) .

Let $T_n : B(0,\varepsilon_o) \to C_\infty$ be defined by

$$T_n(x)(t) = \begin{cases} 0 & \text{if } t \leq \frac{1}{n} \\ \int_o^{t-\frac{1}{n}} k(t,s)f(x,s))ds & \text{if } t \geq \frac{1}{n} \end{cases}$$

Observe that T_n is a compact operator: $\overline{B(0,\varepsilon)} \to C_\infty$, for the same reason ad $K \circ F$.

It is easy to prove that in our hypotheses we have

$$\lim_n T_n = K \circ F$$

uniformly in $B(0,\varepsilon_o)$.

Since $f(t,0) \equiv 0$, the identically zero function is a fixed point of T_n . Since $I - T_n$ are nothing else than the Tonelli approximations, it is well-known that $I - T_n$ is an injective operator for every n . Therefore for the topological degree we have

$$\deg (I - T_n, B(0,\varepsilon), 0) \neq 0$$

by a theorem of Vidossich [3].

Let x_y be an arbitrary solution of (1) existing in a maximal interval J contained in $[0,+\infty[$. Let

$$\alpha = \sup \{t \varepsilon J \mid \|x_y(s)\| \leq \varepsilon_o \quad 0 \leq s \leq t\} \ .$$

Since $x_y(0) = y(0)$, we have $0 \leq \alpha \leq +\infty$.
There are two possible cases:

Case 1. $\alpha = +\infty$
Let n be so large that

$$\|K \circ F(x) - T_n(x)\| < \frac{\eta}{4}, \qquad (x \leq B(0,\varepsilon_o)) .$$

Then for every x with $\|x\|_\infty = \varepsilon$, if we define

$$\tilde{T}_n(x) = T_n(x) - T_n(x_y) + K \circ F(x_y) + y ,$$

we have

$$\|T_n(x) - \tilde{T}_n(x)\|_\infty \leq \frac{1}{2}\eta$$

and

$$\|x - T_n(x)\|_\infty \geq \frac{3}{4}\eta .$$

Since $\frac{3}{4} > \frac{1}{2}$ we can apply a well-known lemma on the topological degree and conclude

$$\deg(I - \tilde{T}_n, B(0,\varepsilon), 0) = \deg(I - T_n, B(0,\varepsilon), 0) \neq 0 .$$

Therefore the equation

$$x - \tilde{T}_n(x) = 0 \tag{3}$$

has at least one solution in $B(0,\varepsilon)$.

x_y being the unique solution of (3) (because there is a unique solution of $x - T_n(x) = T_n(x_y) + K \circ F(x_y) + y)$ it follows that $x_y \in B(0,\varepsilon)$, that is

$$\|x_y\|_\infty < \varepsilon ,$$

as we wanted to prove.

Case 2. $\alpha < +\infty$
This assumption leads to a contradiction with the definition of α .

Remark. By using the variation of constant formula, we may apply these results to systems of ordinary differential equations of the form

$$x' = A(t)x + f(t,x)$$

$$x(0) = x^o$$

where $x^o \varepsilon R^n$, $A(t)$ is a continuous square matrix function and $f : R^+ \times R^n \longrightarrow R^n$ is a continuous function.

For more details, I refer to the proper Ref [2].

REFERENCES

1. Corduneanu, C., Integral Equations and Stability of Feedback Systems, Academic Press, New York, (1973).
2. Marocco, P., A Study of Asymptotic Behaviour and Stability of the Solutions of Volterra Equations Using Topological Degree, submitted.
3. Vidossich, G., Solving Kartsatos' Problem on Nonlinear Boundary Value Problems for Ordinary Differential Equations, submitted.

PERIODIC SOLUTIONS OF SOME NONLINEAR SECOND ORDER DIFFERENTIAL EQUATIONS IN HILBERT SPACES

Jean Mawhin[1]
Michel Willem[1]

Institut Mathématique
Université de Louvain
Louvain-La-Neuve, Belgium

I. INTRODUCTION

There is a vast literature devoted to the existence of periodic solutions of second order vector ordinary differential equations of the form

$$\ddot{u} = f(t,u,\dot{u})$$

when $F: \mathbb{R} \times \mathbb{R}^n \times \mathbb{R}^n \to \mathbb{R}^n$ is continuous. See e.g. [11], [6] for a corresponding bibliography. When one considers the more general case of

$$f: \mathbb{R} \times B \times B \to B$$

with B a (possibly infinite-dimensional) Banach space and replaces the continuity assumption on f by an assumption of

---

[1]Present address: Institut Mathématique, Université de Louvain, B-1348 Louvain-La-Neuve, Belgium.

ISBN 0-12-186280-1

compactness or of k-set contraction, and when one considers, instead of periodic boundary conditions, the Picard conditions of the type

$$u(a) = A, \quad u(b) = B,$$

some extensions of the finite-dimensional results are possible and have been given by Schmitt and Thomson [12], Schmitt and Volkmann [13], Chandra, Lakshmikantham and Mitchell [5]. Basically the two boundary value problems differ in that, abstractly, they correspond to nonlinear perturbations of a linear operator which is invertible in the Picard case and non-invertible in the periodic case. When B is finite-dimensional, the linear operator relative to the periodic case has a finite-dimensional kernel and several existence theorems relative to the Picard problem also hold for the periodic boundary conditions. When B is infinite-dimensional, the same is true for the kernel of the linear operator relative to the periodic case and, as indicated in [12], it is an open problem to know if the Schmitt-Thomson results can be extended to the periodic case, when f is compact.

The aim of this work is to show that, when B is a Hilbert space, a generalization of a result, due in the finite-dimensional case to Hartman [7] for the Picard conditions and to Knobloch [8] for the periodic one, is possible, without compactness assumption on f, at the expense of a monotonicity-type condition. The used technique is essentially a Galerkin type method together with the know classical results in finite-dimensional spaces and the usual limit process involved in the theory of monotone operators.

Finally let us notice that a variant of the main result of [10] can be obtained by the same method of proof without using any result on the associated Cauchy problem.

II. FORMULATION OF THE PROBLEM AND THE EXISTENCE RESULT IN THE FINITE-DIMENSIONAL CASE

Let $\mathcal{H}$ be a real Hilbert space with inner product $(\, , \,)$ and the corresponding norm $|.|$ and $f: \mathbb{R} \times \mathcal{H} \times \mathcal{H} \to \mathcal{H}$ a continuous mapping wich is 1-periodic with respect to the first variable. Consider the following boundary value problem

$$\ddot{u}(t) = f(t,u(t), \dot{u}(t)) \tag{1}$$

$$u(0) - u(1) = \dot{u}(0) - \dot{u}(1) = 0 \tag{P}$$

The following result, which extends to the case of periodic boundary conditions a theorem of Hartman [7] relative to the Picard boundary conditions, was first proved by Knobloch [8] for locally lipschitzian f and extended to continuous f independently by Bebernes and Schmitt [2] and Mawhin [9]. For simple proofs, see [11], [6] or [1].

Theorem 1. *Assume that the following conditions hold.*

(H0) *$\mathcal{H}$ is finite-dimensional*

(H1) *There exists* $R > 0$ *such that*

$$|y|^2 + (f(t,x,y),x) \geq 0$$

for every t, x, y *such that* $|x| = R$ *and* $(x,y) = 0$.

(H2) *There exists a positive continuous nondecreasing function* $\varphi(s)$, $s \in \mathbb{R}^+$, *such that*

$$\frac{\varphi(s)}{s^2} \to 0, \quad s \to \infty$$

and

$$|f(t,x,y)| \leq \varphi(|y|)$$

for every t, x, y *with* $|x| \leq R$.

Then problem (1-P) *has at least one solution* $u(t)$. *Moreover for every* $t \in \mathbb{R}$, $|u(t)| \leq R$ *and* $|\dot{u}(t)| \leq M$ *where* M *is a positive constant depending only on* R *and* φ.

Let H be the Hilbert space $L^2(0,1;\mathcal{H})$ with the usual inner product

$$(u,v)_H = \int_0^1 (u(t),v(t))dt$$

Then problem (1-P) is equivalent to

$$Lu + Nu = 0$$

if we define

$$\text{dom } L = \{u \in H : u \text{ and } \dot{u} \text{ are absolutely continuous, } \ddot{u} \in H \text{ and } (P) \text{ is satisfied}\}$$

$$Lu = -\ddot{u}, \quad u \in \text{dom } L$$

$$\text{dom } N = \mathcal{C}^1(0,1;\mathcal{H})$$

$$Nu = f(.,u,\dot{u}), \quad u \in \text{dom } N$$

If $\mathcal{H}$ is infinite-dimensional it is clear that the kernel of L is infinite-dimensional (since it is isomorphic to $\mathcal{H}$).

Moreover the right-inverse of L is not a compact mapping. It is possible to overcome those difficulties by assuming the following simple analytical hypothese (wich implies that $L + N$ is monotone on dom L).

(H3) $|y-v|^2 + (f(t,x,y) - f(t,u,v), x - u) \geq 0$
for every $t \in \mathbb{R}$ *and for every* $x,y,u,v \in \mathcal{H}$

III. THE EXISTENCE THEOREM IN THE INFINITE-DIMENSIONAL CASE

We now formulate and prove the existence theorem for the infinite-dimensional case.

Theorem 2. *Assume that assumptions* (H1), (H2) *and* (H3) *hold. Then problem* (1-P) *has at least one solution.*

Proof. *First step.* For each finite-dimensional subspace $\mathcal{F} \in \mathcal{H}$ let $P_{\mathcal{F}} : \mathcal{H} \to \mathcal{H}$ be the orthogonal projector onto $\mathcal{F}$. Let us define P_F on H by

$$(P_F u)(t) = P_{\mathcal{F}}(u(t))$$

and let us write $F = \text{Im } P_F$. It is clear that, in F, equation

$$P_F[Lu_F + Nu_F] = 0$$

is equivalent to the problem

$$\ddot{u}_F = P_{\mathcal{F}} f(t, u_F, \dot{u}_F) \qquad (2F)$$

$$u_F(0) - u_F(1) = \dot{u}_F(0) - \dot{u}_F(1) = 0 \qquad (PF)$$

By Theorem 1 and assumptions (H1), (H2) there exists a solution u_F to problem (2F-PF) such that, for all $t \in \mathbb{R}$,

$$|u_F(t)| \leq R \quad \text{and} \quad |\dot{u}_F(t)| \leq M \qquad (3)$$

where M depends only on R and φ.

Second step. Let Λ be the collection of all the subspaces of H consisting in the set of functions whose range is contained in a finite-dimensional subspace of $\mathcal{H}$. For every $F_0 \in \Lambda$ let us write

$$V_{F_0} = \{u_F : F \in \Lambda \quad \text{and} \quad F \supset F_0\}$$

where U_F is a solution to (2F-PF) such that (3) is satisfied. Let us denote by $\bar{V}_{F_0}$ the weak closure of V_{F_0}. Since V_{F_0} is bounded in H, $\bar{V}_{F_0}$ is weakly compact. It is obvious that the family $\bar{V}_{F_0}$, $F_0 \in \Lambda$, has the finite intersection property. Thus there exists $u_0 \in \bigcap_{F_0 \in \Lambda} \bar{V}_{F_0}$.

It follows from (H2), (2F) and (3) that for every $t \in \mathbb{R}$

$$|\ddot{u}_F(t)| \leq |f(t, u_F(t), \dot{u}_F(t))| \leq \varphi(M) \qquad (4)$$

Let F_0 be any element of Λ. Since $u_0 \in \bar{V}_{F_0}$ it follows from a lemma of Kaplansky (for a proof see e.g. [4] p. 81) that there exists a sequence (F_n) in Λ such $F_n \supset F_0$ for every $n \geq 1$ and

$$u_{F_n} \rightharpoonup u_0, \quad n \to \infty$$

in H. By (4), we can assume, going if necessary to a subsequence, that

$$\ddot{u}_{F_n} \rightharpoonup v, \quad n \to \infty$$

in H. The closedness of L implies then that $u_0 \in$ dom L.

Third step. Let us now prove that u_0 is a (classical) solution to (1-P). Let $F_0 \in \Lambda$ and let $f \in F_0 \cap$ dom L. For every $F \in \Lambda$, it follows from (H3) that

$$0 \leq ((L + N)f - (L + N)u_F, f - u_F)_H .$$

If $F \supset F_0$, we have

$$\begin{aligned} 0 &\leq ((L + N)f, f - u_F)_H - (P_F(L + N)u_F, f - u_F)_H \\ &= ((L + N)f, f - u_F)_H . \end{aligned}$$

Thus

$$0 \leq ((L + N)f, f - u_0)_H . \tag{5}$$

Inequality (5) is now proved for every f in dom L whose range is contained in a finite-dimensional subspace of $\mathcal{H}$. But, using Fourier series, it turns out that (5) is true for every $f \in$ dom L. Indeed let (f_n) be the Fourier expansion of $f \in$ dom L. Then $f_n \to f$ and $\dot{f}_n \to \dot{f}$ uniformly on [0,1] when $n \to \infty$. Thus $Nf_n \to Nf$ uniformly on [0,1] when $n \to \infty$. Moreover $Lf_n \to Lf$ in H when $n \to \infty$.

Finally a straighforward use of Minty's trick (see e.g. [3] p. 19) shows that $(L + N)u_0 = 0$, i.e. u_0 is a (classical) solution to problem (1-P).

REFERENCES

1. Bebernes, J., A simple alternative problem for finding periodic solutions of second order ordinary differential systems. *Proc. Amer. Math. Soc. 42*, 121-127 (1974).
2. Bebernes, J., and Schmitt, K., Periodic boundary value problems for systems of second order differential equations. *J. Differential Equations 13*, 32-47 (1973).
3. Browder, F. E., Problèmes Non Linéaires. Séminaire de Mathématiques Supérieures, n. 15. Les Presses de l'Université de Montréal, Montréal, (1966).
4. Browder, F. E., Nonlinear Operators and Nonlinear Equations of Evolution in Banach Spaces. Proc. Sympos. Pure Math. 18, 2, Amer. Math. Soc., Providence, (1976).
5. Chandra, J., Lakshmikantham, V., and Mitchell, A. R., Existence of solutions of boundary value problems for nonlinear second order systems in a Banach space. *Nonlinear Analysis, Theory, Methods and Applications 2*, 157-168 (1978).
6. Gaines, R. E., and Mawhin, J., Coincidence Degree and Nonlinear Differential Equations. Lecture Notes in Math. n. 568, Springer Verlag, Berlin, (1977).
7. Hartman, P., On boundary value problems for systems of ordinary nonlinear second order differential equations. *Trans. Amer. Math. Soc. 96*, 493-509 (1960).
8. Knobloch, H. W., On the existence of periodic solutions for second order vector differential equations. *J. Differential Equations 9*, 67-85 (1971).
9. Mawhin, J., Boundary value problems for nonlinear second order vector differential equations. *J. Differential Equations 6*, 257-269 (1974).
10. Mawhin, J., and Willem, M., Periodic solutions of nonlinear differential equations in Hilbert Spaces, *in* "Equadiff '78, Conv. Intern. Equaz. Diff. Ord. e Equaz. Funz." (R. Conti, G. Sestini, G. Villari, eds.), p. 323-332. Univ. of Firenze, (1978).
11. Rouche, N., and Mawhin, J., Equations Différentielles Ordinaires, 2. Masson, Paris, (1973).
12. Schmitt, K., and Thomson, R., Boundary value problems for infinite systems of second order differential equations.

J. Differential Equations 18, 277-295 (1975).

13. Schmitt, K., and Volkmann, P., Boundary value problems for second order differential equations in convex subsets of a Banach space. *Trans. Amer. Math. Soc. 218*, 397-405 (1976).

OPERATORS OF MONOTONE TYPE AND ALTERNATIVE PROBLEMS WITH INFINITE DIMENSIONAL KERNEL

Jean Mawhin[1]
Michel Willem[1]

Institut Mathématique
Université de Louvain
Louvain-la-Neuve, Belgium

I. INTRODUCTION

Abstract equations of the form

$$Lu + Nu = 0 \tag{1.1}$$

in a Hilbert space, where the linear operator $L : \text{dom } L \subset H \mapsto H$ may have an infinite dimensional kernel have been recently studied by various authors (see e.g. [2], [3], [7], [14], [17]). The aim of this paper is to use an approximation method by nonlinear perturbations of linear operators with a finite dimensional kernel, together with some limit process, to prove a Leray-Schauder's type continuation theorem for (1.1) (Theorem 3.1 of Section 3) when some compactness assumption is made on the composition with N of the generalized inverse of L and

---

[1]Present address: Institut Mathématique, Université de Louvain, B-1348 Louvain-la-Neuve, Belgium.

ISBN 0-12-186280-1

when N is demicontinuous, bounded and of type G_L. The mappings of type G_L are defined in Section 2 and, as indicated there, encompass many well known classes of nonlinear operators of monotone type. It is also shown in Section 3 how the continuation theorem and a variant of it (Theorem 3.1') reduce, for special choices of L, to basic results of nonlinear functional analysis, like the Leray-Schauder and the Browder-Minty theorems. In Section 4, applications are given to existence results of the type of Cesari-Kannan and Brézis-Nirenberg, and an application to the problem of periodic solutions of a nonlinear wave equation is indicated.

Special cases of the results of this paper have been announced in [18] and complete proofs, in a somewhat more general setting, will be given in a subsequent paper.

II. MAPPINGS OF TYPE G_L AND THEIR PROPERTIES

Let H be a real Hilbert space, with inner product $(\, , \,)$ and corresponding norm $| \cdot |$, and let $L : \text{dom } L \subset H \mapsto H$ be a densely defined closed linear operator such that

$$\text{Im } L = (\ker L)^{\perp} ,$$

which, in particular, implies that $\text{Im } L$ is closed in H. Let us denote by $K : \text{Im } L \mapsto \text{dom } L \cap \text{Im } L$ the (continuous) inverse of the restriction of L to $\text{dom } L \cap \text{Im } L$ and by $P : H \mapsto H$ the orthogonal projector onto $\ker L$. If Σ denotes the set of linear operators $L : \text{dom } L \subset H \mapsto H$ satisfying the above properties, one shall notice that $0 \in \Sigma$ and $I \in \Sigma$.

<u>Definition</u>. A mapping $N : H \mapsto H$ is said to be of *type G with respect to a given* $L \in \Sigma$, shortly of type G_L, if, for each sequence (u_n) in H such that, for $n \to \infty$,

$$Pu_n \rightharpoonup v, \quad (I - P)U_n \to w, \quad Nu_n \rightharpoonup z \text{ and } (Nu_n, u_n - (v+w)) \to 0 ,$$

one has

$z = N(v+w)$.

The following remarks are easy consequences of the Definition.

Remarks. 1. A mapping of type G_0 is nothing but a mapping of type G as defined for example in [1], p. 232.

2. If N is of type G_0, then N is of type G_L for every $L \in \Sigma$.

3. If N is of type G_L for some $L \in \Sigma$, then N is of type G_I.

4. If $L \in \Sigma$ and $\dim \ker L < \infty$, then N is of type G_L if and only if N is of type G_I.

Let us now give some useful examples of mappings of type G_L. For the corresponding definitions, see e.g. [13] or [22].

Examples. 1. If N is of type M, then N is of type G_0; in particular, it will occur if N is pseudo-monotone, if N or $-N$ is monotone and hemicontinuous, or if N is weakly sequentially continuous.

2. If N is demicontinuous, then N is of type G_I, and hence of type G_L for every $L \in \Sigma$ such that $\dim \ker L$ is finite.

III. A CONTINUATION THEOREM FOR EQUATIONS INVOLVING SOME PERTURBATIONS OF TYPE G_L OF A LINEAR OPERATOR $L \in \Sigma$

Let H and L be like in Section 2, and let $\Lambda_L = \{\ker L\}$ if $\dim \ker L$ is finite and

$$\Lambda_L = \{F : F \text{ is a finite-dimensional vector subspace of } \ker L\}$$

if $\dim \ker L$ is infinite. If $F \in \Lambda_L$, we shall denote by $P_F : H \mapsto H$ the orthogonal projector onto F, and by H_F the vector subspace

$$H_F = \mathrm{Im}(P_F + I - P) = \mathrm{Im}\, P_F \oplus \ker P = \mathrm{Im}\, P_F \oplus \mathrm{Im}\, L$$

of H. Let $N : H \mapsto H$ be a mapping of type G_L ; we are interested in the existence of solutions for the equation

$$Lu + Nu = 0 . \tag{3.1}$$

Theorem 3.1. *Assume that there exists an open bounded neighborhood* Ω *of* $0 \in H$ *such that the following conditions are satisfied.*

1. N *is demicontinuous on* $\bar{\Omega}$, $N(\bar{\Omega})$ *is bounded and* $K(I - P)N$ *is compact on* $\bar{\Omega}$.

2. *For each* $F \in \Lambda_L$ *and each* $(u,\lambda) \in (\mathrm{dom}\, L \cap \partial\Omega_F) \times]0,1[$, *one has*

$$Lu + (1 - \lambda)P_F u + \lambda N_F u \neq 0 ,$$

where $\Omega_F = \Omega \cap H_F$ *and* $N_F = (P_F + I - P)N$.

Then, equation (3.1) *has at least one solution.*

Proof. We shall only sketch the proof, complete details will appear elsewhere.

Step one. We first show that, for every $F \in \Lambda_L$, the equation

$$Lu + N_F u = 0 \tag{3.2}$$

has at least one solution $u_F \in \mathrm{dom}\, L \cap \bar{\Omega}_F$. In fact, by a known result (see e.g. [10], p. 13-14), for each $\lambda \in]0,1[$, the equation

$$Lu + (1 - \lambda)P_F u + \lambda N_F u = 0 \tag{3.3}$$

is equivalent to the equation

$$u = \lambda M_F u , \tag{3.4}$$

where

$$M_F = P_F - P_F N - K(I - P)N_F = P_F - P_F N - K(I - P)N ,$$

so that M_F is compact on $\overline{\Omega}_F$ using assumption 1. The existence of u_F then follows from assumption 2 and the Leray--Schauder's theorem [12]. Notice that if dim ker L is finite, $\Lambda_L = \{\ker L\}$, $P_F = P$, $N_F = N$ and there is nothing more to prove. Assume thus from now that dim ker L is infinite.

Step two. For each $F_0 \in \Lambda_L$, let us define V_{F_0} by

$$V_{F_0} = \{u_F : F \in \Lambda_L \text{ and } F \supset F_0\}$$

and let us denote by $\bar{V}_{F_0}$ the weak closure of V_{F_0} ; then, there exists $u_0 \in \bigcap_{F_0 \in \Lambda_L} \bar{V}_{F_0}$. Let $f \in H$ and let $F_0 \in \Lambda_L$ be such that $f \in H_{F_0}$ and $u_0 \in H_{F_0}$. By a result in [5], p. 81, there exists a sequence (F_n) in Λ_L such that $F_n \supset F_0$ for every $n \geq 1$ and

$$u_{F_n} \rightharpoonup u_0$$

when $n \to \infty$. Writing, for simplicity,

$$u_n = u_{F_n} , \quad P_n = P_{F_n} ,$$

we deduce from (3.2) that

$$(Nu_n, u_n - u_0) = -(Lu_n, u_n - u_0) .$$

By the boundedness of the sequence (u_n) and the properties of L and N , we also obtain, going if necessary to a subsequence, that

$$(I - P)u_n \to (I - P)u_0 , \quad Lu_n \rightharpoonup Lu_0 , \quad Nu_n \rightharpoonup v ,$$

for some $v \in H$, when $n \to \infty$. Property G_L for N implies then that

$$v = Nu_0 .$$

Now, for each $n \geq 1$, one finds easily that

$$(Nu_n, f) = -(Lu_n, f) ,$$

and hence, going to the limit,

$$(Lu_0 + Nu_0, f) = 0 .$$

f being arbitrary, u_0 is a solution of (3.1).

q.e.d.

The following consequence of Theorem 3.1 replaces assumption (2) by a less general one of coercive type.

Theorem 3.1'. *Assume that condition* (1) *of Theorem* 3.1 *holds together with the following assumption*

2'. $(Lu + Nu, Lu + Pu) \geq 0$

for every $u \in \text{dom } L \cap \partial \Omega$.

Then, equation (3.1) *has at least one solution.*

Proof. We show that condition (2') of Theorem 3.1' implies condition (2) of Theorem 3.1. Suppose that, for some $F \in \Lambda_L$, some $\lambda \in]0,1[$ and some $u \in \text{dom } L \cap \partial \Omega_F$, one has

$$Lu + (1 - \lambda)P_F u + \lambda N_F u = 0$$

i.e.

$$(1 - \lambda)(Lu + P_F u) + \lambda(Lu + N_F u) = 0 .$$

Then, noticing that $Pu = P_F u$, one easily obtains

$$(1 - \lambda)|Lu + Pu|^2 + \lambda(Lu + Nu, Lu + Pu) = 0$$

which is impossible by condition (2') if we notice that $Lu + Pu = 0$ if and only if $u = 0$, so that $u \in \partial \Omega_F$ and $0 \in \Omega_F$ imply $|Lu + Pu| > 0$.

q.e.d.

Remarks. 1. If $L = I$, so that $P = 0$ and $K = I$, the regularity assumptions upon N reduce to the compactness of N on $\overline{\Omega}$ and condition (2) of Theorem 3.1 becomes

$$u + \lambda Nu \neq 0$$

for every $(u,\lambda) \in \partial\Omega \times]0,1[$, so that Theorem 3.1 becomes the usual *Leray-Schauder's fixed point theorem*. Similarly, condition (2') of Theorem 3.1' becomes

$$(u + Nu, u) \geq 0$$

for $u \in \partial\Omega$, so that Theorem 3.1' reduces to the *Krasnosel'skii fixed point theorem* [11].

2. If $\dim \ker L$ is finite, the regularity assumptions upon N reduce to assumption (1) of Theorem 3.1, which implies the L-compactness of N on in the sense of [10], and condition (2) of Theorem 3.1 becomes

$$Lu + (1 - \lambda)Pu + \lambda Nu \neq 0$$

for every $(u,\lambda) \in (\mathrm{dom}\, L \cap \partial\Omega) \times]0,1[$, so that Theorem 3.1 reduces to a special case of Corollary 1 in [15], or Theorem IV.5 in [16]. Similarly, Theorem 3.1' then reduces to a special case of Corollary IV.6 in [16].

3. If $L = 0$, so that $P = I$ and $K = 0$, the regularity assumptions upon N reduce to the demicontinuity of N on $\bar{\Omega}$ and the boundedness of $N(\bar{\Omega})$ and condition (2) of Theorem 3.1 becomes

$$P_F((1 - \lambda)u + \lambda Nu) \neq 0$$

for each finite-dimensional subspace F of H and each $(u, \lambda) \in \partial(\Omega \cap F) \times]0,1[$ and gives a sufficient condition for the solvability of the equation

$$Nu = 0$$

when N is of type G_0. It is unknown to us if, for the general class of mappings N which are of type G_L, condition (2) of Theorem 3.1 could be replaced by the more pleasant one

$$Lu + (1 - \lambda)Pu + \lambda Nu \neq 0$$

for every $(u,\lambda) \in (\mathrm{dom}\, L \cap \partial \Omega) \times]0,1[$. This is true when N is monotone, demicontinuous and $N(\bar{\Omega})$ is bounded, as shown by another approach in [17]. On the other hand, for $L = 0$ and N pseudo-monotone, condition (3.5) can be replaced, without loosing the conclusion, by the simpler one, if $\Omega = B(R)$,

$$(1 - \lambda)u + \lambda Nu \neq 0$$

for every $(u,\lambda) \in \partial B(R) \times]0,1[$, as it follows from a result of Browder [5], Theorem 7.11 and De Figueiredo [9]. On the other hand, for $L = 0$, condition (2') of Theorem 3.1' reduces to

$$(Nu,u) \geq 0$$

for every $u \in \partial \Omega$, which is the extension to the class of demicontinuous and bounded mappings of type G_0 of the basic *theorem of Browder* [4] *and Minty* [19] for monotone operators.

4. The above remarks 1 to 3 above indicate that by varying L from the "best" situation for L with respect to its invertibility, namely $L = I$, to the corresponding "worst" situation $L = 0$, Theorem 3.1 and 3.1' link the Leray-Schauder's type theorems to the Browder-Minty's type theorems, and allow a better localization, with respect to those two corner stones of nonlinear functional analysis, of the more recent results on nonlinear perturbations of noninvertible linear mappings having a finite or a infinite dimensional kernel.

IV. APPLICATIONS

We shall first use Theorem 3.1 to give a simple proof of the generalization, to the case where dim ker L may be infinite, of the Cesari-Kannan abstract formulation of Landesman-Lazer's type conditions (see [6] and [8]) for bounded nonlinear perturbations of some linear operators with finite dimensional kernel. For distinct extensions of this result to the case of infinite dimensional kernel, see the survey paper [7] of Cesari.

Corollary 4.1. *Let* L *and* N *be like in the beginning of Section* 3 *and assume that there exist* $r > 0$ *and* $R > 0$ *such that the following conditions hold.*

a. N *is demicontinuous on* $\bar{B}(r,R)$, $K(I - P)N$ *is compact on* $\bar{B}(r,R)$, $N(\bar{B}(r,R))$ *is bounded and, for every* $x \in \partial B(r,R) \cap \operatorname{dom} L$

$$|K(I - P)Nx| \leq r ,$$

where

$$B(r,R) = \{x \in H : |(I - P)x| < r, \quad |Px| < R\} .$$

b. *One has*

$$(Nx,Px) \geq 0 \quad (\textit{resp.} \quad (Nx,Px) \leq 0)$$

for every $x \in \operatorname{dom} L$ *such that*

$$|Px| = R \quad \textit{and} \quad |(I - P)x| \leq r .$$

The equation (3.1) *has at least one solution.*

Proof. If follows very easily from Theorem 3.1 if we take $\Omega = B(r,R)$ and notice that equation (3.3) is equivalent to the system

$$(I - P_F)x + \lambda K(I - P)Nx = 0 , \quad (1 - \lambda)P_F x + \lambda P_F Nx = 0 ,$$

as follows easily from (3.4).

Remark. When $L = 0$, the conditions of Corollary 4.1 exactly reduce to that of the end of Remark 3 in Section 3, i.e. Corollary 4.1 gives in this case the extension of the Browder-Minty theorem to demicontinuous bounded mappings of type G_0. If $L = I$, the conditions of Corollary 4.1 reduce to the compactness of N on the closed ball $B(r)$ of center 0 and radius r and the Rothe's condition

$$|Nx| \leq r \quad \text{for} \quad |x| = r ,$$

so that Corollary 4.1 becomes when $L = I$ the *Rothe's fixed point theorem* [20] on a ball, which itself contains the *Schauder's fixed point theorem on a ball* [21].

For L satisfying the conditions of Section 2, we shall denote by α, like in [3], the greatest positive constant such that, for every $u \in \operatorname{dom} L$, one has

$$(Lu,u) \geq - \alpha^{-1} |Lu|^2 .$$

One can then deduce, from Theorem 3.1, the following generalization to the case of N of type G_L instead of monotone (Corollary 4.2) and variant (Corollary 4.3) of the fundamental result of Brézis and Nirenberg [3] about the range of mappings of the type $L + N$ with $\dim \ker L$ possibly infinite.

Corollary 4.2. *Let* L *be like in Section* 2 *and let* $B : H \mapsto H$ *be a mapping of type* G_L, *demicontinuous and taking bounded sets into bounded sets, with moreover* $K(I - P)B$ *compact on bounded sets of* H. *If there exists* $\gamma \in]0,\alpha[$ *such that, for all* $u \in H$ *and* $v \in H$, *one has*

$$(Bu - Bv,u) \geq \gamma^{-1} |Bu|^2 - c(y) ,$$

where $c(y)$ *depends only of* y, *then*

$$\operatorname{int}(\operatorname{Im} L + \operatorname{conv} \operatorname{Im} B) \subset \operatorname{Im}(L + B) .$$

Proof. Corollary 4.2 can be deduced from Theorem 3.1 in a way essentially similar to that used in [17] to deduce Corollary 5.2 from Theorem 4.1 and we shall not reproduce the details here.

Corollary 4.3. *Let* L *be like in Section* 2 *and let* $N : N \mapsto H$ *be a mapping of type* G_L, *demicontinuous, taking bounded sets into bounded sets and such that* $K(I - P)N$ *is compact on bounded sets of* H. *Assume that the following assumptions are satisfied.*

(1) *There exists* $\gamma \in \,]0,\alpha[$ *and* $c \geq 0$ *such that, for every* $u \in H$ *, one has*

$$(Nu,u) \geq \gamma^{-1}|Nu|^2 - c \ ;$$

(2) *For every sequence* (u_n) *in* dom L *such that*

$$|Pu_n| \to \infty \text{ if } n \to \infty \ ,$$

one has

$$(Nu_n,u_n) \to +\infty \text{ if } n \to \infty \ .$$

Then equation (3.1) *has at least one solution.*

Proof. We refer to [18] for a sketch of the proof, which consists in proving that, under the assumptions of Corollary 4.3, the set of possible solutions of equation (3.3) is a priori bounded independently of $\lambda \in \,]0,1[$ and $F \in \Lambda_L$, so that one can choose for Ω an open ball of center 0 and sufficiently large radius in theorem 3.1.

Corollary 4.3 can be used, instead of Corollary 4.2 and the theory of maximal monotone operators like in [3], to prove directly the existence result of Brézis and Nirenberg for generalized L^2 solutions of the nonlinear wave equation

$$u_{tt} - u_{xx} + g(t,x,u) = 0$$

which are 2π-periodic in t and verify the boundary conditions

$$u(t,0) = u(t,\pi) = 0 \ , \qquad 0 \leq t \leq 2\pi \ ,$$

when g is measurable in (t,x) , continuous in u, G or $-g$ is nondecreasing as a function of u and satisfies for every $u \in R$ and a.e. $(t,x) \in [0,2\pi] \times [0,\pi]$ the inequalities

$$\pm ug(t,x,u) \geq \eta u^2 - h_1(t,x)|u|, \quad |g(t,x,u)| \leq \gamma|u| + h_2(t,x) \ ,$$

with $\eta > 0$, $\gamma < 1$ and $h_1, h_2 \in L^2([0,2\pi] \times [0,\pi], \mathbb{R}_+)$. We refer to [3] for writing this problem as an abstract equation of type (3.1) in $H = L^2([0,2\pi] \times [0,\pi], \mathbb{R})$.

REFERENCES

1. Berger, M. S., Nonlinearity and Functional Analysis. Lectures on Nonlinear Problems in Mathematical Analysis, Academic Press, New York, (1977).
2. Brézis, H. and Haraux, A., Image d'une somme d'opérateurs monotones et applications, *Israel J. Math. 23*, 165-186 (1976).
3. Brézis, H. and Nirenberg, L., Characterizations of the ranges of some nonlinear operators and applications to boundary value problems, *Ann. Scuola Norm. Sup. Pisa (4) 5*, 225-236 (1978).
4. Browder, F. E., Nonlinear elliptic boundary value problems. II, *Trans. Amer. Math. Soc. 117*, 530-550 (1965).
5. Browder, F. E., Nonlinear Operators and Nonlinear Equations of Evolution in Banach Spaces, Proc. Symp. Pure Math. 18-2, Amer. Math. Soc., Providence, (1976).
6. Cesari, L., Functional analysis, nonlinear differential equations and the alternative method, *in* "Nonlinear Functional Analysis and Differential Equations" (Cesari, Kannan and Schuur, eds.), pp. 1-197. M. Dekker, New York, (1976).
7. Cesari, L., Hyperbolic problems. Existence and applications, Proceedings "Equadiff 78", Florence, (1978).
8. Cesari, L. and Kannan, R., An abstract existence theorem at resonance, *Proc. Amer. Math. Soc. 63*, 221-225 (1977).
9. De Figueiredo, D. G., An existence theorem for pseudo--monotone operator equations in Banach spaces, *J. Math. Anal. Appl. 34*, 151-156 (1971).
10. Gaines, R.E., and Mawhin, J., Coincidence Degree and Nonlilear Differential Equations", Lect. Notes in Math. vol. 568, Springer Verlag, Berlin, (1977).
11. Krasnosel'skii, M. A., On a fixed point principle for completely continuous operators in function spaces (Russian), *Dokl. Akad. Nauk SSSR 74*, 13-15 (1950).

12. Leray, J. and Schauder, J., Topologie et équations fonctionnelles, *Ann. Sci. Ecole Norm. Sup. (3) 51*, 45-78 (1934).

13. Lions, J. L., Quelques méthodes de résolution des problèmes aux limites non linéaires, Dunod, Paris, (1969).

14. Mawhin, J., Contractive mappings and periodically perturbed conservative systems, *Arch. Math. (Brno) 12*, 67-73 (1976).

15. Mawhin, J., Landesman-Lazer's type problems for nonlinear equations, *Confer. Sem. Mat. Univ. Bari n. 147*,(1977).

16. Mawhin, J., Topological Degree Methods in Nonlinear Boundary Value Problems, Lect. Notes in Math., vol. 40, Amer. Math. Soc., Providence, to appear.

17. Mawhin, J. and Willem, M., Range of nonlinear perturbations of linear operators with an infinite dimensional kernel, *in* "Proc. Confer. Ordinary and Partial Differentia Equations", Dundee 978. Lect. Notes in Math., Springer Verlag, Berlin, to appear.

18. Mawhin, J. and Willem, M., Perturbations non linéaires d'opérateurs linéaires à noyau de dimension infinie, *C.R. Acad. Sci. Paris, A, 337* (1978), in press.

19. Minty, G. J., On a "monotonicity" method for the solution of nonlinear equations in Banach spaces, *Proc. Nat. Acad. Sci. USA 50*, 1038-1041 (1963).

20. Rothe, E. H., Zur Theorie der topologischer Ordnung und der Vektorfelder in Banachschen Räumen, *Compositio Math. 5*, 177-197 (1937).

21. Schauder, J., Der Fixpunktsatz in Funktionalräumen, *Studia Math. 2*, 171-180 (1930).

22. Zeidler, E., Vorlesungen über nichtlineare Funktionalanalysis. II. Monotone Operatoren, Teubner, Leipzig, (1977).

STABILITY THEORY FOR COUNTABLY INFINITE SYSTEMS

R. K. Miller[1,2]

Mathematics Department
Iowa State University
Ames, Iowa

A. N. Michel[3,4]

Electrical Engineering Department and
Engineering Research Institute
Iowa State University
Ames, Iowa

I. INTRODUCTION

In the present paper we establish new stability results for a class of countably infinite systems of ordinary differential equations. We consider those systems which may be viewed

---

[1]Supported by N. S. F. Grant ENG77-28446.

[2]Present address: Mathematics Department, Iowa State University, Ames, Iowa 50011.

[3]Supported by N.S.F. Grant ENG77-28446 and by the Engineering Research Institute, Iowa State University.

[4]Present address: Electrical Engineering Department and Engineering Research Institute, Iowa State University, Ames, Iowa 50011.

ISBN 0-12-186280-1

as an interconnection of countably many free or isolated subsystems (which are described by ordinary differential equations defined on finite dimensional spaces). Such systems are often called interconnected systems, composite systems, decentralized systems, large scale systems, and the like (see Michel and Miller [3]). As in [3], our objective will be to analyze interconnected systems in terms of their simpler subsystems and in terms of their interconnecting structure.

For existing results dealing with stability and well posedness of countably infinite systems of differential equations, the reader is referred to Persidskii [4,5], Leung et al. [2], Shaw [7,8] and Deimling [1].

II. COUNTABLY INFINITE SYSTEMS OF DIFFERENTIAL EQUATIONS

A. Initial Value Problem

In this paper we are concerned with the initial value problem

$$\dot{z}_n = h_n(x), \quad z_n(0) = c_n \tag{1}$$

for $n = 1,2,3,\ldots$. Here x is the infinite dimensional column vector $x = (z_1^T, z_2^T, \ldots, z_n^T, \ldots)^T \in R^\omega, z_n \in R^{m_n}$, $D \subset R^\omega$ and $h_n : D \to R^{m_n}$. The infinite product $R^\omega = R^{m_1} \times R^{m_2} \times \ldots \times R^{m_n} \times \ldots$ is given the usual product topology. Since this topology is equivalent to introducing the metric

$$\rho(x,\bar{x}) = \sum_{n=1}^{\infty} \frac{|z_n - \bar{z}_n|}{(1 + |z_n - \bar{z}_n|)2^n}, \tag{2}$$

then R^ω is a convex Frechet space (see e.g., [6]).

A *solution* of (1) is a function $x : [0,b] \to D$ for some $b > 0$ such that $z_n \in C_n^1[0,b]$, $z_n(0) = c_n$ and $\dot{z}_n(t) = h_n(x(t))$ for all $t \in [0,b]$ and for all $n = 1,2,3,\ldots$.

Remark 1. An alternate way to view the above initial value problem is to fix a Banach space X of real sequences and to think of (1)

$$\dot{x} = h(x), \quad x(0) = c \tag{1'}$$

as a problem in X. In this case an X-solution is a function $x : [0,b] \to X$ which is continuously differentiable in the X-sense and satisfies (1'). Clearly, an X-solution for (1') is also a solution in our sense (for (1)). However, in the present paper we shall not concern ourselves with Banach space settings for this initial value problem. Our approach has the advantage that well-posedness (i.e., existence, uniqueness and continuation results for (1)) poses no problems while it may pose serious difficulties in a Banach space setting (for (1')). The disadvantage of our approach is that before applying stability results, it is usually necessary to argue separately that the solution $x(t)$ of (1) remain in a bounded set D.

B. Interconnected Systems

We will regard system (1) as an *interconnected system* or a *composite system* of the form

$$\dot{z}_n = f_n(z_n) + g_n(x), \quad z_n(0) = c_n, \tag{Σ}$$

$n = 1,2,3,\ldots,$ where in the notation of (1),

$$h_n(x) \triangleq f_n(z_n) + g_n(x) .$$

We view (Σ) as an interconnection of countably infinitely many *isolated* or *free subsystems* described by equations of the form

$$\dot{w}_n = f_n(w_n), \tag{$\mathcal{S}_n$}$$

$n = 1,2,\ldots$. The terms $g_n(x)$, $n = 1,2,\ldots,$ comprise the *interconnecting structure* of system (Σ).

C. Well-Posedness

We shall assume that for some constants $r_n > 0$ the set $D = D_k$ for some $k > 0$ where

$$D_k = \{x = (z_1^T, z_2^T, \ldots)^T \in R^\omega : |z_n| \leq kr_n \text{ for } n = 1, 2, \ldots\}.$$

We shall also make the following additional assumptions.

(A-1): $f_n : R^{m_n} \to R^{m_n}$, $f_n(0) = 0$, f_n is Lipschitz continuous with Lipschitz constant L_{n0}.

(A-2): $g_n : R^\omega \to R^{m_n}$, $g_n(0) = 0$, and there exist constants $L_{nj} \geq 0$ such that $\sum_{j=1}^{\infty} L_{nj} r_j < \infty$ and such that

$$|g_n(x) - g_n(x^*)| \leq \sum_{j=1}^{\infty} L_{nj} |z_j - z_j^*|$$

for all $x, x^* \in D_k$, and all $n = 1, 2, \ldots$.

Theorem 1. (see Deimling [1]). If (A-1) a,d (A-2) are true, if $k > 0$, if $|c_n| \leq kr_n$ for all $n \geq 1$ and if

(A-3): there is an $M > 0$ such that for all $n \geq 1$,

$$L_{n0} r_n + \sum_{j=1}^{\infty} L_{nj} r_j \leq Mr_n \tag{3}$$

then (Σ) has a unique solution $x(t,c)$ on $0 \leq t < \infty$ (where $c^T = (c_1^T, c_2^T, \ldots)$ and $x(0,c) = c$).

We will also study *finite approximations* to (Σ) which are of the form

$$\left.\begin{aligned} \dot{z}_n &= f_n(z_n) + g_n((z_1, z_2, \ldots, z_N, 0, \ldots)) \\ z_n(0) &= c_n \end{aligned}\right\} \tag{Σ_N}$$

$n = 1, 2, \ldots, N$, and $z_n(t) \equiv 0$ if $n > N$. In this case we let $x^N(t,c)$ denote the solution of (Σ_N) and let $c^N = (c_1^T, c_2^T, \ldots, c_N^T, 0, 0, \ldots)^T$.

Theorem 2. (See Deimling |1|). Under the hypotheses of Theorem 1

$$\lim_{N \to \infty} x^N(t,c) = x(t,c)$$

uniformly for t on compact subsets.

The derivative of a Lyapunov function along solutions of (Σ) will be computed by finding the derivative w.r.t. (Σ_N) and then letting $N \to \infty$. This can be justified by using Theorem 2 above.

D. Weak-Coupling Conditions

Assume

(A-3): For each free subsystem $(\mathcal{S}_n)$ there is a function $v_n : R^{m_n} \to R^+$, a constant $\sigma_n \in R$, a constant $L_n \geq 0$, and functions $\psi_{jn} \in K$, $j = 1,2,3$, such that

$$\psi_{1n}(|z_n|) \leq v_n(z_n) \leq \psi_{2n}(|z_n|)$$

$$Dv_{n(\mathcal{S}_n)}(z_n) \leq \sigma_n \psi_{3n}(|z_n|)$$

$$|v_n(z_n) - v_n(z_n^*)| \leq L_n |z_n - z_n^*|$$

for all $z_n, z_n^* \in R^{m_n}$.

(A-4): There are constants $a_{nm} \geq 0$ such that $|g_n(x)| \leq \sum_{m=1}^{N} a_{nm}\psi_{3m}(|z_m|)$ whenever $x^T = (z_1^T, z_2^T, \ldots, z_N^T, 0, 0, \ldots)$ for any $N \geq 1$.

In our next result we will employ Lyapunov functions for (Σ) which are of the form

$$v(x) = \sum_{n=1}^{\infty} \lambda_n v_n(z_n)$$

with constants $\lambda_n > 0$ which we will further specify later. We will also utilize $N \times N$ matrices $R_N = [r_{nm}]$ specified by

$$r_{nm} = \begin{cases} -(\sigma_n + L_n a_{nn}) & \text{if} \quad n = m \\ -L_n a_{nm} & \text{if} \quad n \neq m \end{cases}$$

Finally, we will also require the following additional hypothesis:

(A-5): For each N sufficiently large

$$\sum_{j=1}^{N} \lambda_j r_{jn} \geq 0 \quad \text{for} \quad 1 \leq n \leq N.$$

Under the above assumptions we can easily compute

$$\frac{d}{dt} v(t,x^N(t,c)) \leq \sum_{n=1}^{N} \lambda_n \{\sigma_n \psi_{3n}(|z_n|) + \sum_{m=1}^{N} L_n a_{nm} \psi_{3m}(|z_m|)$$

$$= - \Lambda_N R_N \psi_N(|x|)$$

where $\Lambda_N = (\lambda_1, \lambda_2, \ldots, \lambda_N)$ and $\psi_N(|x|) \triangleq (\psi_{31}(|z_1|), \ldots, \psi_{3N}(|z_N|))^T$. By assumption (A-5) it now follows that $\Lambda_N R_N \geq 0$. Thus, $Dv_{(\sum)}(t,x) \leq 0$.
Indeed, we have

$$Dv_{(\sum)}(t,x) \leq - \sum_{n=1}^{\infty} (\sum_{j=1}^{\infty} \lambda_j r_{jn}) \psi_{3n}(|z_n|) = - W(x)$$

Moreover, $W(x)$ is positive definite with respect to the set

$$E^* = \{x = (z_1, z_2, z_3, \ldots): z_n = 0 \text{ when } \sum_{j=1}^{\infty} \lambda_j r_{jn} > 0\}.$$

Summarizing, we now have

Theorem 3. If assumptions (A-1) to (A-5) are true and if $\sum_{n=1}^{\infty} \lambda_n \psi_{2n}(r_n) < \infty$, then

(i) the trivial solution $x = 0$ of system $(\sum)$ is stable w.r.t. D_k for any $k > 0$.

(ii) Let M be the largest invariant subset of E^*. If $c \in D$ and either $x(t,c) \in D_K$ for all $t \geq 0$ and some fixed $k > 1$ or $(\sum)$ is row finite, then $x(t,c) \to M$ as $t \to \infty$.

Example. Consider the problem

$$\dot{z}_1 = -2z_1 + z_2$$
$$\dot{z}_n = z_{n-1} - 2z_n + z_{n+1} \qquad (n \geq 1). \tag{4}$$

For each isolated subsystem (13),

$$\dot{z}_n = -2z_n , \tag{5}$$

$n \geq 1$, we choose Lyapunov functions of the form $v_n(z_n) = |z_n|$. Then $\psi_{1n}(|z_n|) = \psi_{2n}(|z_n|) = |z_n|$, $\psi_{3n}(|z_n|) = |z_n|$, $\sigma_n = -2$ and $L_n = 1$. Thus, (L-3) is satisfied. Hypothesis (L-4) is also satisfied with $a_{n,n-1} = a_{n,n+1} = 1$ for $n \geq 2$, $a_{12} = 1$, and $a_{ij} = 0$ for all other indices. Choose c such that $\sum_{n=1}^{\infty} |c_n| < \infty$ and let $r_n \geq |c_n|$ be such that $\sup_n (r_{n+1} + r_{n-1})/r_n = S < \infty$ and $\sum_{n=1}^{\infty} r_n < \infty$ (e.g. $r_n = \frac{K}{2^n}$). For system (4) we choose

$$v(x) = \sum_{n=1}^{\infty} v_n(z_n) = \sum_{n=1}^{\infty} |z_n| .$$

Then

$$Dv_{(4)}(x) = -2|x_1| + |x_2| + \sum_{n=2}^{\infty} (|x_{n-1}| - 2|x_n| + |x_{n+1}|)$$
$$= -|x_1| \leq 0.$$

Thus $\sup_n \sup_{t \geq 0} |z_n(t,c)| < \infty$. Moreover, $E^* = \{x : z_1 = 0\}$ and if $x(t,c) \in M$, then $z_1(t,c) = 0$ implies that

$$\dot{z}_1 + 2z_1 = z_2 = 0.$$

Similarly,

$$\dot{z}_2 + 2z_2 = z_3 = 0,$$

and so forth. Thus $M = \{(0,0,\ldots,0,\ldots)\}$.

All the hypotheses of Theorem 6 are satisfied. The trivial solution of (4) is asymptotically stable w.r.t. D_k for any $k > 1$.

REFERENCES

1. Deimling, J., Ordinary Differential Equations in Banach Spaces. Lecture Notes in Mathematics No. 596, Springer Verlag, Berlin, (1977).
2. Leung, K. V., Mangeron, D., Oguztorelli, M. N., and Stein, R. B., On the stability and numerical solutions of two neural models. *Utilitas Mathematica 5*, 167-212 (1974).
3. Michel, A.N. and Miller, R. K., Qualitative Analysis of Large Scale Dynamical Systems. Academic Press, New York, (1977).
4. Persidskii, K. P., On stability of solutions of countable systems of differential equations. *Izv. Akad. Nauk. Kazach. SSR. 2*, 3-35 (1948).
5. Persidskii, K. P., Countable systems of differential equations and stability of their solutions III: Fundamental theorems on stability of solutions of countably many differential equations. *Izv. Akad. Nauk. Kazach SSR 9(13)*, 11-14 (1961).
6. Robertson, A. P. and Robertson, W. J., Topological Vector Spaces. Cambridge University Press, No. 53, Cambridge, (1966).
7. Shaw, L., Existence and approximation of solutions to an infinite set of linear time-invariant differential equations. *SIAM J. Appl. Math. 22*, 266-279 (1972).
8. Shaw, L., Solutions for infinite-matrix differential equations. *J. Math. Anal. Appl. 41*, 373-383 (1973).

A NONLINEAR HYPERBOLIC VOLTERRA EQUATION ARISING IN HEAT FLOW

John A. Nohel[1,2]

University of Wisconsin-Madison
Madison, Wisconsin

Abstract. A mathematical model for nonlinear heat flow in a rigid unbounded body of material with memory is analysed by an energy method developed jointly with C. M. Dafermos. Global existence, uniqueness, boundedness and the decay of smooth solutions at $t \to \infty$ are established for sufficiently smooth and "small" data, under physically reasonable assumptions.

I. INTRODUCTION

In this largely expository paper which is based on recent joint work with C. M. Dafermos [1] we use energy methods to discuss the global existence, uniqueness, boundedness, and decay as $t \to \infty$ of smooth (C^2) solutions of the nonlinear Cauchy problem

[1]Research sponsored by the United States Army under Grant No. DAAG29-77-G-0004 and under Contract No. DAAG29-75-C-0024.

[2]Present Address: University of Wisconsin-Madison, Wisconsin 53706, USA.

ISBN 0-12-186280-1

$$(\mathrm{HF})\quad \begin{cases} u_t(t,x) = \int_0^t a(t-s)\sigma(u_x(s,x))_x ds + f(t,x) & (0 < t < \infty,\ x \in \mathbb{R}) \\ u(0,x) = u_0(x) & (x \in \mathbb{R}) \end{cases}$$

for appropriately smooth and "small" data u_0, f. Here $a : [0,\infty) \to \mathbb{R}$, $\sigma : \mathbb{R} \to \mathbb{R}$ $(\sigma(0) = 0)$, $f : [0,\infty) \times \mathbb{R} \to \mathbb{R}$, $u_0 : \mathbb{R} \to \mathbb{R}$ are given functions satisfying assumptions motivated partly by physical considerations sketched below, and partly by the method of analysis; subscripts in (HF) denote partial derivatives. Some comments on closely related initial-boundary value problems are made following the statement of the main result (Theorem 2.1). With appropriate interpretation of the term $\sigma(u_x)_x$, problem (HF) has a valid physical meaning in any number of space dimensions, and we refer to [1, Thm. 7.1] to such a problem in $\mathbb{R}^2$ studied by an extension of this method. An earlier study of (HF) by Mac Camy [6] is based on the method of Riemann invariants and is therefore restricted to a single space dimension. The present method which yields more widely applicable results even in one dimension is more direct and simpler. For a similar approach to a problem in nonlinear viscoelasticity we refer the reader to [1, Theorem 5.1] and to Mac Camy [7] for the Riemann invariant approach.

To motivate the assumptions to be imposed, particularly with regard to the kernel a, we consider briefly the problem of nonlinear heat flow in an unbounded one-dimensional rigid body of a material with memory. Let $u(t,x)$, $\varepsilon(t,x)$, $q(t,x)$, and $h(t,x)$ denote respectively the temperature, the internal energy, the heat flux, and the external heat supply at time t and position x. Following Gurtin and Pipkin [2], and also Mac Camy [6], we assume a model for heat flow in which $\varepsilon(t,x)$ and $q(t,x)$ are respectively the following functionals of the temperature u and of the gradient of temperature u_x:

$$\varepsilon(t,x) = bu(t,x) + \int_{-\infty}^{t} \beta(t-s)u(s,x)ds \qquad (0 < t < \infty,\ x \in \mathbb{R}); \tag{1.1}$$

$$q(t,x) = -\int_{-\infty}^{t} \gamma(t-s)\sigma(u_x(s,x))ds \qquad (0 < t < \infty,\ x \in \mathbb{R}); \tag{1.2}$$

it is assumed that the history $u_0(t,x)$ of u (and hence

also the history $u_{0x}(t,x)$ of the temperature gradient) is prescribed up to $t = 0$ and for $x \in \mathbb{R}$. We can assume without loss of generality that $u_0(t,x) \equiv 0$, $(t < 0, x \in \mathbb{R})$; for if that is not the case, it is easily seen from what follows that this merely alters the forcing term f in (HF). It is reasonable to assume that $\sigma : \mathbb{R} \to \mathbb{R}$ is a smooth function satisfying $\sigma(0) = 0$, $\sigma'(0) > 0$ (in fact, $\sigma'(\xi) \geq \varepsilon > 0$, $(\xi \in \mathbb{R})$ - recall that for linear heat flow $\sigma(\xi) = c\xi$, $c > 0$ a constant). We shall assume that $b > 0$ is a given constant and that the given smooth "memory" functions $\beta, \gamma \in L^1(0,\infty)$; thus $\varepsilon(t,x)$ and $q(t,x)$ are bounded whenever $u(t,x)$ and $u_x(t,x)$ are bounded. It should be noted that in the applied literature β, γ are linear combinations of decaying exponentials with positive coefficients.

If $h(t,x)$ denotes the external heat supply, the balance of heat requires that

$$\varepsilon_t(t,x) = -q_x(t,x) + h(t,x) \qquad (0 < t < \infty,\ x \in \mathbb{R}) \tag{1.3}$$

Substituting (1.1), (1.2) into (1.3), and using the assumption that $u(t,x) \equiv 0$ for $t < 0$, $x \in \mathbb{R}$ yields

$$bu_t(t,x) + \frac{\partial}{\partial t}\int_0^t \beta(s)u(t-s,x)ds = \int_0^t \gamma(t-s)\sigma(u_x(s,x))_x ds + h(t,x) ,$$

or equivalently

$$bu_t(t,x) + \int_0^t \beta(t-s)u_t(t-s,x)ds = \int_0^t \gamma(t-s)\sigma(u_x(s,x))ds + h(t,x) - \beta(t)u(0,x) , \tag{1.4}$$

where we also prescribe the value $u(0,x) = u_0(x)$, $x \in \mathbb{R}$. To reduce (1.4) to (HF) define the resolvent kernel ρ of β by the relation

$$b\rho(t) + (\beta*\rho)(t) = -\frac{\beta(t)}{b} \qquad (0 \leq t \leq \infty) , \tag{ρ}$$

where here and in what follows $*$ denotes the convolution on $[0,t]$. It follows from standard theory of linear Volterra equations that (ρ) has a unique solution $\rho \in C^1[0,\infty)$. If g is a given function on $[0,\infty)$, the solution of the Volterra equation

$$by(t) + (\beta * y)(t) = g(t) \qquad (0 \leq t < \infty) \tag{V}$$

is given by the variation of constants formula

$$y(t) = \frac{g(t)}{b} + (\rho * g)(t) \qquad (0 \leq t < \infty) . \tag{1.5}$$

Applying (1.5) with $y = u_t$ and g the right-hand side of (1.4), one sees that (1.4) is equivalent to (HF) with

$$a(t) = \frac{1}{b}\gamma(t) + (\rho * \gamma)(t) \qquad 0 \leq t < \infty \tag{1.6}$$

$$f(t,x) = \frac{1}{b}(h(t,x) - \beta(t)u_0(x)) + \rho * (h(t,x) - \beta(t)u_0(x)) \qquad (0 \leq t < \infty,\ x \in \mathbb{R}) . \tag{1.7}$$

To motivate the assumptions to be imposed on a, we note from (1.1) and (1.5) that

$$u(t,x) = \frac{\varepsilon(t,x)}{b} + (\rho * \varepsilon)(t,x) \qquad (0 \leq t < \infty,\ x \in \mathbb{R}) ,$$

where ρ is the resolvent of β. If $\rho \in L^1(0,\infty)$ then ε bounded implies u bounded. But since we assumed $\beta \in L^1(0,\infty)$, the Paley-Wiener theorem applied to equation (ρ) yields that $\rho \in L^1(0,\infty)$ if and only if

$$\hat{\beta}(s) \neq -b \quad \text{for} \quad \text{Re } s \geq 0$$

where $\hat{\beta}(s) = \int_0^\infty \exp(-st)\beta(t)dt$. But the internal energy ε is positive and so from (1.1)

$$b + \int_0^\infty \beta(t)dt = b + \hat{\beta}(0) > 0 .$$

Therefore, the Paley-Wiener condition can be modified to the statement: $\rho \in L^1(0,\infty)$ if and only if

$$\text{Re } \hat{\beta}(s) + b > 0 , \qquad (\text{Re } s \geq 0) . \tag{1.8}$$

The assumption $\gamma \in L^1(0,\infty) \cap C^1[0,\infty)$, together with (1.6), then imply that $a \in L^1(0,\infty) \cap C^1[0,\infty)$, and an easy calculation using (ρ) and (1.6) shows that

$$\hat{a}(s) = \frac{\hat{\gamma}(s)}{b + \hat{\beta}(s)} \qquad (\text{Re } s \geq 0) . \tag{1.9}$$

For physical reasons and (1.2) one needs to require $\int_0^\infty \gamma(t)dt > 0$, and so it also follows from (1.9) that

$$\int_0^\infty a(t)dt = \hat{a}(0) = \frac{\int_0^\infty \gamma(t)dt}{b + \int_0^\infty \beta(t)dt} > 0 . \tag{1.10}$$

Physically the function γ represents the heat flux relaxation function, and it is reasonable to assume that $\gamma(0) > 0$. It then follows from (1.6) that $a(0) = \frac{1}{b}\gamma(0) > 0$. If, as is reasonable, it is also assumed that $\gamma'(0) \leq 0$ and $\beta(0) > 0$, one also has from (1.6) that $a'(0) = \frac{1}{b}\gamma'(0) + \rho(0)\gamma(0) = \frac{1}{b}\gamma'(0) - \frac{\beta(0)\gamma(0)}{b} < 0$.

To summarize, the following assumptions concerning the kernel a in (HF) are reasonable for the heat flow problem:

$$\begin{aligned} &a \in C^1[0,\infty) \cap L^1(0,\infty), \quad a(0) > 0, \\ &a'(0) < 0, \quad \int_0^\infty a(t)dt > 0 ; \end{aligned} \tag{1.11}$$

as we shall see below we shall require additional smoothness of a, as well as a positivity "frequency domain" condition involving the Laplace transform of a. The implications of this condition are discussed in Lemma 3.3 and Remarks following it.

In the analysis of (HF) which follows we shall impose other technical assumptions (see assumptions (a), (σ), (f), (u_0) below). To motivate our result for (HF) assume for the moment that $a(t) \equiv a(0) > 0$ for $t \geq 0$. Then (HF) is formally equivalent to the Cauchy problem

$$\begin{aligned} &u_{tt} = a(0)\sigma(u_x)_x + f_t, \quad u(0,x) = u_0(x), \\ &u_t(0,x) = u_1(x) = f(0,x) . \end{aligned} \tag{W}$$

If σ is "genuinely nonlinear" ($\sigma''(\xi) \not\equiv 0$, $\xi \in \mathbb{R}$), Lax [3] has shown that (W) fails to have global smooth solutions in time (even if $f_t \equiv 0$), no matter how smooth one takes the initial data due to the development of "shocks" (the first derivatives of u develop singularities in finite time due to the crossing of characteristics). Note that $a(t) \equiv a(0)$ is excluded by (1.11).

Nishida [8] has shown that for the nonlinear wave equation with frictional damping

$$u_{tt} + \alpha u_t = a(0)\sigma(u_x)_x, \quad u(0,x) = u_0(x), \qquad (W_\alpha)$$
$$u_t(0,x) = u_1(x), \qquad a(0) > 0 ,$$

the dissipation term αu_t, $\alpha > 0$, precludes the development of shocks if the initial data are sufficiently smooth and "small". The proof rests on the concept of Riemann invariant and is restricted to one space dimension. For a generalization of Nishida's method to the forced equation (W_α) we refer to Nohel [9]. As will be seen in (3.3) below, (HF), under physically reasonable assumptions, is equivalent to a variant of (W_α) with an additional memory term which makes our result for (HF) (Theorem 2.1) plausible. The Nishida approach applied to (HF) (necessarily restricted to one space dimension) was studied by Mac Camy [6].

II. STATEMENT OF RESULTS

We make the following assumptions. Concerning σ let

$$\sigma \in C^3(\mathbb{R}), \ \sigma(0) = 0, \quad \sigma'(0) > 0 , \qquad (\sigma)$$

the first for technical reasons, the others on physical grounds (recall that in the linear version of (HF) $\sigma(u_x) = u_x$). Concerning the kernel a assume

(i) $a \in B^{(3)}[0,\infty)$,

(ii) $a(0) > 0, \quad a'(0) < 0$,

(iii) $t^j a^{(m)} \in L^1(0,\infty) \quad (j,m = 0,1,2,3)$, (a)

(iv) letting $\hat{a}(i\eta) = \int_0^\infty e^{-i\eta t} a(t)dt$,

(2.1) $\mathrm{Re}\,\hat{a}(i\eta) > 0 \qquad (\eta \in \mathbb{R})$,

where $B^{(m)}[0,\infty)$ is the set of functions with bounded continuous derivatives up to and including order m. From (1.11) above the conditions $a \in C^1(0,\infty)$, $a \in L^1(0,\infty)$, $a(0) > 0$, $a'(0) < 0$, $\hat{a}(0) = \int_0^\infty a(t)dt > 0$ are reasonable on physical grounds; the remaining ones are needed for technical reasons of the analysis. See additional remarks on alternatives to the frequency domain condition (2.1) following Lemma 3.3 below. Concerning the forcing term f we assume (essentially for technical reasons)

$$f, f_t, f_x, f_{tt}, f_{tx}, f_{xx}, f_{ttt}, f_{ttx}, f_{txx} \in L^2(0,\infty;L^2(\mathbb{R})). \tag{f}$$

The initial datum u_0 is assumed to satisfy

$$u_{0x}, u_{0xx}, u_{0xxx} \in L^2(\mathbb{R}) . \tag{u_0}$$

Note that in (u_0) no assumption is explicitly made about $u_0(x)$; however, for the particular physical problem one would also have to require $u_0 \in L^2(\mathbb{R})$ in order that f defined by (1.7) satisfy (f). Our result concerning (HF) is (see [1; Theorem 4.1]):

Theorem 2.1. *Let the assumptions* (σ), (a), (f), (u_0) *hold. If the* $H^2(\mathbb{R})$ *norm of* u_{0x} *and the* $L^2([0,\infty);L^2(\mathbb{R}))$ *norms of* f *and its derivatives listed in* (f) *are sufficiently small, then the Cauchy problem* (HF) *has a unique solution* $u \in C^2([0,\infty) \times \mathbb{R})$ *with the following properties:*

(i) $u_t, u_x, u_{tt}, u_{tx}, u_{xx}, u_{ttt}, u_{ttx}, u_{txx}, u_{xxx} \in L^\infty([0,\infty);L^2(\mathbb{R}))$,

(ii) $u_t, u_{tt}, u_{tx}, u_{xx}, u_{ttt}, u_{ttx}, u_{txx}, u_{xxx} \in L^2([0,\infty);L^2(\mathbb{R}))$,

(iii) $u_t(t,\cdot), u_{tt}(t,\cdot), u_{tx}(t,\cdot), u_{xx}(t,\cdot) \to 0$ in $L^2(\mathbb{R})$ as $t \to \infty$,

(iv) $u_t(t,x), u_x(t,x), u_{tt}(t,x), u_{tx}(t,x), u_{xx}(t,x) \to 0$

as $t \to \infty$ uniformly on $\mathbb{R}$.

We remark that conclusions (iii), (iv) are easy consequences of (i), (ii). It also follows from the proof that the solution u has a finite speed of propagation.

We also note that the results of Theorem 2.1 hold (with essentially the same proofs) for the following two problems of heat flow in a body on the interval $[0,1]$ (see [1, Theorem 6.1]):

(i) (HF) on $(0,\infty) \times (0,1)$ with homogeneous Neumann boundary conditions at $x = 0$ and $x = 1$, and with $u_0(x)$ prescribed on $[0,1]$;

(ii) (HF) on $(0,\infty) \times (0,1)$ with homogeneous Dirichlet boundary conditions at $x = 0$ and $x = 1$, and with $u_0(x)$ prescribed on $[0,1]$.

In both problems assumptions (σ) and (a) are unchanged while assumptions (f) and (u_0) hold in $L^2[(0,1) : L^2(\mathbb{R})]$ and in $L^2(0,1)$ respectively. For problem (ii) one adds the assumption $f(t,0) = f(t,1) = 0$.

For a version of (HF) in two space dimensions and with a similar but technically more involved proof we refer the reader to [1, Theorem 7.1].

We note also that (HF) is of the abstract form

$$\begin{aligned} &u'(t) + \int_0^t a(t-\tau)Au(\tau)d\tau = f(t) \quad (0 < t < \infty) \\ &u(0) = u_0 \quad , \end{aligned} \tag{A}$$

where A is the nonlinear operator $Au = -\frac{\partial}{\partial x}\sigma(u_x)$ plus appropriate conditions at $\pm\infty$ or suitable boundary conditions at $x = 0$ and 1. Such abstract problems have been recently studied by Londen [4],[5] for a class of kernels $a(\cdot)$ which are positive, decreasing, convex on $[0,\infty)$ and which satisfy the condition $a'(0+) = -\infty$ which is crucial for his technique. In addition, the solution obtained by Londen is not sufficiently regular, and no comparable decay results are obtained.

Finally, we observe that a comparison of Theorem 2.1 and

of its proof with the results and method of proof by Mac Camy [6] shows that our approach is more direct, not restricted to one space dimension, and yields more general results (see additional remarks following Lemma 3.3).

III. OUTLINE OF PROOF OF THEOREM 2.1

To simplify the exposition we shall assume that $f \equiv 0$ in (HF), and we refer the reader to [1] for the technically more involved treatment resulting from $f \not\equiv 0$; the method is unaltered by this simplification.

a. Transformation of (HF). Differentiation of (HF) with respect to t brings it to the form

$$\begin{aligned} u_{tt}(t,x) &= a(0)\sigma(u_x)_x(t,x) + (a' * \sigma(u_x)_x)(t,x) \\ u(0,x) &= u_0(x), \quad u_t(0,x) \equiv 0 \qquad (x \in \mathbb{R}) . \end{aligned} \tag{3.1}$$

We transform (3.1) to an equivalent form by observing that this equation is linear in $y = \sigma(u_x)_x$. Define the resolvent kernel k of a' by the equation

$$a(0)k(t) + (a' * k)(t) = -\frac{a'(t)}{a(0)} \qquad (0 \leq t < \infty) ; \tag{k}$$

since $a(0) > 0$, assumptions a(i) imply that k is uniquely defined and $k \in C^2[0,\infty)$ (k has other properties - see Lemma 3.3 below). By the variation of constants formula for linear Volterra equations one has

$$a(0)y + a' * y = \quad \Longleftrightarrow \; y = \varphi/a(0) + k * \varphi$$

for any given function φ. Applying this to (3.1) one sees that if u is a classical solution of (HF) with $f \equiv 0$, then u satisfies the equation

$$u_{tt} + a(0)k * u_{tt} = a(0)\sigma(u_x)_x .$$

Performing an integration by parts and using $u_t(0,x) \equiv 0$

shows that (HF) with $f \equiv 0$ is equivalent to the Cauchy problem

$$u_{tt}(t,x) + a(0)\frac{\partial}{\partial t}(k*u_t)(t,x) = a(0)\sigma(u_x(t,x))_x \qquad (0 < t < \infty,\ x \in \mathbb{R}) \tag{3.2}$$
$$u(0,x) = u_0(x),\ u_t(0,x) \equiv 0 \qquad (x \in \mathbb{R}) .$$

Another important equivalent form of (HF) with $f \equiv 0$ resulting from (3.2) is

$$u_{tt}(t,x) + a(0)k(0)u_t(t,x) = a(0)\sigma(u_x(t,x))_x - a(0)(k'*u_t)(t,x) \qquad (0 < t < \infty,\ x \in \mathbb{R}) \tag{3.3}$$
$$u(0,x) = u_0(x),\ u_t(0,x) \equiv 0 \qquad (x \in \mathbb{R}) .$$

Since $a(0)k(0) = -a'(0) > 0$, (3.3) suggests the dissipative mechanism induced by the memory term in (HF) and the relationship with the damped wave equation (W_α). We remark that if $f \not\equiv 0$ in (HF) one adds the forcing term $\Phi(t,x) = f_t(t,x)/a(0) + (k*f_t)(t,x) + a(0)k(t)f(0,x)$ to (3.2) and (3.3), and one replaces the zero initial condition by $u_t(0,x) = f(0,x)$.

The proof of Theorem 2.1 is carried out in two stages: (i) A suitable local existence and uniqueness result is established. (ii) A priori estimates are established to continue the local solution; these will at the same time yield conclusion (i), (ii) of Theorem 2.1.

b. Local Theory. We shall make the temporary additional assumption concerning σ:

$$\text{there exists } p_0 > 0 \text{ such that } \sigma'(\xi) \geq p_0 > 0 \qquad (\xi \in \mathbb{R}) \tag{σ^*}$$

Proposition 3.1. *Let the assumptions* (σ), (σ^*), (u_0) *hold, and let* $k', k'' \in C[0,\infty) \cap L^1(0,\ )$. *Then the Cauchy problem* (3.2) (resp. (3.3)) *has a unique solution* $u \in C^2([0,T_0) \times \mathbb{R})$ *on a maximal interval* $[0,T_0) \times \mathbb{R}$, $T_0 \leq +\infty$, *such that for* $T \in [0,T_0)$ *one has*

(i) *all derivatives of* u *of orders one to three inclusive* $\in L^\infty([0,T]; L^2(\mathbb{R}))$;

(ii) *if* $T_0 < \infty$, *then*

$$\int_{-\infty}^{\infty} [u_t^2(t,x) + u_x^2(t,x) + u_{tt}^2(t,x) + \dots + u_{xxx}^2(t,x)]dx \to \infty \quad \text{as} \quad t \to T_0^- .$$

We remark that the property of finite speed of propagation of solutions of (HF) is an easy consequence of the proof of Proposition 3.1.

The proof uses the Banach fixed point theorem. Let $X(M,T)$ be the set of functions $u \in C^2([0,T] \times \mathbb{R})$ for any $T > 0$ such that $u(0,x) = u_0(x)$, $u_t(0,x) = 0$ and such that

(i) $u_t, u_x, u_{tt}, \dots, u_{xxx} \in L^\infty([0,T]; L^2(\mathbb{R}))$ and

(ii) $\sup_{[0,T]} \int_{-\infty}^{\infty} [u_t^2(t,x) + u_x^2(t,x) + u_{tt}^2(t,x) + \dots + u_{xxx}^2(t,x)]dx \leq M^2$.

Note that $X(M,T)$ is not empty if M is sufficiently large, and that if $u \in X(M,T)$, then

(iii) $\sup_{[0,T]\times\mathbb{R}} \{|u_t(t,x)|, |u_x(t,x)|, |u_{tt}(t,x)|, |u_{tx}(t,x)|, |u_{xx}(t,x)|\} \leq M$.

Let S be the map: $X(M,T) \to C^2([0,T] \times \mathbb{R})$ which carries a function $v \in X(M,T)$ into the solution of the linear Cauchy problem (see (3.3) for motivation)

$$\begin{aligned} u_{tt}(t,x) + k(0)u_t(t,x) &= a(0)[\sigma'(v_x(t,x))u_{xx}(t,x) \\ &\quad - (k' * v_t)(t,x)] \qquad (0 < t < T,\ x \in \mathbb{R}) \end{aligned} \tag{3.4}$$

$$u(0,x) = u_0(x),\ u_t(0,x) = 0 .$$

Clearly a fixed point of S will be a solution of (3.2) (respectively (3.3)). To apply the Banach fixed point theorem to the map S one first shows (by an energy argument, for details see [1, Lemma 3.1]) that if M is sufficiently large and if T is sufficiently small, then S maps $X(M,T)$ into itself. One next equips $X(M,T)$ with the metric

$$\rho(u,\bar{u}) = \max_{[0,T]} \{\int_{-\infty}^{\infty} |(u_t(t,x) - \bar{u}_t(t,x))^2 + (u_x(t,x) - \bar{u}_x(t,x))^2]dx\}^{1/2} .$$

By the lower semicontinuity of norms under weak convergence in Banach space, $X(M,T)$ becomes a complete metric space. One then shows that for M sufficiently large and T sufficiently small the map S is a strict contraction of $X(M,T)$ and the proof of Proposition 3.1 is completed in a standard manner (for details see [1, Lemma 3.2].

If σ, k, u_0, u_1 are smoother, the solution becomes smoother. A precise regularity result which is needed for the a priori estimates is

Proposition 3.2. *Let the assumptions of Proposition 3.1 be satisfied. In addition, assume that*

$$\sigma \in C^4(\mathbb{R}) , \; u_{0xxxx} \in L^2(\mathbb{R}) \quad . \tag{3.5}$$

Then the solution u *of Proposition 3.1 has the addition property*

$$u_{tttt}, u_{tttx}, u_{ttxx}, u_{txxx}, u_{xxxx} \in L^\infty([0,T];L^2(\mathbb{R})) , \tag{3.6}$$

for every $T < T_0$, *where* $[0,T_0) \times \mathbb{R}$ *is the maximal interval of existence.*

For the proof see [1, Theorem 3.2].

c. A Priori Estimates and Continuation. We wish to show that the maximal interval $[0,T_0)$ of Proposition 3.1 is in fact $[0,\infty)$. Recall that the local theory assumes that (σ^*) is satisfied; this assumption will be removed. The a priori estimates will be deduced from equation (3.2) above. We shall restrict the range of $u_x(t,x)$ for a local solution u to the set on which $\sigma'(\cdot) > 0$; choose $c_0 > 0$ such that

$$\sigma'(w) \geq p_0 > 0, \quad w \in [-c_0,c_0] \quad . \tag{3.7}$$

We wish to show that there exists a constant $\mu > 0$, $\mu < c_0$, depending on p_0, $\int_0^\infty |k'(t)|dt$, $\max_{[-c_0,c_0]} \{|\sigma'(\cdot)|, |\sigma''(\cdot)|, |\sigma'''(\cdot)|\}$, but not on $T > 0$ such that if the local solution u of (3.1) satisfies

$$\sup_{0\le t<T, x\in\mathbb{R}} \{|u_t(t,x)|, |u_x(t,x)|, |u_{tx}(t,x)|, |u_{xx}(t,x)|\} \le \mu , \qquad (\mu^*)$$

then certain functionals of the solution u are *controllably small* (i.e. these functionals can be made arbitrarily small by choosing the initial data sufficiently small in the appropriate H norms). More precisely, the result of the a priori estimates which follow is that if the assumptions of Theorem 2.1 hold (with $f \equiv 0$) and if the $H^2(\mathbb{R})$ norm of u_{0x} is sufficiently small, then for as long as the local solution u of (3.2) satisfies the condition (μ^*) for $\mu > 0$ sufficiently small, the condition

$$\int_{-\infty}^{\infty} [u_t^2(s,x) + u_x^2(s,x) + u_{tt}^2(s,x) + \dots + u_{xxx}^2(s,x)]dx + \int_0^s \int_{-\infty}^{\infty} [u_t^2(t,x) + u_{tt}^2(t,x) + \dots + u_{xxx}^2(t,x)]dxdt \le \mu^2 \qquad (0 \le s < T) \qquad (\mu^{**})$$

is satisfied. The inequality (μ^{**}) in turn implies that condition (μ^*) holds and the cycle closes in a standard manner using Proposition 3.1. Thus the maximal interval of existence of the solution $u(t,x)$ is $[0,\infty) \times \mathbb{R}$ and (μ^{**}) holds for $0 \le s < \infty$. This proves properties (i) and (ii) and Theorem 2.1.

To establish the a priori inequality (μ^{**}) we shall work with the equivalent form (3.2) of (HF) and we require some additional properties of the resolvent kernel k.

Lemma 3.3. *Let assumptions* (a) *be satisfied and let* k *be the resolvent kernel of* a' *defined by equation* (k). *Then*

(i) $k \in B^2[0,\infty)$, $k(0) = -\dfrac{a'(0)}{[a(0)]^2} > 0$,

(ii) $k(t) = k_\infty + K(t)$, $k_\infty = \dfrac{1}{\hat{a}(0)} > 0$; $K^{(m)} \in L^1(0,\infty)$, $m = 0,1,2$,

(iii) *for every* $T > 0$ *and for every* $v \in L^2[0,T]$ *there exists a number* $\gamma > 0$ *such that*

$$\int_0^T v(t) \frac{d}{dt} (k*v)(t)dt \ge \gamma \int_0^T v^2(t)dt . \qquad (3.8)$$

Remarks on Lemma 3.3. We refer to [6, Lemma 3.1], for the proof of Lemma 3.3 and to [1, Lemma 2.1] for some comments and corrections of that proof. Here we make some additional comments concerning the energy inequality (3.8) which is of independent interest. If, as is the case here $k' \in L^1(0,\infty)$, the inequality (3.8) is derived by the following simple argument (see the method of [10, Theorem 1]. Extend k' evenly for $t < 0$, and let

$$v_T(t) = \begin{cases} v(t) & \text{if } t \in [0,T] \\ 0 & \text{otherwise .} \end{cases}$$

Then

$$\int_0^T v(t)\,\frac{d}{dt}\,(k*v)(t)dt = k(0)\int_0^T v^2(t)dt + \int_0^T v(t)(k'*v)(t)dt$$

$$= k(0)\int_0^T v^2(t)dt + \frac{1}{2}\int_0^T v(t)\int_0^T k'(t-\tau)v(\tau)d\tau dt$$

$$= k(0)\int_{-\infty}^{\infty} v_T^2(t)dt + \frac{1}{2}\int_{-\infty}^{\infty} v_T(t)\int_{-\infty}^{\infty} k'(t-\tau)v_T(\tau)d\tau dt \ .$$

Letting $\tilde{v}_T(\eta) = \int_{-\infty}^{\infty} e^{-i\eta t} v_T(t)dt$, $(\eta \in \mathbb{R})$, the Parseval and convolution theorems give

$$\int_0^T v(t)\,\frac{d}{dt}\,(k*v)(t)dt = \frac{k(0)}{2\pi}\int_{-\infty}^{\infty} |\tilde{v}_T(\eta)|^2 d\eta$$

$$+ \frac{1}{4\pi}\int_{-\infty}^{\infty} |\tilde{v}_T(\eta)|^2 \tilde{k}'(\eta)d\eta \ .$$

But $\tilde{k}'(\eta) = 2\mathrm{Re}\,\widehat{k'(i\eta)}$, where $\hat{}$ is the Laplace transform, and $\mathrm{Re}\,\widehat{k'(i\eta)} = \mathrm{Re}[i\eta\hat{k}(i\eta) - k(0)]$. Therefore,

$$\int_0^T v(t)\,\frac{d}{dt}\,(k*v)(t)dt = \frac{1}{2\pi}\int_{-\infty}^{\infty} |\tilde{v}_T(\eta)|^2 \mathrm{Re}[i\eta\hat{k}(i\eta)]d\eta \ .$$

Now an easy calculation from equation (k) yields

$$\mathrm{Re}\; i\eta\hat{k}(i\eta) = \mathrm{Re}\,\frac{1}{\hat{a}(i\eta)} = \frac{\mathrm{Re}\;\hat{a}(i\eta)}{|\hat{a}(i\eta)|^2} \ .$$

Thus, to prove the inequality (3.8) it suffices to establish the existence of $\gamma > 0$ such that

$$\frac{\operatorname{Re} \hat{a}(i\eta)}{|\hat{a}(i\eta)|^2} \geq \gamma > 0 \qquad (\eta \in \mathbb{R}) .$$

This is precisely what is done with the aid of assumptions (a) in [6, Lemma 3.1]), although Mac Camy's derivation of (3.8) is different from the above and unnecessarily complicated. The above suggests that the frequency domain condition (2.1) in assumptions (a) should be replaced by the condition

$$\text{there exists } \alpha > 0 \text{ such that} \quad \alpha \operatorname{Re} \hat{a}(i\eta) \geq |\hat{a}(i\eta)|^2 \quad (\eta \in \mathbb{R}) . \tag{S}$$

The importance of condition (S) was first recognized by O. J. Staffans [11] in a different context. He showed [11, Theorem 2] that condition (S) is satisfied for at least two classes of kernels of importance for the problem (HF)

(i) $a \in L^1(0,\infty) \cap BV[0,\infty)$ and a *strongly positive* on $[0,\infty)$, i.e. there exists an $\varepsilon > 0$ such that $\operatorname{Re} \hat{a}(i\eta) \geq \varepsilon(1 + \eta^2)^{-1}$,

and

(ii) $a \in L^1(0,\infty)$ and a and $-a'$ are nonnegative and convex on $(0,\infty)$ (here $a(0+) = -a'(0+) = +\infty$ are allowed).

Staffans also gives an example of a kernel which is a positive definite measure μ satisfying (S), but such that μ is *not* strictly positive $(\operatorname{Re} \hat{\mu}(i\eta) > 0)$.

Incidentally, it is not hard to show that if $a \in L^1(0,\infty)$, $a(0) > 0$, and a is either strongly positive on $[0,\infty)$ or a satisfies condition (S), then $a'(0) < 0$. It is also important to notice that if a satisfies (S) and $\hat{a}(i\eta)$ is defined as a function (e.g. if $a \in L^1(0,\infty)$), then $\hat{a}(i\eta)$ can vanish at most on a set of measure zero on the imaginary axis.

The above considerations suggest that the energy inequality (3.8) is true under other useful conditions which are much more general than assumptions (a), and such results are now being obtained.

The remainder of this section is devoted to the derivation of the a priori estimates which imply (μ^{**}). Define

$$W(w) = \int_0^w \sigma(\xi)\,d\xi \geq \frac{p_0}{2} w^2 \qquad w \in [-c_0, c_0] \quad , \tag{3.9}$$

where the inequality follows from (3.7). Let u be a local solution of (3.2) satisfying (μ^*) for some $T > 0$ and $0 < \mu < c_0$. Multiply (3.2) by u_t and integrate over $[0,s] \times \mathbb{R}$. Using (3.9) and Lemma 3.3 (iii) one obtains the estimate (recall we are doing the special case $f \equiv 0$ in (HF) so that $u_t(0,x) \equiv 0$)

$$\frac{1}{2}\int_{-\infty}^{\infty} u_t^2(s,x)\,dx + a(0)\int_{-\infty}^{\infty} W(u_x(s,x))dx + \gamma\int_0^s\int_{-\infty}^{\infty} u_t^2(t,x)\,dxdt$$

$$\leq a(0)\int_{-\infty}^{\infty} W(u_x(0,x))\,dx \qquad (0 \leq s < T) \ , \tag{3.10}$$

where an integration by parts with respect to x was carried out in the term

$$\int_0^s\int_{-\infty}^{\infty} \sigma(u_x(t,x))_x u_t(t,x)\,dxdt \ ,$$

followed by an application of (3.9). It follows from (3.9), (3.10) that

$$\int_{-\infty}^{\infty} u_t^2(s,x)\,dx + a(0)p_0\int_{-\infty}^{\infty} u_x^2(s,x)\,dx + 2\gamma\int_0^s\int_{-\infty}^{\infty} u_t^2(t,x)\,dxdt$$

$$\leq 2a(0)\int_{-\infty}^{\infty} W(u_{0x}(x))\,dx \qquad (0 \leq s < T) \ . \tag{3.11}$$

Thus, for as long as (μ^*) holds with $\mu < c_0$, the quantities

$$\int_{-\infty}^{\infty} u_t^2(s,x)\,dx, \ \int_{-\infty}^{\infty} u_x^2(s,x)\,dx, \ \int_0^s\int_{-\infty}^{\infty} u_t^2(t,x)\,dxdt$$

are controllably small, uniformly on $[0,T)$.

We now derive two additional estimates, the first by dif

ferentiating (3.2) with respect to t, multiplying by $u_{tt}(t,x)$ and integrating over $[0,s] \times \mathbb{R}$, the second by differentiating (3.2) with respect to x, multiplying by $u_{tx}(t,x)$ and integrating over $[0,s] \times \mathbb{R}$. Following the procedure in obtaining (3.10), and noting that since $u_t(0,x) \equiv 0$ one now has

$$\frac{\partial^2}{\partial t^2}(k*u_t)(t,x) = \frac{\partial}{\partial t}(k*u_{tt})(t,x) ,$$

we obtain

$$\begin{aligned}
&\frac{1}{2}\int_{-\infty}^{\infty} u_{tt}^2(s,x)\,dx + \frac{a(0)}{2}\int_{-\infty}^{\infty} \sigma'(u_x(s,x))u_{tx}^2(s,x)\,dx \\
&\quad + a(0)\gamma\int_0^s\int_{-\infty}^{\infty} u_{tt}^2(t,x)\,dxdt \le \frac{1}{2}\int_{-\infty}^{\infty} u_{tt}^2(0,x)\,dx \\
&\quad + \frac{a(0)}{2}\int_0^s\int_{-\infty}^{\infty} \sigma''(u_x(t,x))u_{tx}^3(t,x)\,dxdt \qquad (0 \le s < T), \qquad (3.12)
\end{aligned}$$

and

$$\begin{aligned}
&\frac{1}{2}\int_{-\infty}^{\infty} u_{tx}^2(s,x)\,dx + \frac{a(0)}{2}\int_{-\infty}^{\infty} \sigma'(u_x(s,x)u_{xx}^2(s,x)\,dx \\
&\quad + a(0)\gamma\int_0^s\int_{-\infty}^{\infty} u_{tx}^2(t,x)\,dxdt \le \frac{a(0)}{2}\int_{-\infty}^{\infty} \sigma'(u_{0x}(x))u_{0xx}^2(x)\,dx \\
&\quad + \frac{a(0)}{2}\int_0^s\int_{-\infty}^{\infty} \sigma''(u_x(t,x))u_{tx}(t,x)u_{xx}^2(t,x)\,dxdt \qquad (3.13) \\
&\qquad (0 \le s < T) .
\end{aligned}$$

We add up (3.12), (3.13) and we claim that in the resulting inequality, and as long as (μ^*) holds with μ sufficiently small, each term on the right-hand side is either controllably small or can be majorized by the sum of such a quantity and a quantity that is dominated by one of the dissipation terms. Indeed, since from (3.3)

$$u_{tt}(0,x) = a(0)\sigma(u_{0x})_x , \qquad (3.14)$$

the $L^2(\mathbb{R})$ norm of $u_{tt}(0,x)$ is controllably small. The two space-time integrals in (3.12), (3.13) are majorized as follows

$$|\frac{1}{2}\int_0^s\int_{-\infty}^{\infty}\sigma''(u_x(t,x))u_{tx}^3(t,x)\,dxdt|$$

$$\leq \frac{\mu}{2}\max_{[-c_0,c_0]}|\sigma''(\cdot)|\int_0^s\int_{-\infty}^{\infty}u_{tx}^2(t,x)\,dxdt\ , \tag{3.15}$$

$$|\frac{1}{2}\int_0^s\int_{-\infty}^{\infty}\sigma''(u_x(t,x))u_{tx}(t,x)u_{xx}^2(t,x)\,dxdt|$$

$$\leq \frac{\mu}{2}\max_{[-c_0,c_0]}|\sigma''(\cdot)|\int_0^s\int_{-\infty}^{\infty}u_{xx}^2(t,x)\,dxdt\ . \tag{3.16}$$

To estimate the integral on the right-hand side of (3.16) we have from (3.3)

$$a(0)\sigma'(u_x(t,x))u_{xx}(t,x) = u_{tt}(t,x)$$

$$+\ a(0)k(0)u_t(t,x) + a(0)(k'*u_t)(t,x)\ , \tag{3.17}$$

and this yields from (3.7) and standard estimates

$$a^2(0)p_0^2\int_0^s\int_{-\infty}^{\infty}u_{xx}^2(t,x)\,dxdt \leq 4\int_0^s\int_{-\infty}^{\infty}u_{tt}^2(t,x)\,dxdt$$

$$+\ 4a^2(0)k^2(0)\int_0^s\int_{-\infty}^{\infty}u^2(t,x)\,dxdt$$

$$+\ 4a^2(0)(\int_0^{\infty}|k'(t)|dt)^2\int_0^s\int_{-\infty}^{\infty}u_t^2(t,x)\,dxdt\ . \tag{3.18}$$

The restrictions imposed on μ are expressed in terms of parameters fixed a priori. For example, the estimate (3.15) and the desire to absorb this term into the corresponding dissipation term in (3.13), impose the restriction $\mu\max_{[-c_0,c_0]}|\sigma''(\cdot)| \leq \frac{\gamma a(0)}{4}$. The combination of (3.12)-(3.18) then yields that the quantities $\int_{-\infty}^{\infty}u_{tt}^2(s,x)\,dx$, $\int_{-\infty}^{\infty}u_{tx}^2(s,x)\,dx$, $\int_{-\infty}^{\infty}u_{xx}^2(s,x)\,dx$, $\int_0^s\int_{-\infty}^{\infty}u_{tt}^2(t,x)\,dxdt$, $\int_0^s\int_{-\infty}^{\infty}u_{tx}^2(t,x)\,dxdt$, and $\int_0^s\int_{-\infty}^{\infty}u_{xx}^2(t,x)\,dxdt$ are controllably small uniformly on $[0,T)$, provided (μ^*) holds for μ sufficiently small.

To get the final set of estimates one assumes temporarily that the additional smoothness assumption (3.5) of Proposition 3.2 is satisfied, so that u satisfies (3.6). We form the second derivative of equation (3.2) first with respect t, multiply by $u_{ttt}(t,x)$ and then integrate over $[0,s] \times \mathbb{R}$, secondly with respect to t and x, multiply by $u_{ttx}(t,x)$, and then integrate over $[0,s] \times \mathbb{R}$. Following the above procedure and using equation (3.3) to compute $u_{ttt}(0,x)$ and $u_{ttx}(0,x)$ a tedious but straightforward calculation (see [1], section 4) shows that the quantities $\int_{-\infty}^{\infty} u_{ttt}^2(s,x)ds$,

$\int_{-\infty}^{\infty} u_{ttx}^2(s,x)dx$, $\int_{-\infty}^{\infty} u_{txx}^2(s,x)dx$, $\int_{-\infty}^{\infty} u_{xxx}^2(s,x)dx$,

$\int_0^s\int_{-\infty}^{\infty} u_{ttt}^2 dxdt$, $\int_0^s\int_{-\infty}^{\infty} u_{ttx}^2 dxdt$, $\int_0^s\int_{-\infty}^{\infty} u_{txx}^2 dxdt$,

$\int_0^s\int_{-\infty}^{\infty} u_{xxx}^2 dxdt$ are controllably small uniformly on $[0,T)$, provided (μ^*) holds for μ sufficiently small. Moreover, the detailed estimates show they depend solely on parameters which do not involve the additional assumption (3.5). Therefore, a simple density argument can be used to remove the extraneous assumption (3.5).

Combining all the controllable estimates obtained above yields the inequality (μ^{**}), for any local solution u satisfying (μ^*) for $\mu > 0$ sufficiently small. This completes the sketch of the proof of Theorem 2.1.

REFERENCES

1. Dafermos, C. M. and Nohel, J. A., Energy methods for nonlinear hyperbolic Volterra integrodifferential equations, *Comm. in P.D.E.* (to appear).
2. Gurtin, M.E. and Pipkin, A. C., A general theory of heat conduction with finite wave speeds, *Arch. Rat. Mech. Anal.* *31*, 113-126 (1968).
3. Lax, P.D., Development of singularities of solutions of nonlinear hyperbolic partial differential equations, *J. Math. Phys.* *5*, 611-613 (1964).

4. Londen, S. O., An existence result for a Volterra equation in Banach space, *Trans. Amer. Math. Soc.* *235*, 285-305 (1978).

5. Londen, S. O., An integrodifferential Volterra equation with a maximal monotone mapping, *J. Differential Equations* (to appear).

6. Mac Camy, R. C., An integro-differential equation with applications in heat flow, *Q. Appl. Math.* *35*, 1-19 (1977).

7. Mac Camy, R. C., A model for one-dimensional nonlinear viscoelasticity, Ibid *35*, 21-33 (1977).

8. Nishida, T., Global smooth solutions of the second-order quasilinear wave equations with the first-order dissipation (unpublished).

9. Nohel, J. A., A forced quasilinear wave equation with dissipation, Proceedings of EQUADIFF 4, Lecture Notes, Springer Verlag (to appear).

10. Nohel, J. A. and Shea, D. F., Frequency domain methods for Volterra equations, *Advances in Math.* *22*, 278-304 (1976).

11. Staffans, O. J., An inequality for positive definite kernels, *Proc. Amer. Math. Soc.* *58*, 205-210 (1976).

LINEARITY AND NONLINEARITY IN THE THEORY OF G-CONVERGENCE

Livio Clemente Piccinini[1,2]

Istituto di Matematica
Facoltà di Scienze Statistiche
Università di Padova, Padova, Italy

0. INTRODUCTION

In this paper we discuss the main lines of the theory of G-convergence for ordinary differential equations. We deal in particular with the various definitions of G-convergence, with the equivalence properties and with the phenomena of non-linearity that can arise in the theory. An informal exposition of a part of this paper has been object of a lecture held at the Colloquio Internazionale sui metodi recenti di analisi non lineare e applicazioni, Roma, May 1978.

Even if the theory, as it was stated originally in [4],[5] and [6] (and also in [3]), does not depend on De Giorgi's abstract definition (see [1]), it seems better to unify the different approachs starting from De Giorgi's abstract definition. This paper is divided in the following sections: 1 and

---

[1]Present address: Facoltà di Scienze Statistiche, Via VIII Febbraio 9, Padova.

[2]Lavoro eseguito nell'ambito del G.N.A.F.A. del C.N.R.

ISBN 0-12-186280-1

2 for the definitions of G-convergence, 3 for an analysis of the problems that arise in the theory, 4 to 6 for the discussions of the single problems, 7 for an extension toward a general theory, 8 for a short discussion of the results of the general theory.

I. ABSTRACT DEFINITION OF G-CONVERGENCE

Instead of using the actual definition of [1], that requires as a prerequisite the knowledge of Γ-convergence, we give the following equivalent direct definition:

Abstract definition of G-convergence. *Suppose* (U,τ) *and* (V,ϑ) *are two topological spaces. We attach to* $U \times V$ *the product topology* $\tau \times \vartheta$. *We say that a sequence of correspondences* $\mathcal{C}_n \subset U \times V$ *G-converges to a correspondence* $\mathcal{C} \subset U \times V$ *if both the upper and the lower Kuratowski limit of* $\mathcal{C}_n$ *according to the topology* $\tau \times \vartheta$, *coincide and are equal to* $\mathcal{C}$.

Alternatively we can say, for all the spaces we use in this paper, that G-convergence holds if the two following properties hold:

$$\text{If } n_k \to \infty, \quad u_{n_k} \xrightarrow{\tau} u, \quad v_{n_k} \xrightarrow{\vartheta} v, \quad (u_{n_k}, v_{n_k}) \in \mathcal{C}_{n_k}, \quad \text{then} \quad (u,v) \in \mathcal{C}; \tag{1.1}$$

If $(u,v) \in \mathcal{C}$ then there exists two sequences

$$u_n \xrightarrow{\tau} u, \quad v_n \xrightarrow{\vartheta} v \quad \text{such that} \quad (u_n, v_n) \in \mathcal{C}_n. \tag{1.2}$$

Usually we refer to (1.1) as *limiting property* and to (1.2) as *approximability property*.

Remarks. In terms of Γ-convergence the definition given above corresponds to stating that the characteristic functions of $\mathcal{C}_n$ $\Gamma(N^{\pm}, \tau^{+}, \vartheta^{+})$-converge to the characteristic function of $\mathcal{C}$. For this point see [1] and [8].

The abstract definition of G-convergence is general in the sense that it is not bounded to differential equations. It is not the most general definition, because it corresponds only to a special case of Γ-convergence; it is possible to consider other signs for Γ-convergence (e.g. $\Gamma(N^+,\tau^+,\vartheta^+)$, $\Gamma(N^-,\tau^+,\vartheta^+)$) and the corresponding G-convergences; up to now this problem has not been particularly investigated.

II. G-CONVERGENCE FOR ORDINARY DIFFERENTIAL EQUATIONS

First of all we put some bounds on the class of the differential equations that are allowed in section 2 to 6. We fix a strip $[a,b]\times\mathbb{R}^s\subset\mathbb{R}^{s+1}$. Let $f : [a,b] \times \mathbb{R}^s \mapsto \mathbb{R}^s$ be a function satisfying Caratheodory's condition. We require also that there exists a function $C(t) \in L^\alpha([a,b])$ (1), for some $\alpha > 1$, such that

$$\|f(t,y)\| \leq C(t)\ (1 + \|y\|)\ . \tag{2.1}$$

The associated differential operator is

$$A[y] = y' - f(t,y)\ .$$

The conditions stated, as it is well known, ensure the existence of solutions to any Cauchy problem for the equation $A[y] = 0$ on the whole $[a,b]$. Of course condition (2.1) is not necessary to develop a theory of G-convergence, but in this case some local definitions are required; the same problem arises if a, b are allowed to be unbounded. Generalizations in these cases are treated in [3].

We consider only equations of normal type; some hints to the problems that arise in case of non normal equations are given in section 8.

In order to use the abstract definition of G-convergence we must now associate to a differential operator two topological spaces, roughly speaking a space U of tests and a space V of solutions, and a correspondence between the two spaces.

---

(1) Also any Morrey space different from $L^1([a,b])$ is suitable.

Is is easy to build a correspondence between U and V induced by the equation; the nontrivial part of the construction lies in the fact that the knowledge of the correspondence between U and V ought to be sufficient to determine uniquely (in general almost everywhere) the coefficient $f(t,y)$ of the vector field associated to the operator A. It is possible to renounce to this requirement, generating thus equivalence classes of differential equations, but up to now this generalization has not been studied.

Many cases of different choices of U and V are considered in [8]. Here we treat only the cases that are meaningful for all the equations satisfying condition (2.1).

Case of Cauchy data: we denote as usual by $v(t;u) \equiv v(t;t_o;y_o)$ a solution (or the unique solution if there is uniqueness) of the Cauchy problem $A[v] = 0$, $v(t_o) = y_o$, $(t_o,y_o) \equiv u$. We let $U = [a,b] \times \mathbb{R}^s$, $V = \{C_o([a,b])\}^s$ and we define the correspondence as

$$\mathcal{C}_n = \{(u_n,v_n) \in U \times V : v_n = v_n(t;u_n)\}, \quad n \in \bar{N}.$$

Case of perturbations: we let $U = \{L^\infty([a,b])\}^s$, $V = \{C_o([a,b])\}^s$ and

$$\mathcal{C}_n = \{(u_n,v_n) \in U \times V : A_n[v_n] = u_n\}, \quad n \in \bar{N}.$$

A reduction of the previous case is given by the

Case of parameters: $U = \mathbb{R}^s$, V as before, and

$$\mathcal{C}_n = \{(\lambda_n,v_n) \in U \times V : A_n[v_n] = \lambda_n\}.$$

The greatest reduction is obtained in the

Elementary case: $U = \{0\}$, V as before:

$$\mathcal{C}_n = \{(0,v_n) \in U \times V : A_n[v_n] = 0\}.$$

Some choices of the topologies are listed in the table below:

U	τ	V	ϑ	
$[a,b] \times \mathbb{R}^s$	discrete	$\{C_o([a,b])\}^s$	$\{L^\infty\}^s$	(2.3)
"	Euclidean	"	"	(2.4)
$\{L^\infty([a,b])\}^s$	$\{H^{-1,\infty}\}^s$	"	"	(2.5)
"	$\{L^\infty\}^s$	"	"	(2.6)
"	discrete	"	"	(2.7)
$\mathbb{R}^s$	Euclidean	"	"	(2.8)
"	discrete	"	"	(2.9)
$\{0\}$		"	"	(2.10)

The most interesting cases are (2.3) and (2.6). We recall explicitely the definition of G-convergence (2.6); when we omit the specification of the G-convergence we refer to this one:

Definition of G-convergence.

i) *Suppose* $n_k \to \infty$, $u_{n_k} \xrightarrow{(L^\infty)^s} u$, $v_{n_k} \xrightarrow{(L^\infty)^s} v$ *and* $A_{n_k}[v_{n_k}] = u_{n_k}$, *then* $A[v] = u$.

ii) *Suppose* $A[v] = u$, *then there exist two sequences* $u_n \xrightarrow{(L^\infty)^s} u$, $v_n \xrightarrow{(L^\infty)^s} v$ *such that* $A_n[v_n] = u_n$.

If conditions i) *and* ii) *hold we say that* A_n *G-converges to* A *and we write* $A_n \xrightarrow{G} A$.

The case (2.3) is of interest only when all the equations A_n and A have a unique solution for the Cauchy problem, that is A_n and A are regular. In this case definition (2.3) is equivalent to the following:

Definition of Cauchy G-convergence.

Suppose A_n, A *are regular differential operators in the interval* $[a,b]$. *We say that* A_n *Cauchy-G-converges to* A ($A_n \xrightarrow{C-G} A$) *if for any Cauchy data* $u \in [a,b] \times \mathbb{R}^s$, $v_n(t;u) \to v(t;u)$ *uniformly on* $[a,b]$.

In fact C-G-convergence implies (2.3): if $v_{n_k} \xrightarrow{(L^\infty)^s} \bar{v}$ and $v_{n_k} = v_{n_k}(t;u)$ then by C-G-convergence $v_{n_k}(t;u) \to v(t;u)$; hence $\bar{v} = v$, $\bar{v} = v(t;u)$. For property ii): let $\bar{v} = \bar{v}(t;u)$; by C-G-convergence $v_n(t;u) \xrightarrow{(L^\infty)^s} v(t,u)$. Since there is uniqueness of solutions $v(t;u) = \bar{v}(t;u)$. Conversely $v(t;u)$ by ii) is the limit of $v_n(t;u)$; since all of them are uniquely defined the statement follows.

Remark. We have not considered here boundary value problems for second order equations, which should correspond to the original definition of G-convergence given for elliptic equations (see [2]). Actually such an approach seems too restrictive for ordinary differential equations. Compare anyhow [8] on this point.

We are dealing here with systems in which no variable has a leading role; anyhow when a system represents an higher order single equation some variables (often half of them) have a leading role and the topology on V can be $(\tau_1)^{s/2} \times (\tau_2)^{s/2}$, where τ_2 is coarser than τ_1, often the trivial topology, that is convergence is required only for the first group of variables.

The reason why we do not attach to V L^p- topologies is that in compactness theorems (that are not treated in this paper) we wish to avoid degeneracy into transmission equations. The generalization to the case of transmission equations ought anyhow to be straightforward.

III. MAIN PROBLEMS

Four problems seem of some interest in this theory. The first is to find equivalences between the different definitions and to find simple criteria for G-convergence. To this problem we dedicate section 4 and part of section 5.

Anothere relevant problem leads to compactness: if a sequence of solutions of different equations converges in some topology, is the limit still a solution of a differential equation of the same kind? We do not deal in this paper with compactness problems; for results in this sense see [3], [4], [5], [8].

An interesting problem, not yet solved in general, lies in homogeneization. The non trivial known cases, that is cases in which homogeneization takes place with respect to the dependent variable (called non-linear homogeneization), are referred to in section 5 and 6. For the proofs see also [4] and [6].

The problem of G-approximating non regular equations by regular ones becomes trivial using the results of section 4. The general problem of G-approximating differential inequalities by regular equations requires on the contrary the general theory developped in section 7. Section 6 essentially points out the limits of the theory of section 2 with respect to this problem. The complication is caused by the fact that non-linear homogeneization and Peano phaenomenon must be dealt with at the same time. In section 8 we sketch a possible solution of the problem of approximation.

IV. G-CONVERGENCE AND CONVERGENCE OF COEFFICIENTS

The examples given in the next section show that the different definitions of G-convergence are not always equivalent between each other. In particular it is important to remark that in general G-convergence does not imply convergence of coefficients, nor a convergence of coefficients implies G-convergence to the corresponding operator. In this section we study those cases in which G-convergence is equivalent to the convergence of the coefficients, and some cases in which different definitions of G-convergence are equivalent.

We shall use often the following

Lemma. *If* $A_n \xrightarrow{G} A$ *according to any definition from* (2.5) *to* (2.10) *on the interval* $[a,b]$ *and* (2.1) *holds uniformly, then* $A_n \xrightarrow{G} A$ *according the same definition also on any subinterval* $[c,d]$ *of* $[a,b]$.

Proof. Property ii) of G-convergence is automatically satisfied.

As for property i) suppose $y_{n_k} \xrightarrow{\vartheta([c,d])} y$, $u_{n_k} \xrightarrow{\tau([c,d])} u$ and $A_{n_k}[y_{n_k}] = u_{n_k}$ on $[c,d]$. Take any solution of the problem

$$\begin{cases} A_{n_k}[y_{n_k}] = u_{n_k}(c) & ^{(1)} \\ \tilde{y}_{n_k}(c) = y_{n_k}(c) \end{cases} \quad a \le t \le c \qquad \begin{cases} A_{n_k}[\tilde{y}_{n_k}] = u_{n_k}(d) & ^{(1)} \\ \tilde{y}_{n_k}(d) = y_{n_k}(d) \end{cases} \quad d \le t \le b$$

and define it as $\tilde{y}_{n_k}(t) = y_{n_k}(t)$ for $c \le t \le d$.

By (2.1) $\tilde{y}_{n_k}$ is a sequence of uniformly bounded equicontinuous functions, thus there is a subsequence that converges uniformly on $[a,b]$. The right hand sides obviously converge according τ in $[a,b]$. Hence property i) of G-convergence on $[a,b]$ ensures the result. Q.E.D.

We say that a sequence of coefficients $f_n(t,y)$ *weakly converges* to a coefficient $f(t,y)$ if for any $z \in R^s$

$$f_n(\cdot,z) \xrightarrow{[H^{-1,\infty}]^s} f(\cdot,z),$$

that is for any $\varepsilon > 0$, there exists n_ε such that for $n > n_\varepsilon$ and any $t_1, t_2 \in [a,b]$ it holds

$$\left|\int_{t_1}^{t_2} [f_n(t,z) - f(t,z)]\right| dt < \varepsilon.$$

The following theorem states a relation between G-convergence and weak convergence of coefficients:

Theorem 1. *Suppose the following properties of equicontinuity and local boundedness are satisfied:*

there exists a function $C(t)$, $\int_a^b |C(t)|^\alpha dt = C^\alpha$ *for some* $\alpha > 1$, *and a function* $\psi(z)$ *increasing with respect to* z *such that*

$$\lim_{z \downarrow 0} \psi(z) = 0 \tag{4.1}$$

for which it holds for $n \in \bar{N}$

---

(1) In case (2.5) instead of $u_{n_k}(c)$, $u_{n_k}(d)$ let 0.

$$\|f_n(t,y_2) - f_n(t,y_1)\| \leq \psi(\|y_2 - y_1\|)$$
$$\forall\, t \in [a,b],\ \forall\ y_1, y_2 \in \mathbb{R}^s \tag{4.2}$$

$$\|f_n(t,0)\| \leq C(t) \qquad \forall\, t \in [a,b]. \tag{4.3}$$

Then the following statements are equivalent:

i) A_n *G-converges to* A

ii) A_n *converges according to* (2.5) *to* A

iii) *The corresponding coefficients weakly converge to the coefficient of* A .

Proof. First we prove that i) or ii) imply iii). Fix $z \in \mathbb{R}^s$. Suppose there is a constant M such that $\|\varphi_n\|_\tau \leq M$, $\varphi_n \in U$. Here τ denotes one of the topologies induced by $(H^{-1,\infty})^s$, $(L^\infty)^s$. Consider now a solution of the Cauchy problem

$$\begin{cases} y_n' = f_n(t,y_n) + \varphi_n(t) & n \in \bar{N} \\ y_n(t_o) = z & t_o \in [a,b]. \end{cases} \tag{4.4}$$

By (4.2) and (4.3) it is possible to determine a constant $M_1 = M_1(z,M)$ such that

$$\|y_n(t;t_o,z)\| \leq M_1 \qquad \forall\ t_o \in [a,b], \quad n \in \bar{N}. \tag{4.5}$$

For any $\varepsilon > 0$ it is possible to find η_ε such that

$$\psi(\eta_\varepsilon) < \varepsilon/(b-a). \tag{4.6}$$

It is also possible to find $\delta_o \equiv \delta_o(\varepsilon,z,M)$, $0 < \delta_o \leq 1$, such that

$$[\psi(M_1)\delta_o + C\delta_o^{\frac{\alpha-1}{\alpha}}] \leq \min\left(\frac{\eta_\varepsilon}{2},\varepsilon\right) . \tag{4.7}$$

Then for a solution of problem (4.4) with $\|\varphi_n\|_\tau < \frac{1}{2}\eta_\varepsilon$ it holds for $t_o \leq t \leq t_o + \delta$, $n \in \bar{N}$, $\delta \leq \delta_o$

$$\|y_n(t;t_o,z) - z\| \leq \int_{t_o}^{t} \|f_n(\tau,y_n(\tau))\| \, d\tau$$

$$+ \|\int_{t_o}^{t} \varphi_n(\tau)d\tau\| \leq \int_{t_o}^{t} \|f_n(\tau,y_n(\tau)) - f_n(\tau,0)\|d\tau$$

$$+ \int_{t_o}^{t} \|f_n(\tau,0)\| \, d\tau + \frac{1}{2}\eta_\varepsilon \leq \psi(M_1)\,\delta_o$$

$$+ C\delta^{\frac{\alpha-1}{\alpha}} + \frac{1}{2}\eta_\varepsilon \leq \eta_\varepsilon \quad . \tag{4.8}$$

We divide now the interval $[a,b]$ in N_ε subintervals $[t_{i-1},t_i]$, $i = 1,2,\ldots,N_\varepsilon$ of length $\delta < \delta_o$. We have thus for $n \in \bar{N}$, $a \leq t' < t'' \leq b$

$$\|\int_{t'}^{t''} [f_n(\tau,z) - f(\tau,z)]d\tau\| \leq \|\int_{t'}^{t_r}[\ldots]\, d\tau\|$$

$$+ \sum_{j=p+1}^{q} \|\int_{t_{j-1}}^{t_j} [\ldots]d\tau\| + \|\int_{t_q}^{t''} [\ldots]d\tau\| \tag{4.9}$$

The first and the last integral on the right hand side are less than 2ε by (4.7). Now let $y_j(t)$ be a solution of the Cauchy problem $y_j'(t) = f(t,y_j)$, $y_j(t_{j-1}) = z$. By the hypothesis, using the lemma, there exist two sequences $y_{jn} \xrightarrow{(L^\infty)\,s} y_j$ and $\varphi_{jn} \xrightarrow{\tau} 0$ such that $y_{jn}' = f_n(t,y_{jn}) + \varphi_{jn}$. Then it is possible to choose n_ε such that for $j = 1,2,\ldots,N_\varepsilon$ and $n > n_\varepsilon$ it holds

$$\|\varphi_{jn}\|_\tau < \min\left(\frac{\eta_\varepsilon}{2}, \frac{\varepsilon}{N_\varepsilon}\right) \tag{4.10}$$

$$\|y_{jn}-y_j\|_{L^\infty} < \min\left(\eta_\varepsilon, \frac{\varepsilon}{N_\varepsilon}\right) . \tag{4.11}$$

So we have

$$\|\Sigma \int_{t_{j-1}}^{t_j} [f_n(\tau,z) - f(\tau,z)]d\tau\| \leq \|\Sigma \int_{t_{j-1}}^{t_j} [f_n(\tau,z)$$

$$- f_n(\tau,y_j(\tau))]d\tau\| + \|\Sigma\int_{t_{j-1}}^{t_j} [f_n(\tau,y_j(\tau))$$

$$- f_n(\tau,y_{jn}(\tau))]d\tau\| + \|\Sigma\int_{t_{j-1}}^{t_j} [f_n(\tau,y_{jn}(\tau))$$

$$- f(\tau,y_j(\tau))]d\tau\| + \| \Sigma \int_{t_{j-1}}^{t_j} [f(\tau,y_j(\tau)) - f(\tau,z)]d\tau\| .$$

Since y_j satisfies (4.8) the first and the last summation are estimated by (4.6). The second summation is estimated by (4.11). The integrals that appear in the third summation satisfy the relation

$$\| \int_{t_{j-1}}^{t_j} [f_n(\tau,y_{jn}(\tau)) - f(\tau,y_j(\tau))]d\tau \|$$

$$\leq \| y_{jn}(t_j) - y_j(t_j) \| + \| y_{jn}(t_{j-1}) - y_j(t_{j-1}) \|$$

$$+ \| \int_{t_{j-1}}^{t_j} \varphi_{nj}(\tau)d\tau \|$$

and are estimated, using (4.10) and (4.11), by $2\varepsilon/N_\varepsilon$. Hence the whole third summation is estimated by 2ε. Collecting the estimates we have stated it follows i) or ii) implies iii).

We prove now the converse. Suppose that $n_k \to \infty$, $y_{n_k} \xrightarrow{(L^\infty)^s} y$, $\varphi_{n_k} \xrightarrow{\tau} \varphi$ and $A_{n_k}[y_{n_k}] = \varphi_{n_k}$. In order to prove that also $A[y] = \varphi$ we start from the equalities

$$y_{n_k}(t) = y_{n_k}(t_o) + \int_{t_o}^{t} f_{n_k}(\tau,y_{n_k}(\tau))d\tau + \int_{t_o}^{t} \varphi_{n_k}(\tau)d\tau .$$

Let $M_1 = \sup_{k \in N} \|y_{n_k}\|_{L^\infty}$; for any $\varepsilon > 0$ we determine η_ε, $\delta_o > 0$ in order that (4.6) and (4.7) are satisfied and moreover $\max \|\varphi\| \delta_o < \frac{1}{2}\eta_\varepsilon$. We divide $[a,b]$ in subintervals of length $\delta < \delta_o$.

Let now $z_j = y(t_{j-1})$, $j = 1,\dots,N_\varepsilon$. We choose also n_ε such that for $n_k > n_\varepsilon$ it holds (4.11) and

$$\|\varphi_{n_k} - \varphi\|_\tau < \min\left(\tfrac{1}{2}\eta_\varepsilon,\varepsilon\right) \tag{4.10'}$$

$$\| y_{n_k} - y \|_{L^\infty} < \min(\eta_\varepsilon,\varepsilon) \tag{4.11'}$$

$$\|\int_{t'}^{t''} [f_{n_k}(\tau,z_j) - f(\tau,z_j)]d\tau\| < \varepsilon/N_\varepsilon \tag{4.12}$$

for $a \leq t' < t'' \leq b$.

Then we have

$$\| y(t) - y(t_o) - \int_{t_o}^{t} f(\tau,y(\tau))d\tau - \int_{t_o}^{t} \varphi(\tau)d\tau \|$$

$$\leq \| y_{n_k}(t) - y_{n_k}(t_o) - \int_{t_o}^{t} f_{n_k}(\tau,y_{n_k}(\tau))d\tau - \int_{t_o}^{t} \varphi_{n_k}(\tau)d\tau \|$$

$$+ \| y(t) - y_{n_k}(t) \| + \| y(t_o) - y_{n_k}(t_o) \|$$

$$+ \| \int_{t_o}^{t} [f_{n_k}(\tau,y_{n_k}(\tau)) - f(\tau,y(\tau))]d\tau \|$$

$$+ \| \int_{t_o}^{t} [\varphi_k(\tau) - \varphi(\tau)]d\tau \|$$

$$\leq 0 + 4\varepsilon + \| \sum_{j=p+1}^{q} \int_{t_{j-1}}^{t_j} [f_{n_k}(\tau,y_{n_k}(\tau)) - f(\tau,y(\tau))]d\tau \| + \varepsilon ,$$

by (4.10)', (4.11)', (4.7). Splitting the last term we get then

$$\| \sum_j \int_{t_{j-1}}^{t_j} [f_{n_k}(\tau,y_{n_k}(\tau)) - f(\tau,y(\tau))]d\tau \|$$

$$\leq \| \sum_j \int_{t_{j-1}}^{t_j} [f_{n_k}(\tau,y_{n_k}(\tau)) - f_{n_k}(\tau,y(\tau))]d\tau \|$$

$$+ \| \sum_j \int_{t_{j-1}}^{t_j} [f_{n_k}(\tau,y(\tau)) - f_{n_k}(\tau,z_j)]d\tau \|$$

$$+ \| \sum_j \int_{t_{j-1}}^{t_j} [f_{n_k}(\tau,z_j) - f(\tau,z_j)]d\tau \|$$

$$+ \| \sum_j \int_{t_{j-1}}^{t_j} [f(\tau,z_j) - f(\tau,y(\tau))]d\tau \| .$$

From the definition of δ it follows, by (4.7),

$$\| y(\tau) - z_j \| \leq \eta_\varepsilon . \qquad (4.13)$$

So the first, the second and the fourth summation are estimated by (4.11)', (4.13) and (4.6), while the third summation is estimated by (4.12). This achieves the proof that $A[y] = \varphi$.

We prove now the approximability property of G-convergence. We remark that (4.8) ensures us that for any function $\varphi(t)$, $\|\varphi\|_{L^\infty} < K$ and for any y_o there exists a constant M such that any solution of the problem

$$\begin{cases} y_n' = f_n(t,y_n) + \varphi(t) & t \geq t_o \\ y_n(t_o) = y_o \end{cases}$$

satisfies $\|y_n\| \leq M = M(y_o,K)$. Furthermore it holds a modified version of (4.8)

$$\|y_n(t) - y_o\| < (\psi(2M) + K)|t-t_o| + C\,|t-t_o|^{\frac{\alpha-1}{\alpha}} \ . \tag{4.14}$$

Suppose now we must approximate a solution of the equation

$$y' = f(t,y) + \varphi(t) \ .$$

Let $M = 2 \max \|y(t)\|$. For a fixed $\varepsilon > 0$ $(\varepsilon < \frac{1}{2})$ we choose η such that

$$\psi(\eta) < \varepsilon \tag{4.15}$$

and δ_o such that $\delta_o > 0$ and

$$[\psi(2M) + 1]\delta_o + C\,\delta_o^{\frac{\alpha-1}{\alpha}} < \min(\eta, \tfrac{\varepsilon}{2}) \ . \tag{4.16}$$

We divide $[a,b]$ in subintervals of lenght $\delta < \delta_o$. Let $y_j = y(t_j)$. On each subinterval we construct functions $h_n(t)$ such that the following functions y_n satisfy the conditions $y_n(t_j) = y_j$ and the equations $y_n' = f_n(t,y_n) + \varphi(t) + h_n(t)$. We discuss the construction on a single subinterval, say $[t_o,t_1]$. We can determine a constant H such that (calling $f(t,y) + \varphi(t) = g(t,y)$)

$$z_o' = \frac{1}{\delta}\int_{t_o}^{t_1} g(\sigma,y_o)\,d\sigma + H_o, \quad z_o(t_o) = y(t_o), \quad z_o(t_1) = y(t_1).$$

It holds then, by (4.14) (with $K = 0$), (4.16), (4.15)

$$\|H_o\| \cdot \delta \leq \int_{t_o}^{t_1} \|g(\tau, y(\tau)) - g(\tau, y_o)\| \, d\tau \leq \psi(\eta) \cdot \delta < \varepsilon \cdot \delta \ ,$$

that is $\|H_o\| < \varepsilon$.

We can determine a sequence of constants H_n such that

$$z_n' = \frac{1}{\delta} \int_{t_o}^{t_1} g_n(\sigma, y_o) d\sigma + H_n; \qquad z_n(t_o) = y(t_o), \qquad z_n(t_1) = y(t_1),$$

and we get the estimate

$$\|H_n\| \delta \leq \|H_o\| \delta + \|\int_{t_o}^{t_1} [g(\sigma, y_o) - g_n(\sigma, y_o)] d\sigma\| \ .$$

Since δ is fixed we can find, by the weak convergence of the coefficients, n_ε such that for $n > n_\varepsilon$ it holds

$$\|\int_{t_o}^{t_1} [g(\sigma, y_o) - g_n(\sigma, y_o)] d\sigma\| \leq \varepsilon \delta \ ,$$

so that for $n > n_\varepsilon$ $\|H_n\| \leq 2\varepsilon$. Finally we let y_n to be the solution of the problem

$$\begin{cases} y_n' = f_n(t, y_o) + \varphi(t) + H_n \\ y_n(t_o) = y(t_o) \ . \end{cases}$$

Of course it holds also $y_n(t_1) = y(t_1)$, since $y_n(t_1) = z_n(t_1)$. It holds also, by (4.14) with $K = 2\varepsilon$ and (4.16), $\|y_n(t) - y_o\| < \min(\frac{\varepsilon}{2}, \eta)$; hence $\|y_n(t) - y(t)\| < \varepsilon$. Now $y_n' = g_n(t, y_n(t)) + g_n(t, y_o) - g(t, y_n(t)) + H_n$ and $\|g_n(t, y_o) - g(t, y_n(t))\| \leq \psi(\|y_o - y_n(t)\|) < \psi(\eta) < \varepsilon$.
So $y_n' = g_n(t, y_n(t)) + \varphi_n(t)$, where $\|\varphi_n\|_{L^\infty} < \varepsilon + \|H_n\| < 3\varepsilon$.
This achieves approximability in case (2.6). Case (2.5) is much easier: let $y(t)$ be a solution of $A[y] = \varphi$; with no loss of generality we can suppose $\varphi = 0$, since also $f_n(t, y) - \varphi(t)$ satisfies iii). We consider the constant se-

quence $y_n(t) \equiv y(t)$, satisfying the equation $A_n[y_n] = \varphi_n$. We get the estimate

$$\left\| \int_{t'}^{t''} \varphi_n(t)\, dt \right\| \leq \left\| \int_{t'}^{t''} [f_n(\tau, y(\tau)) - f(\tau, y(\tau))] d\tau \right\| ,$$

and from (4.15) we get ii). Q.E.D.

Under stronger assumptions a sharper form of the theorem holds:

Theorem 2. *Suppose that* $f_n(t,y)$ *are uniformly Lipschitz continuous with respect to* y, *namely*

$$\| f_n(t,z_1) - f_n(t,z_2) \| \leq L(t) \cdot \| z_2 - z_1 \| \tag{4.17}$$

$$\int_a^b L(t)\, dt = L$$

and it holds

$$\| f_n(t,0) \| \leq C(t) \qquad C(t) \in L^{\alpha}([a,b]), \quad \alpha > 1 .$$

Then all the definitions of G-*convergence given in section 2 are equivalent to the weak convergence of the coeffficients.*

Proof. The proof that G-convergence implies the weak convergence of the coefficients is the same of Theorem 1. For the converse the proof of property i) of G-convergence is essentially the same. Approximability property in this case is quite obvious, since it is enough to consider the *unique* solution to the Cauchy problem having the same initial data and then to apply Gronwall's lemma. Q.E.D.

Corollary 3. *If the hypothesis of theorem* 1 *or* 2 *hold, in order that* G-*convergences* (2.5) *or* (2.6) *be satisfied it is sufficient that properties* i) *and* ii) *of* G-*convergence hold for a particular* φ .

Proof. In order to prove the weak convergence of the coefficients only G-convergence for a particular φ has been required ($\varphi = 0$ in the actual proof, but there is no loss of generality). The second part of theorem 1 allows to complete the proof. Q.E.D.

Remarks. Theorem 1 and theorem 2 allow to state also obvious theorems about approximation of differential operators by regular differential operators or by piecewise autonomous differential operators. A result derived from theorem 2 is the following theorem related to the linear homogeneization (for the proof see [4] and [6]).

Theorem 4. *Suppose* $f(t,y)$ *is periodic of period* L *with respect to* t *and is Lipschitz continuous with respect to the* y-*variables. Then the sequence of differential operators*

$$A_n[y] = y' - f(nt,y) \tag{4.18}$$

G-*converges according to any of the given definitions to the equation* $y' = g(y)$ *where* $g(y) = L^{-1}\int_0^L f(\tau,y)\,d\tau$.

V. MORE ABOUT CAUCHY G-CONVERGENCE. NON-LINEARITY

From the results of the last section it could appear that G-convergence (or convergence (2.5)) is more general than C-G-convergence, inasmuch it is suitable to deal with equations that present Peano phaenomenon, while C-G-convergence can be used (essentially) only for regular equations. In this section we report some counterexamples which show that C-G-convergence is suitable to deal with non-linear homogeneization, where G-convergence fails and convergence (2.5) leads to improper results.

Theorem 5. (*Non linear homogeneization*). *Let* $f : [a,b] \times \mathbb{R} \to \mathbb{R}$ *be periodic of period* L *with respect to* y *and* $f \in \mathcal{C}^1$. *Then the sequence of equations*

$$y' - f(t,ny) = 0 \tag{5.1}$$

C-G-*converges to the equation*

$$y' - g(t) = 0 ; \tag{5.2}$$

$$g(t) = \begin{cases} \{L^{-1}\int_0^L f^{-1}(t,\eta)d\eta\}^{-1} & \textit{if} \quad \int_0^L |f^{-1}(t,\eta)|d\eta < +\infty \\ 0 & \textit{otherwise.} \end{cases}$$

We remark that by theorem 9 (see below) the sequence (5.1) converges to (5.2) also according (2.4) and (2.10). On the contrary it *does not* G-converge. It is enough to check that the limiting property of G-convergence does not hold: consider in fact $A_n[y_n] = \varphi_n$, $\varphi_n(t) \equiv \lambda$, and a sequence of solutions of the problem

$$A_n[y_n] = \lambda ; \qquad y_n(a) = y_o .$$

The sequence of solutions converges, according to the statement of theorem 5, to y, solution of the problem

$$y' = \{L^{-1}\int_0^L [f(t,\eta) + \lambda]^{-1}d\eta\}^{-1}, \quad y(a) = y_o$$

and not to a solution of the equation

$$y' = \{L^{-1}\int_0^L f(t,\eta)^{-1}d\eta\}^{-1} + \lambda .$$

This depends on the fact that whenever a non trivial non linear homogeneization takes place, condition (4.15) and even (4.2) are systematically violated, so that corollary 3 cannot be applied. Actually in this case, in general, there is not even weak convergence of the coefficients.

It is striking to remark that convergence (2.5) holds, but the limit is different from (5.2); the result is specified in the following:

Theorem 6. *Under the same hypothesis of theorem* 5 *the limit of the sequence of differential equations*

$$y' - f(t,ny) = \varphi(t) \tag{5.3}$$

according to convergence (2.5) *is the differential inequality*

$$\varphi(t) + \min_{0\le\eta\le L} f(t,\eta) \le y' \le \varphi(t) + \max_{0\le\eta\le L} f(t,\eta) . \qquad (5.4)$$

Proof. Property i) of G-convergence is obvious, since for the approximating functions it holds for $\varepsilon > 0$, provided h is small enough,

$$h^{-1}\int_t^{t+h} \varphi(\tau)d\tau + \min_{0\le\eta\le L} f(t,\eta) - \varepsilon \le \frac{y_n(t+h) - y_n(t)}{h}$$

$$\le h^{-1}\int_t^{t+h} \varphi(\tau)d\tau + \max_{0\le\eta\le L} f(t,\eta) + \varepsilon .$$

Taking the limit it holds almost everywhere (5.4), since y is absolutely continuous as a uniform limit of absolutely continuous functions.

We prove property ii) of G-convergence only for strictly positive functions f, φ, leaving the complete proof to the reader. So we assume $f(t,\eta) > 0$, $\varphi(t) > 0$. For any $\varepsilon > 0$ there exists $\delta_o > 0$ such that $|f(t',\eta) - f(t'',\eta)| < \varepsilon$ provided $|t'' - t'| < \delta_o$, for each $\eta \in [0,L]$. It can also be assumed that $\delta_o\{\sup_{a\le t\le b} \varphi(t) + \max f(t,y)\} < \varepsilon$. We divide $[a,b]$ in subinterval of lenght $\delta \le \delta_o$. Let y be a solution of (5.4). We have then

$$\frac{1}{\delta}\int_{t_i}^{t_{i+1}} \varphi(\sigma)d\sigma + \min_{0\le\eta\le L} f(t_i,\eta) - \varepsilon \le \frac{y(t_{i+1}) - y(t_i)}{\delta}$$

$$\le \frac{1}{\delta}\int_{t_i}^{t_{i+1}} \varphi(\sigma)d\sigma + \max_{0\le\eta\le L} f(t_i,\eta) + \varepsilon ,$$

hence for some interval $[\eta_1,\eta_2] \subset [-L,L]$, of lenght L/s (s integer) it holds

$$\left|\frac{y(t_{i+1}) - y(t_i)}{\delta} - (C_i + f(t_i,\eta))\right| < 2\varepsilon , \qquad (5.5)$$

$$rL + \eta_1 \le \eta \le rL + \eta_2, \quad r \in Z,$$

where $C_i = \frac{1}{\delta}\int_{t_i}^{t_{i-1}} \varphi(\sigma)d\sigma$. We remark that on each interval

$[t_i, t_{i+1}]$ the solutions of the piecewise autonomous equation

$$y' = f(t_i, y) + c_i$$

are solutions of the equation

$$y' = f(t,y) + \varphi(t) + \psi(t) ,$$

where $\|\psi\|_{H^{-1,\infty}([a,b])} \leq \max [\varepsilon, \varepsilon(b-a)]$. So it is enough to prove approximability for autonomous equations, provided $y_n(t_i) = y(t_i)$.

Proof for the autonomous equation, with $\dfrac{y(b) - y(a)}{b-a} = f(\bar{\eta})$.

We need to find a function $\psi_n(t)$ such that $y_n' = f(ny_n) + \psi_n(t)$ $y_n(a) = y(a)$, $y_n(b) = y(b)$ and $\|\psi_n\|_{L^\infty([a,b])} \leq K n^{-1}$

First we determine η_1, η_2 such $\eta_1 < \bar{\eta} < \eta_2$ and for $rL + \eta_1 < \eta < rL + \eta_2$ it holds $|f(\eta) - f(\bar{\eta})| < n^{-1}$. We can choose η_1, η_2 such that $\eta_2 - \eta_1 = L/s$, where s is an integer.

We let $t_{1,1} = a + (b-a)/n$, $t_\infty = b - (b-a)/n$ and determine by recurrence

$$t_{j,2i} - t_{j,2i-1} = L\ (sn\ f(\bar{\eta}))^{-1} \qquad i = 1,2,\ldots,s$$

$$t_{j,2i+1} - t_{j,2i} = L\ (sn^2)^{-1} \qquad i = 1,2,\ldots,s$$

$$t_{j,1} - t_{j-1,2s} = (s-1)L\ (sn^2)^{-1} \qquad i = 2,\ldots,h ,$$

where h satisfies

$$h = \mathrm{Int}\left[(b-a)(1-2/n)(n/L)(f(\bar{\eta})^{-1} + (2s-1)(sn)^{-1})^{-1}\right] - 1$$

$$\sim \mathrm{Int}[n(b-a)\ f(\bar{\eta})/L] .$$

We consider the solution of the problem so defined: $y_n(a) = y(a)$; let r be the least integer such that $rL/n+\eta_1 > y(a)$, then

$$y_n'(t) = \begin{cases} (rL/n + \eta_1 - y(a))n/(b-a) & \text{for } a<t<t_{1,1} \\ f(\bar{\eta}) & \text{for } t_{j,2i-1}<t<t_{j,2i} \\ -n & \text{for } t_{j,2i}<t<t_{j,2i+1} \\ n & \text{for } t_{j,2s}<t<t_{j+1,1} \\ (y(b)-y(t_\infty))n/(b-a) & \text{for } t_\infty<t<b \end{cases} \tag{5.6}$$

Remark also that $y(t_\infty) > \text{Int}\left[(b-a)(1-2/n)\left(\frac{1}{f(\bar{\eta})} + \frac{2s-1}{sn}\right)^{-1} \frac{L}{n}\right]\frac{n}{L} + y(a)$, hence
$y(b)-y(t_\infty) < f(\bar{\eta})(b-a) - \{\text{Int}\left[(b-a)(1-2/n)\left(\frac{1}{f(\bar{\eta})} + \frac{2s-1}{sn}\right)^{-1} \frac{L}{n}\right]\frac{n}{L}\}$
$\sim 0(n^{-1})$, and similarly $rL/n + \eta_1 - y(a) \sim 0(n^{-1})$.

It is only left to estimate ψ_n; we have

$$\psi_n(t) \sim \begin{cases} 0(n^{-1})n+0(1) = 0(1) & \text{for } a<t<a+(b-a)/n = t_{1,1} \\ 0(n^{-2}) & \text{for } t_{j,2i-1}<t<t_{j,2i} \\ -n + 0(1) & \text{for } t_{j,2i}<t<t_{j,2i+1} \\ n + 0(1) & \text{for } t_{j,2s}<t<t_{j+1,1} \\ 0(1) & \text{for } t_\infty<t<b \end{cases} \tag{5.7}$$

Then $\|\psi\|_{H^{-1}\,\infty([a,b])} \sim 0(n^{-1})$, since the terms containing n are all balanced except for the last index. This result achieves the proof, since it is enough to choose $n_\varepsilon > (b-a)/(\varepsilon\delta)$. Remark that the proof of the approximability can be shortened allowing approximation with solutions of transmission equations.

Q.E.D.

The following theorems complete the equivalence theory related to C-G-convergence.

Theorem 7. *G-convergence according* (2.8) *is equivalent to G-convergence according* (2.10).

Proof. The only non trivial point is to prove in the part (2.10) (2.8) that if $t_{o,n} \to t_o$, $y_{o,n} \to y_o$,

$y_n(t;t_{o,n},y_{o,n}) \to y(t)$ then $y(t) = y(t;t_o,y_o)$. Suppose it is false, namely $y(t_o) = y_1$, $\|y_1-y_o\| = \delta$. Then, since $f(t,y)$ satisfies (2.1) it holds (4.14) in the form $\|y(t) - y_1\| \leq K(|t-t_o| + |t-t_o|^{\frac{\alpha-1}{\alpha}})$ (remark that Caratheodory's condition implies, locally, (4.1) for a single function). This fact contradicts the uniform convergence of y_n to y.

Q.E.D.

Theorem 8. *If* (2.8) *holds,* $\|f_n(t,y)\| \leq C(1+\|y\|)$ *and the limiting operator* $A[y]$ *has unique solutions for the Cauchy problem, then there is* C-G-*convergence, and* C-G-*convergence is equivalent to* (2.9).

Proof. It is obvious that there is C-G-convergence; it is left to prove the approximability property, in order to have convergence (2.9). Consider $y(t;t_o,y_o)$ and a choice of $y_n(t;t_o,y_o)$. By our hypothesis y_n satisfy the conditions of Ascoli Arzelà lemma, hence there is a converging subsequence; by convergence (2.8) the limit is a solution of $A[y] = 0$. By the uniqueness this limit is actually $y(t;t_o,y_o)$. Since any convergent subsequence converges to the same limit and the set is compact, the whole sequence converges to $y(t;t_o,y_o)$.

Q.E.D.

Theorem 9. *For ordinary differential equations if all of them are regular, then* C-G-*convergence implies convergence* (2.8).

First we show by a counterexample that without supplementary requirements the theorem does not hold for systems.

$$\begin{cases} x' = f_n(x,y) \\ y' = 0 \end{cases} \quad \text{where} \quad \begin{cases} f_n = 1 & \text{outside} \quad (y - \frac{2}{n})^2 + x^2 = \frac{1}{n^2} \\ f_n = n^2\left[(y - \frac{2}{n})^2 + x^2\right] & \text{inside} . \end{cases}$$

Then the sequence C-G-converges to $x' = 1$, $y' = 0$, but the sequence of solutions of the Cauchy problems of initial data $x_n(0) = 0$, $y_n(0) = 2/n$, that is $x(t) = 0$, $y(t) = 2/n$ converges to $x(t) = 0$, $y(t) = 0$, while the solution of the Cauchy problem with the limiting data is $x = t$, $y = 0$.

Proof of theorem 9. Suppose $t_{o_n} \to t_o$, $y_{o_n} \to y_o$, $y_n \xrightarrow{L^\infty} y$ and $y_n = y_n(t;t_{o_n},y_{o_n})$. Suppose by contradiction $y(t) \not\equiv y(t;t_o,y_o)$. Let

$$\varepsilon = \max |y(t) - y(t;t_o,y_o)| . \tag{5.8}$$

Since uniqueness of solutions implies continuous dependence on initial data, there exists δ_ε such that $|y(t;t_o,y_o \pm \delta_\varepsilon) - y(t;t_o,y_o)| < \varepsilon/4$ for $t \in [a,b]$. By C-G-convergence there exists n_ε such that for $n > n_\varepsilon$, $|y_n(t;t_o,y_o \pm \delta_\varepsilon) - y(t;t_o,y_o)| < \varepsilon/2$. Hence by uniqueness for any $\bar{y} \in [y_o - \delta_\varepsilon, y_o + \delta_\varepsilon]$, for $n > n_\varepsilon$

$$|y_n(t;t_o,\bar{y}) - y(t;t_o,y_o)| < \varepsilon/2 . \tag{5.9}$$

Since y_n converges uniformly to y and y is continuous, it follows that for $n > N_\varepsilon$, $y_n(t_o) \in [y_o - \delta_\varepsilon, y_o + \delta_\varepsilon]$ and

$$|y_n(t;t_{o_n},y_{o_n}) - y(t)| < \varepsilon/2 . \tag{5.10}$$

But (5.9) and (5.10) contradict (5.8).

The approximability property is obvious. Q.E.D.

We end this section with an example taken from [7], where homogeneization is performed with respect to two variables and there is weak convergence of the coefficients, but to a different limit. Consider the sequence of equations $y' = \cos(n(t-y))$. This sequence C-G-converges to the equation $y' = 1$. Using the results of [7] it is immediate to check that the sequence does not G-converge to $y' = 1$, in fact the solutions of the equations $y' = \cos(n(t-y)) + \lambda$, $0 \leq \lambda \leq 2$, converge all to a solution of the equation $y' = 1$, instead of $y' = 1 + \lambda$ as it would be required by G-convergence. Remark also that in this case the coefficients weakly converge to $f = 0$, that is not the coefficient of the C-G-limit. For more details on the problem of homogeneization with respect to more than one variable compare [4], chapter 3, [6], [8]. For a summary of the results [7].

VI. LIMITS OF THE STANDARD THEORY OF G-CONVERGENCE

Whe have introduced in the last section examples of non linear homogeneization for which C-G-convergence is a suitable tool, while G-convergence cannot be used. When Peano phaenomenon arises, on the contrary, G-convergence can be used and C-G-convergence usually does not hold (and not even modified versions of C-G-convergence, like (2.8) or (2.10)). At first sight it seems just to be the problem of a correct choice of the definition that fits our needs. We show below an example in which both situations arise at the same time, hence it does not enter in any of the definitions developped till now.

Let $f(t,y)$, periodic in both the variables of period 4, be so defined

$$\begin{cases} f(t,y) = 2\sqrt{y} & 0 \le t < 2 \quad 0 \le y < 1 \\ f(t,y) = 2\sqrt{2-y} & 0 \le t < 2 \quad 1 \le y < 2 \\ f(t,y) = -f(4-t,y) & 2 \le t < 4 \quad 0 \le y < 2 \\ f(t,y) = -f(t,-y) & 0 \le t < 4 \quad -2 \le y < 0 \end{cases} \tag{6.1}$$

Then the sequence of differential equations

$$y' = f(nt,ny) \qquad a \le t < b \tag{6.2}$$

C-G-converges to the inequality $|y'| \le 1$. For a sketch of the proof see [8]. Consider now the following sequence of regular equations: f_n is periodic of period 4 with respect to both variables:

$$f_n(t,y) = \begin{cases} f(t,y) & 0 \le t < 2 \quad n^{-1} \le y < 2-n^{-1} \\ 2\sqrt{n}\, y & 0 \le t < 2 \quad 0 \le y < n^{-1} \\ 2\sqrt{n}\,(2-y) & 0 \le t < 2 \quad 2-n^{-1} \le y < 2 \\ -f_n(4-t,y) & 2 \le t < 4 \quad 0 \le y < 2 \\ -f_n(t,-y) & 0 \le t < 4 \quad -2 \le y < 2 \end{cases} \tag{6.3}$$

Then the sequence of equations

$$y' = f_n(t,y) \qquad a \le t \le b \tag{6.4}$$

G-converges to the equation $y' = f(t,y)$ according to theorem 1. Obviously it does not C-G-converge because of the arising of Peano phaenomenon. It would be expected that combining the homogeneization of (6.2) and the approximation (6.4) by regular equations, namely using the sequence

$$y' = f_n(nt,ny) \tag{6.5}$$

it were possible to obtain as a G-limit of regular equations the inequality $|y'| \le 1$. Unfortunately (6.5) converges only according to (2.5), but to a different limit, namely, for each $\varphi(t)$,

$$\varphi(t) - 2 \le y' \le \varphi(t) + 2 ,$$

that is to the inequality $|y'| \le 2$. This result is due to the fact that for each t, $\max_y f_n(t,y) = 2$, $\min_y f_n(t,y) = -2$.
Of course it is possible to modify (6.3) in order that the G-limit is, as expected, $|y'| \le 1$. In fact define f_n starting from $\frac{n}{n-1} y^{1/n}$ instead of $2\sqrt{y}$, that is, on the single period:

$$f_n(t,y) = \begin{cases} \frac{n}{n-1} n^{(n-1)/n} y & \text{if } 0 \le t < 2 \quad 0 \le y < n^{-1} \\ \frac{n}{n-1} y^{1/n} & \text{"} \quad n^{-1} \le y < 1 \\ \frac{n}{n-1} (2-y)^{1/n} & \text{"} \quad 1 \le y < 2-n^{-1} \\ \frac{n}{n-1} n^{(n-1)/n} (2-y) & \text{"} \quad 2-n^{-1} \le y < 2 \\ -f_n(4-t,y) & 2 \le t < 4 \quad 0 \le y < 2 \\ -f_n(t,-y) & 0 \le t < 4 \quad -2 \le y < 0 \end{cases} \tag{6.6}$$

Then the sequence $y' = f_n(nt,ny)$ converges according (2.5) to $|y'| \leq 1$. It still does not converge according to any other definition; in particular is does not G-converge because using a test $\varphi = 0$ we get inconsistent results. This suggest to give a reduced definition of G-convergence, namely:

DEFINITION OF SINGLE-POINT-G-CONVERGENCE (S-P-G-convergence).

A sequence of differential equations $A_n[y] = y' - f_n(t,y) = 0$ S-P-G-*converges to the equation* $A[y] = y' - f(t,y) = 0$ *if the following properties hold:*

i) *if* $n_k \to \infty$, $y_{n_k} \xrightarrow{(L^\infty)^s} y$, $\varphi_{n_k} \xrightarrow{(L^\infty)^s} 0$ *and* $A_{n_k}[y_{n_k}] = \varphi_{n_k}$, *then* $A[y] = 0$.

ii) *If* $A[y] = 0$ *there exist two sequences* $y_n \xrightarrow{(L^\infty)^s} y$, $\varphi_n \xrightarrow{(L^\infty)^s} 0$, *such that* $A_n[y_n] = \varphi_n$.

This is the simplest definition that allows to overcome the difficulties met up to now. It has anyhow the disadvantage of not being, properly a G-convergence according to the abstract definition. In fact it is a G-convergence only with respect to $\varphi = 0$. In the next sections we give some ideas for a general theory of G-convergence, that holds provided we consider parametric differential equations, and overcomes this problem of definitions. Anyhow we remark that in all the cases considered up to now (also case (6.5)) S-P-G-convergence holds and yields the expected results.

VII. GENERAL THEORY OF G-CONVERGENCE FOR PARAMETRIC DIFFERENTIAL EQUATIONS

Since in this section we use multivalued functions we introduce some notations and definitions. Let $A, B \subset \mathbb{R}^s$, we let $\mathrm{DIST}(A,B) = \sup_{a \in A} \{ \inf_{b \in B} |a-b| \}$.

If $f : X \to \mathcal{P}(\mathbb{R})$ and if for each x ε $f(x)$ is a closed interval we let $f^+(x) = \sup\{\xi : \xi \in f(x)\}$, $f^-(x) = \inf\{\xi : \xi \in f(x)\}$. By the notation $f(x) + g(x)$ we mean $\{\xi + \eta : \xi \in f(x), \eta \in g(x)\}$. In general we shall deal with functions $f : [a,b] \times \mathbb{R}^s \to \mathcal{P}(\mathbb{R}^q)$. A solution of the equation $y' \in f(t,y)$ is an absolutely continuous function such that almost everywhere it holds $y'(t) \in f(t,y(y))$. In the theory developped up to now we have considered equations of the form $y' \in f(t,y) + \varphi(t)$. In this section we allow a parametric dependence on the function φ, namely we consider

$$f : [a,b] \times \mathbb{R}^s \times (C_o([a,b]))^q \to \mathcal{P}(\mathbb{R}^s) ,$$

and the equations

$$y' \in f(t,y(t);\varphi) . \tag{7.1}$$

We call an expression like (7.1) parametric differential equation. We make the usual assumptions that ensure the existence of solutions on the whole $[a,b]$, namely boundedness:

$$\sup \{|\xi| : \xi \in f(t,y;\varphi)\} \leq C(t)(1 + \|y\| + \|\varphi\|_{L^\infty}) \tag{7.2}$$

$$\int_a^b |C(t)|^\alpha dt = C^\alpha, \quad \alpha > 1$$

and continuity with respect to y, that becomes in this case: for each K there exists $\psi_K(t,z)$ increasing with z, such that $\lim_{z \downarrow 0} \int_a^b |\psi_K(t,z)| dt = 0$ and for each $|y'|, |y''| < K$, $\|\varphi\|_{L^\infty} < K$ it holds for any $t \in [a,b]$

$$\text{DIST}\,(f(t,y'';\varphi), f(t,y';\varphi)) < \psi_K(t, \|y''-y'\|) .$$

DEFINITION OF GENERALIZED G-CONVERGENCE.

Let A_n, A *be parametric differential equations. The sequence* A_n G-*converges if the two following properties are satisfied:*

i) *If* $n_k \to \infty$, $y_{n_k} \xrightarrow{(L^\infty)^s} y$, $\varphi_{n_k} \xrightarrow{(L^\infty)^q} \varphi$ *and* y_{n_k} *are solutions of* $y'_{n_k} \in f_{n_k}(t,y_{n_k};\varphi_{n_k})$, *then also* y *is a solution of* $y' \in f(t,y;\varphi)$.

ii) *If* y *is a solution of* $y' \in f(t,y;\varphi)$, *then there exist two sequences* $y_n \xrightarrow{(L^\infty)^q} y$, $\varphi_n \xrightarrow{(L^\infty)^q} \varphi$ *such that*

y_n *are solutions of* $y'_n \in f_n(t, y_n(t); \varphi)$.

Sometimes the space of parameters is given only by constants, then we can apply definition (2.8) of G-convergence (or (2.9)).

Also when parameters are taken from $\{C_0([a,b])\}^q$ we can use G-convergence (2.8), considering thus a restricted convergence, much easier to handle. Theorem 12 shows an important case in which restricted G-convergence implies G-convergence.

We call *linear parametric equations* those for which $f(t,y;\varphi) = f(t,y;0) + \varphi(t)$. Theorem 1 states that under assumptions of equicontinuity of the coefficients also the G-limit, if it exists, is a linear parametric equation. Theorem 5 shows that in general this property does not hold.

It seems advisable to restrict our attention only to those parametric equations that are consistent with G-convergence, that is equations that G-converge to themselves. This property is satisfied under the very weak assumption of the next theorem; we only remark that it is not automatic as it is shown by the trivial counterexample:

$$y' = \operatorname{sgn}\,[\varphi(t)] \qquad 0 \le t \le 1 . \tag{7.3}$$

Take $\varphi(t) = n^{-1}$, $y_n(t) = t$, where $y_n(t) \to t$, $\varphi_n(t) \to 0$ and $\frac{dt}{dt} \neq \operatorname{sgn}(0)$.

Theorem 10. *Suppose that the following continuity property holds: for each* $K > 0$ *there exists a function* $\psi_K(t,z)$ *increasing with respect to* z,

$$\lim_{z \downarrow 0} \int_a^b |\psi_K(t,z)|\,dt = 0 , \tag{7.4}$$

such that for any φ', φ'', *bounded in* L^∞ *by* K, *for* $\|y'\|$, $\|y''\| < K$ *it holds*

$$\mathrm{DIST}(f(t,y'';\varphi''),\ f(t,y';\varphi'))$$

$$< \psi_K(t, \max\{\|y''-y'\|, \|\varphi''-\varphi'\|_{L^\infty}\}) . \tag{7.5}$$

Then $y' \in f(t,y;\varphi)$ *G-converges to itself.*

Proof. Property ii) of G-convergence does not require to be proved since the very solution approximates itself. In order to prove i) suppose $y_{n_k} \to y$, $\varphi_{n_k} \to \varphi$, $y'_{n_k} \in f(t, y_{n_k}; \varphi_{n_k})$. There exists a constant K such that $\|y_{n_k}\|, \|y\|, \|\varphi_{n_k}\|$, $\|\varphi\| < K$. We let

$$H(t) = y(t) - y(a) - \int_a^t f(\tau, y(\tau), \varphi) d\tau .$$

Then it holds

$$H(t) = y(t) - y_{n_k}(t) - y(a) + y_{n_k}(a)$$
$$- \int_a^t [f(\tau, y(\tau); \varphi) - f(\tau, y_{n_k}(\tau); \varphi_{n_k})] d\tau .$$

For any $\varepsilon > 0$ let $\delta \leq \varepsilon$ be such that $\int_a^b \psi_K(t, \delta) dt < \varepsilon$; there exists k_ε such that for $k > k_\varepsilon$, $\|y_{n_k} - y\|$, $\|\varphi_{n_k} - \varphi\| < \delta$, then

$$\|H\|_{L^\infty} < 2\varepsilon + \int_a^b \psi_K(t, \max(\|y_{n_k} - y\|_{L^\infty}, \|\varphi_{n_k} - \varphi\|_{L^\infty})) dt \leq 3\varepsilon .$$

Since H does not depend on n_k, $H \equiv 0$. Q.E.D.

The same theorem holds when parameters are taken from $\mathbb{R}^q$.

Another property of interest is the local dependence on the parameters; *we say that a parametric differential equation is local when, if* $\varphi'' = \varphi'$ *on a subinterval* $[c,d]$ *of* $[a,b]$, *for each solution of the equation* $y_1' \in f(t, y_1(t); \varphi')$ *there exists a solution* y_2 *of the equation* $y_2' \in f(t, y_2(t); \varphi'')$ *such that* $y_1 = y_2$ *on* $[c,d]$.

Trivial examples of non locality are given by:

$$y' = \sup_{a \leq t \leq b} \|\varphi(t)\| \quad \text{or} \quad y' = \int_a^b \varphi(\tau) \, d\tau .$$

A condition sufficient for the locality is the following: the equation is consistent, (7.2) holds; there exists a function $\bar{f}(t, y; \lambda)$, $\bar{f} : [a,b] \times \mathbb{R}^s \times \mathbb{R}^q \to \mathcal{P}(\mathbb{R}^s)$ such that $f : [a,b] \times \mathbb{R}^s \times (C_o)^q \to \mathcal{P}(\mathbb{R}^s)$ can be factorized into $\bar{f} \circ P$, where

$$P(t,y;\varphi) = (t,y,\varphi(t)) \in [a,b] \times \mathbb{R}^s \times \mathbb{R}^q \ . \tag{7.6}$$

Remark that consistency can get lost taking the G-limit; locality on the contrary is preserved; it holds:

Theorem 11. *Let* $f_n(t,y,\varphi)$ *satisfy* (7.2) *uniformly. Let the equations* $y' \in f_n(t,y;\varphi)$ *be local and* G-*converge to* $y' \in f(t,y;\varphi)$; *then also the limit equation is local.*

Proof. Suppose $\varphi_1 = \varphi_2$ on $[c,d]$ and y_1 is a solution of $y_1' \in f(t,y_1,\varphi_1)$. Then by property ii) of G-convergence there exist two sequences $y_{1n} \xrightarrow{(L^\infty)^s} y_1$, $\varphi_{1n} \xrightarrow{(L^\infty)^q} \varphi_1$, $y_{1n}' \in f_n(t,y_{1n};\varphi_{1n})$. Let now define the sequence

$$\varphi_{2n}(t) = \begin{cases} \varphi_{1n}(t) & \text{on } [c,d] \\ \varphi_2(t) + \varphi_{1n}(c) - \varphi_1(c) & \text{on } [a,c] \\ \varphi_2(t) + \varphi_{1n}(d) - \varphi_1(d) & \text{on } [d,b] \ , \end{cases}$$

so that φ_{2n} uniformly converges to φ_2. Then by the locality of the approximating equations, there exist solutions y_{2n} of $y_{2n}' \in f_n(t,y_{2n};\varphi_{2n})$ which coincide with y_{1n} on $[c,d]$. By (7.2) they are equicontinuous and uniformly bounded, so there is a converging subsequence. The limit y_2, by property i) of G-convergence is a solution of $y_2' \in f(t,y_2;\varphi_2)$ which coincides with y_i on $[c,d]$. Q.E.D.

The next theorem provides a tool for reducing the proof of G-convergence for the non linear homogeneization of local parametric equations to the simpler convergence (2.8).

We call regular a real multivalued function $f : X \to \mathcal{P}(\mathbb{R})$ if

i) $f(t)$ is the closed interval $[f^-(t), f^+(t)]$ for any $t \in X$

ii) $f^+(t)$ is upper semicontinuous, $f^-(t)$ is lower semicontinuous.

We say that an equation is regular if the equation is local, $\bar{f}(t,y,.)$ is uniformly regular when t and y vary in a compact set and boundedness and continuity property hold.

Theorem 12. *Let* A_n *be a sequence of regular parametric differential equations; suppose*

i) $f_n(t,y;.)$ *is non decreasing with respect to* φ *, that is if* $\varphi' \geq \varphi''$, *then* $f_n(t,y;\varphi') \geq f_n(t,y,\varphi'')$.

ii) *for any* H *there is a constant* C_H, *and there is a sequence* $\varepsilon_n \downarrow 0$ *such that if* $\|\varphi\| < H$, $\|y\| < H$, $y' \in f_n(t,y;\varphi)$, *then there exists* y_1 *solution of* $y_1' \in f_n(t,y_1;\varphi)$ *such that for all* t', t'' *it holds*

$$|y(t'') - y_1(t'')| \leq (|y(t')-y_1(t')|+C_H\varepsilon_n)\exp(C_H|t''-t'|) \qquad (7.7)$$

iii) *The sequence* $y' \in \bar{f}_n(t,y;\lambda)$ *converges according convergence* (2.8) *to a regular parametric equation* $y' \in \bar{f}(t,y;\lambda)$.

Then the sequence $y' \in f_n(t,y;\varphi)$ G-*converges to* $y' \in f(t,y;\varphi)$, *where* $f = \bar{f} \circ P$.

Remark that we do not require uniform equicontinuity with respect to n .

Proof. Recall that if $\bar{f}_n$ converges on $[a,b]$, because of the locality, converges also on $[t_1,t_2] \subset [a,b]$.
In order to prove property i) of G-convergence, let $y_k \xrightarrow{L^\infty} y$, $\varphi_k \xrightarrow{L^\infty} \varphi$, $y_k' \in f_k(t,y_k;\varphi_k)$. Consider $t_1,t_2 \in [a,b]$, and solutions of the equations

$$\begin{cases} Z_k' \in f_k(t,Z_k;\max\limits_{[t_1,t_2]}\varphi_k(\tau)) \\ Z_k(t_1) = y_k(t_1) \end{cases} \qquad \begin{cases} z_k' \in f_k(t,z_k;\min\limits_{|t_1,t_2|}\varphi_k(\tau)) \\ z_k(t_1) = y_k(t_1) \end{cases} .$$

By hypothesis i) it is possible to find solutions such that $z_k(t_2) \leq y(t_2) \leq Z_k(t_2)$. By (7.2) $Z_k(t)$ and $z_k(t)$ form two sequences of uniformly bounded equicontinuous functions, hence there are two subsequences that converge uniformly to two functions $z(t)$, $Z(t)$ such that $z(t_1) = Z(t_1) = y(t_1)$ and $z(t) \leq y(t) \leq Z(t)$. On the interval $[t_1,t_2]$ it holds also, by hypothesis iii) $z'(t) \in \bar{f}(t,z;\min\limits_{[t_1,t_2]}\varphi(\tau))$,
$Z'(t) \in \bar{f}(t,Z;\max\limits_{[t_1,t_2]}\varphi(\tau))$, hence there exist

$\eta_1(t) \in \bar{f}(t,z; \underset{[t_1,t_2]}{-}\varphi(\tau))$, $\eta_2(t) \in \bar{f}(t,Z; \underset{[t_1,t_2]}{}\varphi(\tau))$ such that

$$\frac{1}{t_2-t_1}\int_{t_1}^{t_2}\eta_1(t)dt \leq \frac{y(t_2)-y(t_1)}{t_2-t_1} \leq \frac{1}{t_2-t_1}\int_{t_1}^{t_2}\eta_2(t)dt \; .$$

For each $\eta > 0$, by the monotonicity of $\bar{f}(t,y;.)$ and the continuity of φ is is possible to determine δ such that for $|t_2-t_1| < \delta$

$$\eta_1(t) \in \bigcup_{0\leq\sigma<\eta} \bar{f}(t,z;\varphi(t_1)-\sigma) \qquad \eta_2(t) \in \bigcup_{0\leq\sigma<\eta} \bar{f}(t,Z;\varphi(t_1)+\sigma);$$

for each $\varepsilon > 0$ by the regularity of the equation it is possible to find $\eta > 0$ such that

$$\inf \bigcup_{0\leq\sigma<\eta} \bar{f}(t,z;\varphi(t_1)-\sigma) > \bar{f}^-(t,z,\varphi(t_1)) - \varepsilon$$

$$\sup \bigcup_{0\leq\sigma<\eta} \bar{f}(t,Z;\varphi(t_1)+\sigma) < \bar{f}^+(t,Z;\varphi(t_1)) + \varepsilon \; .$$

Then

$$\frac{1}{t_2-t_1}\int_{t_1}^{t_2} \bar{f}^-((t,z;\varphi(t_1))\, dt - \varepsilon \leq \frac{y(t_2)-y(t_1)}{t_2-t_1}$$

$$\leq \frac{1}{t_2-t_1}\int_{t_1}^{t_2} \bar{f}^+(t,Z;\varphi(t_1))\, dt + \varepsilon \; .$$

The limits, as t_2 tends to t_1, of the three elements of the inequality exist almost everywhere, since y is absolutely continuous, and satisfy almost everywhere

$$\bar{f}^-(t_1,z(t_1);\varphi(t_1)) - \varepsilon \leq y'(t_1)$$

$$\leq \bar{f}^+(t_1,Z(t_1);\varphi(t_1)) + \varepsilon \; ,$$

that is, by the definition of z, Z, that almost everywhere

$$y' \in \bar{f}(t,y;\varphi(t)) = f(t,y;\varphi) \; .$$

We prove now property ii) of G-convergence. Suppose $y' \in \bar{f}(t,y;\varphi(t))$. For any $\varepsilon > 0$ there is δ_o such that $\int_t^{t+\delta_o} |\bar{f}(\tau, \|y\|; \|\varphi\|_{L^\infty})|d\tau < \varepsilon$, what is possible by (7.2).

We divide $[a,b]$ in N_ε subintervals of length $\delta \leq \delta_o$. By the monotonicity of the approximating functions, and by the continuity it is possible to find in each interval a point $\bar{t}_i$, such that a solution of the problem $\bar{y}' \in f(t,\bar{y},\varphi(\bar{t}_i))$, $\bar{y}(t_i) = y(t_i)$ satisfies $\bar{y}(t_{i+1}) = y(t_{i+1})$. This is possible since the solution of $\bar{y}' \in f(t,\bar{y};\min\ )$, $\bar{y}(t_i) = y(t_i)$ satisfies $\bar{y}(t_{i+1}) \leq y(t_{i+1})$ and conversely for the upper solution. We find a sequence $\lambda_{in} \to \varphi(\bar{t}_i)$ and a sequence $y_{in} \xrightarrow{L^\infty([t_i,t_{i+1}])} \bar{y}$ such that $y'_{in} \in f_n(t,y_{in};\lambda_{in})$. Let

$$H = \max\left[\sup|\lambda_{in}|,\ 2\|\bar{y}\|_{L^\infty([a,b])}\right] ;$$

let η and n_1 be such that for $n > n_1$ it holds

$$2\eta \exp(C_H \delta N_\varepsilon) + \frac{\exp(C_H\delta)(2\eta + C_H\varepsilon_n)}{\exp(C_H\delta) - 1}(\exp(C_H \delta N_\varepsilon) - 1) < \varepsilon. \tag{7.8}$$

Let $n_2 \geq n_1$ be such that for $n > n_2$ it holds

$$\|y_{in} - \bar{y}\|_{L^\infty} < \eta ; \tag{7.9}$$

consider now a solution of the problems

$$\begin{cases} \bar{y}'_{in} \in f(t,\bar{y}_{in};\lambda_{in}) & t_i \leq t \leq t_{i+1} \\ \bar{y}_{in}(a) = y_{in}(a) & \\ \bar{y}_{in}(t_i) = \bar{y}_{i-1,n}(t_i) & i > 1. \end{cases}$$

By (7.7), (7.8) and (7.9) it holds then $|\bar{y}_{in}(t) - \bar{y}(t)| < \varepsilon$; defining $\bar{y}_n(t) = \bar{y}_{in}(t)$, $t_i \leq t < t_{i+1}$, $\bar{y}_n(t)$ is now a global solution of the equation $\bar{y}'_n(t) \in \bar{f}_n(t,\bar{y}_n(t);\varphi_n(t))$ where $\varphi_n(t) = \lambda_{in}$, $t_i \leq t < t_{i+1}$ and it holds by the definition of δ_o

$$|\bar{y}_n(t) - y(t)| \leq |\bar{y}_n(t) - \bar{y}(t)| + |\bar{y}(t) - y(t)| < 2\varepsilon .$$

By the monotonicity and by (7.2) it is possible now to regularize $\varphi_n(t)$ into continuous functions such that the corresponding $\bar{y}_n$ still satisfy $|\bar{y}_n - y| < 3\varepsilon$ for $n > n_2$.

Q.E.D.

VIII. SOME RESULTS IN GENERALIZED HOMOGENEIZATION

First we come back to theorem 5. In generalized G-convergence theory it becomes

Theorem 13. *Let* $f : [a,b] \times \mathbb{R} \to \mathcal{P}(\mathbb{R})$ *be regular and periodic of period* L *with respect to* y; *let* f^- *and* f^+ *be of class* C^1. *Then the sequence of linear parametric equations*

$$y' \in f(t,ny) + \varphi(t) \tag{8.1}$$

G-*converges to the local non linear parametric equation*

$$y' \in g(t;\varphi) , \tag{8.2}$$

where $g(t;\varphi) = \bar{g}(t, \varphi(t))$ *and*

$$\bar{g} \pm t,\lambda) = \begin{cases} \{L^{-1}\int_0^L (f\pm(\eta)+\lambda)^{-1}d\eta\}^{-1} & \\ & \textit{if } \int_0^L |f\pm(\eta)+\lambda|^{-1}d\eta < +\infty \\ 0 & \textit{otherwise.} \end{cases} \tag{8.3}$$

Proof. A_n is a sequence of regular parametric equations; A is also regular. G-convergence (2.8) follows from theorem 5 considering the greatest and the least solution starting from any Cauchy data for a fixed λ. Theorem 9 allows to pass from discrete topology on the Cauchy data to the Euclidean topology, hence to G-convergence (2.10) for each λ. The explicit formula (8.3) allows to check that we can use Euclidean topology on $\mathbb{R}$ as requested by the G-convergence restricted to parameters.

Hypothesis i) of theorem 12 is obviously satisfied.

From the periodicity of $f_n(t,y) + \lambda = f(t,ny) + \lambda$, of period L/n, it follows ii) with $\varepsilon_n = L/n$. So it is possible to apply theorem 12 that yields the thesis. Q.E.D.

As an easy computable example consider the sequence of equations

$$y' = \sin(ny) + \varphi(t) \ ,$$

then according to theorem 13 it G-converges to

$$y' = \bar{g}(t;\varphi(t)) \ ,$$

where $\bar{g}(t;\lambda) = \text{sgn}(\lambda) \cdot [\max(0,\lambda^2-1)]^{1/2}$.

We come now to the examples given in section 6. Let $f(t,y)$ be defined by (6.1), then the sequence of equations

$$y' = f(nt,ny) + \varphi(t)$$

G-converges to the non linear parametric equation

$$y' = \bar{g}(\varphi(t))$$

where $\bar{g}(\lambda)$ is defined as follows

$$\bar{g}(\lambda) = \begin{cases} \{p(\lambda)\} & \lambda > 2 \\ \{1\} & 0 < \lambda \leq 2 \\ [-1,1] & \lambda = 0 \\ \{-1\} & -2 \leq \lambda < 0 \\ \{-p(-\lambda)\} & \lambda < -2 \end{cases}$$

$\frac{p(\lambda)}{\lambda}$ increasing to 1 as λ tends to ∞, $p(2) = 1$.

In order to check that it holds convergence (2.8) one uses the same technique of theorem 4 of [6]. All the hypothesis of theorem 12 are satisfied; in particular ii) because of the periodicity, so we get the result.

The same limit is obtained taking $f_n(t,y)$ defined in (6.3) and considering the sequence of regular differential equations

$$y' = f_n(nt,ny) + \varphi(t) .$$

On the contrary the G-limit of the sequence of non linear parametric equations $y' = \bar{f}(nt,n(y-\varphi(t)t) + \varphi(t)$ where $\bar{f}$ is given by (6.1) G-converges to the linear parametric equation $y' \in \bar{g}(\quad(t))$, $\bar{g}(\lambda) = \{|\xi| \leq 1\} + \lambda$.
In order to check this result it is sufficient to check G-convergence (2.8). But in this case the equation

$$y' = \bar{f}(nt,n(y-\lambda t)) + \lambda$$

can be transformed in the form $z' = \bar{f}(nt,nz)$ for any λ, where $z = y - \lambda t$. The last sequence C-G-converges to the inequality $|z'| \leq 1$. Substituting back $y = z + \lambda t$ we get the result. The same results are obtained starting from the sequence f_n given in (6.3).

For what the approximation theory is concerned the previous examples are the starting point for proving that any regular parametric differential equation is the G-limit of regular parametric differential equations which for each value of φ are *regular ordinary differential equations.*

REFERENCES

1. De Giorgi, E., Γ-convergenza e G-convergenza, *Boll. Uni. Mat. It. (5), 14-A*, 213-220 (1977).
2. De Giorgi, E., Spagnolo, S., Sulla convergenza degli integrali dell'energia per operatori ellittici del secondo ordine, *Boll. Un. Mat. It., 8*, 391-411 (1973).
3. Patuzzo, P., G-convergenza per le equazioni differenziali ordinarie su domini illimitati, to appear on *Boll. Un. Mat. It.*
4. Piccinini, L.C., Stampacchia, G., Vidossich, G., Equazioni

differenziali ordinarie, Liguori, Napoli, (1978).

5. Piccinini, L.C., G-convergence for ordinary differential equations with Peano phaenomenon, *Rend. Sem. Mat. Padova*, (1977).
6. Piccinini, L. C., Homogeneization problems for ordinary differential equations, *Rend. Circ. Mat. Palermo*, (1978).
7. Piccinini, L.C., G-convergence for ordinary differential equations. Lecture notes of the Colloquio internazionale su metodi recenti di analisi non lineare e applicazioni. Roma, 8-12 maggio 1978.
8. Piccinini, L.C., G and Γ-convergence, Mimeographed lecture notes at I.T.C.P. Trieste-Miramare, reference SMR/42/7 and SMR/42/17, (1977).

PATH INTEGRALS AND PARTIAL DIFFERENTIAL EQUATIONS

A. Pliś[1]

Mathematical Institute
PAN, Kraków

Consider an equation of the form

$$\frac{du}{dt} = A\,u, \tag{1}$$

where u belongs to a Banach space B and A is a linear (not necessarily bounded) operator defined on a dense subspace D of space B. The non-relativistic quantum mechanics can be founded on equation (1) for

$$A\,u = i\,H\,u = i\,\left(\sum_{k=1}^{3} \frac{\partial^2 u}{\partial x_k^2} + v(x)\right) u \tag{2}$$

(Schrödinger equation), and in fact the majority of calculations in solid state theory and in quantum chemistry is based on this equation. Nevertheless in some circumstances the Feynman formulation of quantum mechanics using integration over functional spaces, so called path integral, is very useful.

In this note we give a theorem (Theorem 1) on approximations of evolution operator of (1) for $A = A_1 + A_2$ by compositions of evolution operators for $A = A_1$ and for $A = A_2$. We reinterpret these approximations for A from (2) as approximations of a path integral (Theorem 2).

---

[1] Present address: Mathematical Institute, Polish Academy of Sciences, Krakow, Poland.

ISBN 0-12-186280-1

Now we recall a working definition of evolution operator. We say that an operator $E_A(t)$ is an evolution operator for A on D if, for any $t \geq 0$, $E_A(t)$ is an operator $D \to D$, $E_A(t)$ depends continuously on t for $v \in D$, $E_A(0)\, v = v$ for $v \in D$ and $E_A(t-\tau)v$ satisfies (1) for $t > \tau$, $v \in D$ (in the definition of $\frac{du}{dt}$ the topology of B is used).

<u>Theorem 1</u>. Let B be a Banach space with norm $|\ |_B$ and D a dense subspace of B. Suppose that

$$\text{linear operators } A_j : C_j \to B, \quad j = 1,2, \quad D \subset C_1, \quad C_2 \subset B \text{ are closed,} \tag{3}$$

on D there is defined another norm $|\ |_D$ ($|\ |_D$ need not be defined on the whole space B),

$$\text{there exist evolution operators } E_{A_1}(t),\ E_{A_2}(t) \text{ on } D \tag{4}$$

for $0 \leq t \leq 1$ defined and continuous on $B \times [0,1]$, and satisfy the following inequalities

$$(1-qt)\,|u|_D \leq |E_{A_j}(t)\, u|_D \leq (1+qt)\,|u|_D, \tag{5}$$

$$|E_{A_j}(t_1)\, u - E_{A_j}(t_2)\, u|_B \leq h\,|t_1 - t_2|\,|u|_D, \tag{6}$$

$$|A_j E_{A_l}(t_1)\, u - A_j E_{A_l}(t_2)\, u|_B \leq h\,|t_1 - t_2|\,|u|_D, \tag{7}$$

$$|E_{A_2}(t_1)\, A_1\, u - E_{A_2}(t_2)\, A_1\, u|_B \leq h\,|t_1 - t_2|\,|u|_D, \tag{8}$$

for $u \in D$, $0 \leq t_1, \ t_2 \leq 1$, $j,l = 1,2$, where q,h are positive numbers,

$$\text{the sets } S_r,\ A_1 S_r,\ A_2 S_r \text{ are compact in the norm } |\ |_B, \tag{9}$$

where $S_r = \{u \in D : |u|_D \leq r\}$, $r > 0$,

$$u(t) = 0 \tag{10}$$

for $t \geq 0$ is the unique solution of equation

$$\frac{du}{dt} = (A_1 + A_2)\ u, \tag{11}$$

satisfying initial condition $u(0) = 0$ and inclusion $u(t) \in D$.

Under these assumptions there exists an evolution operator $E_A(t)$ on D and

$$E_A(s) = \lim_{k\to\infty} F_k(s),$$

$$F_k(s) = E_{A_2}(\tfrac{s}{k})\ E_{A_1}(\tfrac{s}{k})\ \ldots\ E_{A_2}(\tfrac{s}{k})\ E_{A_1}(\tfrac{s}{k}) \quad \text{(k times)}.$$

Proof. Let us fix a positive number s and an element $w \in D$. Consider a sequence of functions $[0,s] \to D$: $f_k(0,w) = w$

$$f_k(t,w) = E_{A_2}(t - \tfrac{js}{k})\ E_{A_1}(t - \tfrac{js}{k})\ f_k(\tfrac{js}{k},w), \tag{12}$$

for $\frac{js}{k} < t \leq \frac{(j+1)s}{k}$, $j = 0,1,\ldots,k-1$. In virtue of (4) f_k are defined for large k.

It is easy to see, that $F_k(s)\ w = f_k(s,w)$.

In virtue of (5) we have

$$|f_k(t,w)|_D \leq c\ |w|_D, \tag{13}$$

(here and in the following we denote by c any constant independent of w, k, t, t_1, t_2).

From (6), (7) it follows that

$$|f_k(t_1,w) - f_k(t_2,w)|_B \leq c\ |t_1 - t_2|\ |w|_D \tag{14}$$

$$|A_j\ f_k(t_1,w) - A_j\ f_k(t_2,w)| \leq c\ |t_1 - t_2|\ |w|_D, \tag{15}$$

for $j = 1,2$, $k = 1,2,\ldots$.

Properties (9), (13) imply that

values $f_k, A_1 f_k, A_2 f_k$ for $0 \leq t \leq s$ belong to a compact, $|\ |_B$, set. (16)

In virtue of (13)-(16), for any sequence $j'(k)$ there exists a subsequence $l(k)$ that

$$f_{1(k)} \text{ together with } A_1 f_{1(k)} \text{ and } A_2 f_{1(k)} \text{ is uniformly convergent.} \tag{17}$$

We claim that the limit f of $f_{1(k)}$ satisfies (11). Indeed, from (12) it follows for $t \neq 0, s/k, 2s/k, \ldots s$ that the derivative $f'_{1(k)}$ satisfies the equality

$$f'_{1(k)}(t) = A_2 E_{A_2}(t-r) E_{A_1}(t-r) u + E_{A_2}(t-r) A_1 E_{A_1}(t-r) u,$$

where $r = \frac{js}{k}$, $u = f_k(\frac{js}{k}, w)$ and $|t-r| < \frac{1}{k}$. Putting in (8) $t_1 = 0$, $t_2 = t = r$, $u = E_{A_1}(t-r) u$ we get

$$|A_1 E_{A_1}(t-r) u - E_{A_2}(t-r) A_1 E_{A_1}(t-r) u|_B \leq h \frac{1}{k} |E_{A_1}(t-r) u|_D \leq c \frac{1}{k} |w|_D .$$

Analogously using (7) we get

$$|A_1 E_{A_1}(t-r) u - A_1 E_{A_2}(t-r) E_{A_1}(t-r) u|_B \leq c \frac{1}{k} |w|_D.$$

Therefore $|f'_{1(k)}(t,w) - (A_1+A_2) f_{1(k)}(t,w)|_B \leq c/k |w|_D$. By integration we obtain that the Lipschitz-continuous functions $f_{1(k)}$ satisfy the inequality

$$|f_{1(k)}(t,w) - w - \int_0^t (A_1+A_2) f_{1(k)}(r,w)\, dr|_B \leq \frac{c}{k} |w|_D,$$

for $0 \leq t \leq s$, and therefore by (17) and (3)

$$f(t,w) = w + \int_0^t (A_1+A_2) f(r,w)\, dr,$$

or $f'(t,w) = (A_1+A_2) f(t,w)$ and $f(0,w) = w$. In virtue of (10) $f(t,w)$ is the unique solution of this problem. Therefore the original sequence $f_k(t,w)$ is convergent to $f(t,w)$. The proof of Theorem 1 is thus complete.

If we take

$$A_1 = i\,T = i \sum_{k=1}^{3} \frac{\partial^2 u}{\partial x_k^2}, \quad A_2 = i\,V,$$

where V denotes the operator of multiplication by a function

$v(x)$ satisfying certain conditions, we can apply Theorem 1 to the Schrödinger equation $u' = i\,(T+V)\,u$, where T corresponds to kinetic energy of a particle (the arbitrary mass and other constants can be introduced easily) and $v(x)$ to its potential energy. Hamiltonian $H = T+V$ corresponds to total energy.

The evolution operators for A_1, A_2 can be given by the following formulas

$$w(x) = E_{A_1}(t)\;u(-)\;(x) = $$
$$= (\pi\, i\, t)^{-3/2} \int_{R^3} \exp(i(x-y)^2/4t)u(y)\;dy$$

$$w(x) = E_{A_2}(t)\;u(-)\;(x) = u(x)\;\exp(itv(x)).$$

We suppose that $v(x)$ is of class C^4 on R^3 with derivatives bounded up to the order four. We take the norms

$$|u|_B = \left(\int_{R^3}|u(y)|^2\;dy\right)^{\frac{1}{2}},\quad |u|_D = \max\left(\int_{R^3}|Xu(y)|^2 dy\right)^{\frac{1}{2}},$$

where maximum is taken over all derivatives X up to the order four (including of order zero). As B we take Hilbert space and as D we take complete (Sobolev) space with norm $|u|_D$.

One can prove that the assumptions of Theorem 1 with the exception of property (9) are satisfied. We can remove the difficulty with (9) using topology defined by seminorms

$$|u|_{S(r)} = \int_{S^r}|u(y)|^2\;dy$$

instead of Hilbert space topology defined by $|u|_B$. Therefore the evolution operator E_{iH} for Schrödinger equation is equal to the limit $\lim_{k\to\infty} F_k(s)$, where $F_k(s) = E_{iV}(s/k)E_{iT}(s/k)\;\ldots\; E_{iV}(s/k)E_{iT}(s/k)$, (k times); using expressions given above we get for $E_{iV}(t)E_{iT}(t)$ the formula

$$u \to w = (\pi\, i\, t)^{-3/2} \int_{R^3} \exp(i((x-y)^2/4t + tv(x)))\;u(y)\;dy;$$

using it we get formula with k-fold integral for $F_k(s)$.

We shall write it in the following

Theorem 2. Let function $v(x)$ be of class C^4, bounded together with its derivatives up to the order four, complex valued, defined on R^3.

Under these assumptions the evolution operator $E_{iH}(s)$ for Schrödinger equation

$$\frac{\partial u}{\partial t} = i \left(\sum_{j=1}^{3} \frac{\partial^2 u}{\partial x_j^2} + v(x)\, u \right)$$

is given by the limit

$$\lim_{k\to\infty} (\pi i s/k)^{-3k/2} \overbrace{\int_{R^3} \cdots \int_{R^3}}^{k} \exp\Big(i\Big(\sum_{j=1}^{k} (x_j - x_{j-1})\, k/4s + s v(x_j)/k\Big)\Big) u(x_0)\, d\,x_0 \ldots d\,x_{k-1}, \quad x_k = s.$$

Remark. The k-fold integral is a classical approximation for Feynman integral. Strictly speaking this expression, with meaning as in Fourier integral, can be used to define path integral.

ON PERIODIC SOLUTIONS OF HAMILTONIAN SYSTEMS OF ORDINARY DIFFERENTIAL EQUATIONS

Paul H. Rabinowitz[1,2]

Department of Mathematics
University of Wisconsin
Madison, Wisconsin

In this note we will survey some recent results about the existence of periodic solutions of Hamiltonian systems of ordinary differential equations. Thus let $p, q \in \mathbb{R}^n$, $H = H(p,q) : \mathbb{R}^{2n} \to \mathbb{R}$ and consider the Hamiltonian system

$$\dot{p} = -H_q(p,q), \quad \dot{q} = H_p(p,q) \tag{1}$$

where $\cdot$ denotes $\frac{d}{dt}$ and H_p, H_q denote the partial derivatives of H with respect to p and q. Setting $z = (p,q)$ and $\mathcal{J} = \begin{pmatrix} 0 & -I \\ I & 0 \end{pmatrix}$ where I is the identity matrix in $\mathbb{R}^n$, we can rewrite (1) as

$$z = \mathcal{J}H_z(z) . \tag{2}$$

---

[1]Sponsored in part by the Office of Naval Research under Contract No. N00014-76-C-0300. Reproduction in whole or in part is permitted for any purpose of the United States Government.

[2]Present address: Department of Mathematics, University of Wisconsin, Madison, Wisconsin 53706.

ISBN 0-12-186280-1

The study of the existence of periodic solutions of such general Hamiltonian systems has mainly focused on two questions: (i) solutions having a prescribed energy, (ii) solutions having a prescribed period. We begin by discussing the fixed energy case. The first work we know of is due to Seifert [1] who essentially treated

$$H(p,q) = \sum_{i,j=1}^{n} a_{ij}(q)\, p_i\, p_j + V(q)$$

where the kinetic energy term is positive definite in p for $q \in D$, a domain in $\mathbb{R}^n$ with $\bar{D}$ diffeomorphic to the closed unit ball $\overline{B_1(0)} \subset \mathbb{R}^n$, the coefficients a_{ij} and V real analytic in D, $V < b$ (a positive constant) in D, $V = b$ on ∂D, and $V_q \neq 0$ on ∂D. He then showed that under these hypotheses (2) has a periodic solution on $H^{-1}(b)$. His proof used arguments from differential geometry, obtaining the solution as a geodesic for the Riemannian metric associated with the kinetic energy term.

Seifert's result was extended by A. Weinstein [2] who proved:

<u>Theorem 3</u>. Suppose $H(p,q) = K(p,q) + V(q)$ where $K \in C^2(\mathbb{R}^{2n}, \mathbb{R})$, $V \in C^2(\mathbb{R}^n, \mathbb{R})$ and for some $b > 0$, K and V satisfy

(V_1) $D = \{q \in \mathbb{R}^n \mid V(q) < b\}$ is C^2 diffeomorphic to $B_1(0)$ in $\mathbb{R}^n$.

(V_2) $V_q \neq$ on ∂D.

(V_3) For each $q \in \bar{D}$ and $p \in \mathbb{R}^n$, $K(0,q) = 0$ and $K(p,q)$ is even and convex in p.

(K_2) For each $q \in \bar{D}$ and $p \in S^{n-1}$, $\lim_{\alpha\to\infty} K(\alpha p,q) > b - V(q)$.

Then there is a $T > 0$ and a solution $z = (p,q)$ of (2) such that $q(0), q(T) \in \partial D$ and $q(t) \in D$ for $t \in (0,T)$.

<u>Remark 4</u>. Weinstein observes that since $H(z(t)) \equiv b$ for this solution, $p(0) = p(T) = 0$. Thus if p is extended as an odd function and q as an even function about 0 and T, (K_1) and the form of H show the resulting extension of z is a $2T$ periodic solution of (2). In fact Seifert used similar reasoning for his case. Weinstein's proof was based in part

on that of [1] replacing the Riemannian metric by a Finsler metric.

Another fixed energy result was obtained independently by Weinstein [2] and this author [3,4].

Theorem 5. Suppose $H \in C^1(\mathbb{R}^{2n}, \mathbb{R})$ and $H^{-1}(b)$ is a manifold which is diffeomorphic to S^{2n-1} under radial projection. Then (2) has a periodic solution which lies on $H^{-1}(b)$.

Remark 6. Actually Weinstein assumed $H \in C^2(\mathbb{R}^{2n}, \mathbb{R})$ and that $H^{-1}(b)$ bounds a convex region. By a clever device he reduced the proof of the theorem to an application of Theorem 3. The methods used in [3]-[4] are rather different and will be described later.

Remark 7. An interesting open question in connection with Theorem 5 is whether better lower bounds obtain for the number of periodic solutions of (2) on $H^{-1}(b)$. For the special case of $H(z) = Q(z) + R(z)$ where Q is a positive definite quadratic form and $R(z) = o(|z|^2)$ at $z = 0$, Weinstein [5] showed for each small $b > 0$, (2) possesses at least n geometrically distinct periodic orbits on $H^{-1}(b)$. A simpler proof has been given by Moser [6]. An analogue of these bifurcation results for a fixed period situation was obtained by Fadell and this author [7].

Remark 8. Quite recently, Clarke [8] studies the convex case of Theorem 5 assuming only that H is locally Lipschitz continuous. Hence, since H is not necessarily C^1, (2) is replaced by an inclusion, i.e., a set valued differential equation. Using methods from convex analysis, he obtains a periodic solution of this problem on $H^{-1}(b)$.

Motivated by [2], we have recently proved a variant of Theorem 3 under a weaker condition than (K_1). Below $(\cdot,\cdot)_{\mathbb{R}^j}$ denotes the usual inner product in $\mathbb{R}^j$.

Theorem 9 [9]. Suppose $H(p,q) = K(p,q) + V(q)$ where $K \in C^2(\mathbb{R}^{2n}, \mathbb{R})$, $V \in C^2(\mathbb{R}^n, \mathbb{R})$, and (V_1)-(V_2), (K_2) are satisfied. If also

$(\overline{K_1})$ $K(0,q) = 0$ for $q \in \bar{D}$ and $(p, K_p(p,q))_{\mathbb{R}^n} > 0$ for $0 \neq p \in \mathbb{R}$ and $q \in \bar{D}$

holds, then (2) possesses a periodic solution on $H^{-1}(b)$.

The proofs given in [3] for Theorem 5 and in [9] for Theorem 9 are based on the same method which will be sketched next.

Observe first that the period T of any periodic solution is not known a priori. It is convenient to introduce T explicitly into the problem by scaling time via $t \to 2\pi t T^{-1} \equiv \lambda^{-1}$ so that (2) becomes

$$\dot{z} = \lambda J H_z . \tag{10}$$

We now seek a 2π periodic function z and a nonzero scalar λ such that the pair (λ, z) satisfies (6). Consider the functional

$$A(z) = \int_0^{2\pi} (p, \dot{q})_{\mathbb{R}^n} \, dt \tag{11}$$

where $z = (p, q)$ is a 2π periodic function subjected to the constraint

$$\frac{1}{2} \int_0^{2\pi} H(z(t))dt = b . \tag{12}$$

It is easy to check that formally any critical point of A subjected to (12) and having a nonzero Lagrange multiplier satisfies (10), λ essentially being the multiplier. Since any solution of (10) satisfies $H(z(t)) \equiv$ constant, (12) then implies that $z(t)$ lies on $H^{-1}(b)$. We do not know how to find critical points of this variational problem directly, the main difficulty being that A subjected to (12) is not bounded from above or below, even modulo a finite dimensional submanifold. Instead we use a finite dimensional approximation argument. Note that the functionals appearing in (11) and (12) are invariant under the translations $z(t) \to z(t+\theta)$ for any $\theta \in \mathbb{R}$. This symmetry allows an index theory developed in [7] to be applied to our variational problem. With the aid of this index theory, by restricting A to appropriate finite dimensional sub-manifolds of (12), minimax methods can be applied to get approximate critical points of (11)-(12). Furthermore a variational characterization given for the corresponding critical values yields bounds in turn for the critical values, Lagrange multipliers, and critical points. These estimates enable us to pass to a limit and get a solution of (10) on $H^{-1}(b)$. For more details we refer the reader to [3] and [9]. Another proof of Theorem 5 using Theorem 13 below has been given in [4].

We now turn to the case where the period, T, rather than the energy is prescribed. Several results for this situation have been obtained in [3]. In particular it was shown that

Theorem 13. Suppose that $H \in C^1(\mathbb{R}^{2n}, \mathbb{R})$ and satisfies

(H_1) $H(z) \geq 0$ for all $z \in \mathbb{R}^{2n}$

(H_2) $H(z) = o(|z|^2)$ at $z = 0$

(H_3) There is a $\theta \in (0, \frac{1}{2})$ and an $M > 0$ such that

$$0 < H(z) \leq \theta\,(z, H_z(z))_{\mathbb{R}^{2n}}$$

for $|z| \geq M$.

Then for any $T > 0$, (2) possesses a nonconstant T periodic solution.

Remark 14. If H depends explicitly on t in a T periodic fashion, then (2) possesses a nonconsistant T periodic solution provided that $(H_1)-(H_3)$ hold and e.g. $|H_z(z)| \leq a_1 |(z, H_z(z))_{\mathbb{R}^{2n}}| + a_2$ for all $z \in \mathbb{R}^{2n}$.

Remark 15. On integration, (H_3) implies that $H(z) \geq a_3|z|^{\frac{1}{\theta}} - a_4$ for all $z \in \mathbb{R}^{2n}$, i.e. H grows at a "superquadratic" rate. Conditions $(H_1)-(H_2)$ are not essential to the conclusion of Theorem 13. In fact we have the following stronger result:

Theorem 16. Suppose $H \in C^1(\mathbb{R}^{2n}, \mathbb{R})$ and satisfies (H_3). Then for any T, $r > 0$, (2) possesses a T periodic solution z with $\|z\|_{L^\infty} > r$.

Remark 17. Theorem 13 and 16 make no assertion about the solution obtained for (2) having T as a primitive or minimal period. We suspect this is the case in the setting of Theorem 13. However simple examples (see e.g. [3]) show the period need not be minimal if only (H_3) is satisfied.

The proofs of Theorem 13 and 16 follow essentially the same pattern as that sketched for Theorems 5 and 9. Scaling time as earlier, we seek a 2π periodic solution of (10) where λ is now fixed. Setting

$$I(z) = A(z) - \lambda \int_0^{2\pi} H(z)\,dt\,, \tag{18}$$

a formal calculation shows that any critical point of (18) gives a solution of (10). A direct existence proof of critical points of I encounters the same difficulties indicated earlier with the further complication that $z \equiv 0$ is a known critical point which must be avoided. However we can successfully employ an approximation argument involving (a) finding critical points of I restricted to appropriate finite dimensional subspaces of periodic functions via minimax methods, and (b) getting strong enough estimates for these approximate solutions to pass to a limit. These two steps must be supplemented by (c) an argument to show the solution thus obtained is nonconstant. The existence proof for critical points of the approximate problem for Theorem 16 again uses the index theory of [7]; however for Theorem 13 the additional structure provided by $(H_1)-(H_2)$ permits a simpler proof.

Remark 19. By using some elementary tricks, it was shown in [4] that Theorem 5 follows from Theorem 13. However the original proof in [3] yields better upper and lower bounds for the unknown period than does the simpler argument.

Remark 20. In a recent paper, Clarke and Ekeland [10] have studied the fixed period problem for (2) for Hamiltonians which are globally convex and appropriately "subquadratic" at 0 and ∞, e.g. $|z|^{-2}H(z) \to \infty$ as $|z| \to 0$ and $\to 0$ as $|z| \to \infty$, just the contrary situation to $(H_2)-(H_3)$ of Theorem 13. Using methods from convex analysis as in [8], they find solutions of (2) which have minimal period T for any $T > 0$.

REFERENCES

1. Seifert, H., Periodische Bewegungen Mechanischer Systeme, *Math. Z. 51*, 197-216 (1948).
2. Weinstein, A., Periodic orbits for convex Hamiltonian systems. To appear *Annals of Math.*
3. Rabinowitz, P. H., Periodic solutions of Hamiltonian systems, *Comm. Pure Appl. Math. 31*, 157-184 (1978).
4. Rabinowitz, P. H., A variational method for finding periodic solutions of differential equations. To appear in "Proc. Sym. on Nonlinear Evolution Eq." (M.G. Crandall, ed.).

5. Weinstein, A., Normal modes for nonlinear Hamiltonian systems, *Inv. Math. 20*, 47-57 (1973).
6. Moser, J., Periodic orbits near an equilibrium and a theorem of Alan Weinstein, *Comm. Pure Appl. Math. 29*, 727-747 (1976).
7. Fadell, E. R. and Rabinowitz, P. H., Generalized cohomological index theories for Lie group actions with an application to bifurcation questions for Hamiltonian systems, *Inv. Math. 45*, 139-174 (1978).
8. Clarke, F. H., Periodic solutions to Hamiltonian inclusions, preprint.
9. Rabinowitz, P. H., Periodic solutions of a Hamiltonian system on a prescribed energy surface. To appear *J.Diff. Eq.*
10. Clarke, F. H. and Ekeland, I., Hamiltonian trajectories having prescribed minimal period, preprint.

SOME RESULTS IN FUNCTIONAL INTEGRAL EQUATIONS IN A BANACH SPACE

D. R. K. Rao[1]

Faculty of Sciences
Razi University
Kermanshah, Iran

The study of abstract integral equations and abstract differential equations has drawn the attention of many recent researchers. The aim of the present paper is to investigate some problems of qualitative nature concerning functional integral equations of the form

$$x(t) = \gamma(\sigma,\varphi) + g(t,x_t) + \int_\sigma^t k(t,s)f(s,x_s)ds + \int_\sigma^t h(s)ds \tag{1}$$

with the initial value $x_\sigma = \varphi$ where $x_t(\theta) = x(t+\theta)$, $-r \leq \theta \leq 0$, in a Banach Space. As has been pointed out by Lakshmikantham, in the case of functional integral equations and functional differential equations in Banach Spaces, there are some difficulties in imposing the assumptions since, in this case, the domain and the range of the functions involved in the functional equations are not in the same Banach Space.

---

[1]Present address: Faculty of Sciences, Razi University, Kermanshah, Iran.

ISBN 0-12-186280-1

Following the idea of Lakshmikantham we overcome this difficulty by imposing conditions over a subset of the domain in a suitable way and obtain several results, namely, existence, uniqueness, boundedness, stability and asymptotic behaviour of the solutions fo equation (1). Thereby we extend the results of Nohel [6] and also of Corduneanu [2] and R. K. Miller, Nohel and James Wong [5] on perturbation problems.

Let $r > 0$ be a given number and let E be a Banach Space with $\|\cdot\|$. Let $E_o = C([-r,0],E)$ denote the Banach Space of continuous functions with the norm given by $\|\varphi\|_o = \max_{-r\le s\le 0} \|\varphi(s)\|$. If $\sigma \in R^+$ and $x \in C([\sigma-r,\infty),E)$ then for any $t \in [\sigma,\infty)$, $x_t \in E_o$ implies

$$x_t(s) = x(t+s), \quad -r \le s \le 0$$

Theorem 1. We need the following assumptions with respect to the initial value problem (1)

a) $h(t) \in C(R^+,E)$

b) $g(t,x_t),\ f(t,x_t) \in C(R^+ \times E_o,E)$

c) $k(t,s) = \{k_{ij}(t,s)\}$, $i = 1,2,\ldots,n$, $j = 1,2,\ldots,m$, is a continuous matrix for $0 \le s < t < \infty$, with the assumption that the Banach Space $E \subset R^n$

d) there exist two continuous non-negative functions $a(t)$ and $b(t)$ defined on R^+ such that

$$\|f(t,x_t)\| \le b(t) \quad \text{and} \quad \|g(t,x_t)\| \le b(t) \quad \text{for}$$
$$\|x(t)\| \le a(t), \quad t \in R^+ \tag{2}$$

$$\|\gamma(\sigma,\varphi)\| + \|b(t)\| + \int_\sigma^t \|k(t,s)\|\, b(s)\,ds \le a(t),$$
$$t \in R^+, \quad t > \sigma > 0 \tag{3}$$

Then there exists at least one solution $x(t)$ of the equation (1) in the Banach Space E such that

$$\|x(t)\| \le a(t), \quad t \in R^+$$

Proof. We prove the theorem by using Schauder's fixed point theorem.

In the space $C([\sigma-r,\infty),E)$ let us consider the set

$$S = \{x(t);\ \|x(t)\| \le a(t),\quad t \in [\sigma-r,\infty)\}$$

The set S is a closed convex set.

For any $x(t) \in S$, let

$$(Tx)(t) = \gamma(\sigma,\varphi) + g(t,x_t) + \int_\sigma^t k(t,s)f(s,x_s)ds + \int_\sigma^t h(s)ds,\quad t \in R^+,\quad t > \sigma > 0 \tag{4}$$

From the assumptions it follows that the operator T is continuous operator from S to $C([\sigma-r,\infty),E)$. Also for every $x(t) \in S$, we have

$$\|(Tx)(t)\| = \|\gamma(\sigma,\varphi)\| + \|g(t,x_t)\| + \int_\sigma^t \|k(t,s)\|\,\|f(s,x_s)\|\,ds + \int_\sigma^t \|h(s)\|\,ds \le a(t),\quad t \in R^+,\quad t > \sigma > 0$$

which implies that $TS \subset S$ (5)

Next we prove that TS is a relatively compact set in $C([\sigma-r,\infty),E)$

$C([\sigma-r,\infty),E)$ is metrizable space.

$\|(Tx)(t)\| \le a(t)$, $t \in [\sigma-r,\infty)$ for every $x(t) \in S$ and the continuity of $a(t)$ imply the uniform boundedness of the functions belonging to TS on every bounded interval $[a,b]$ $[\sigma-r,\infty)$. Also

$$\|(Tx)(t) - (Tx)(t')\| \le \|g(t,x_t) - g(t',x_{t'})\| + \int_\sigma^t \|k(t,s) - k(t',s)\|b(s)\,ds + \int_{t'}^t \|k(t,s)\|\,b(s)ds + \int_{t'}^t \|h(s)\|\,ds$$

which implies that the functions of TS are equicontinuous on every compact interval of $[\sigma-r,\infty)$.

Using Ascoli's theorem we get that TS is a relatively compact set in $C([\sigma-r,\infty),E)$.

Let S_1 be the closure of TS in $C([\sigma-r,\infty),E)$, that is

$$TS \subset S_1 \subset S$$

S_1 is compact. Hence applying the fixed point theorem of Schauder we obtain the required result.

Remarks. On differentiating equation (1) we will get neutral functional differential equation. That is, neutral functional differential equations and functional difference equations are included as special cases of equation (1).

In this section we consider two functional integral equations

$$x(t) = \gamma_1(\sigma,\varphi) + g_1(t,s_t) + \int_\sigma^t k(t,s) f_1(s,x_s)ds + \int_\sigma^t h_1(s)ds \tag{6}$$

with the initial value $x_\sigma = \varphi$ and

$$y(t) = \gamma_2(\sigma,\psi) + g_2(t,y_t) + \int_\sigma^t k(t,s) f_2(s,y_s)ds + \int_\sigma^t h_2(s)ds \tag{7}$$

with the initial function $y_\sigma = \psi$ and obtain some interesting properties for solutions of (6) and (7) by using the well known comparison technique.

Theorem 2. We assume the following conditions

a) $g_i(t,x_t)$, $f_i(t,x_t) \in C(R^+ \times E_o,E)$, $i = 1,2$

b) $h_i(t) \in C(R^+,E)$, $i = 1,2$

c) Suppose there exists continuous scalar functionals $G(t,r_t)$, and $W(t,r_t)$ and scalar functions $K(t,s)$, for $0 \leq s < t < \infty$, $H(t)$ which are all non-negative defined on R^+ and $\gamma(\sigma,u)$ such that

i) $\|\gamma_1(\sigma,\varphi) - \gamma_2(\sigma,\psi)\| \leq \gamma(\sigma,\|\varphi-\psi\|)$

ii) $\|g_1(t,x_t) - g_2(t,y_t)\| \leq G(t, \|x_t-y_t\|)$

iii) $\|k(t,s)\| \leq K(t,s)$

iv) $\|f_1(t,x_t) - f_2(t,y_t)\| \leq W(t, \|x_t-y_t\|)$

v) $\|h_1(t) - h_2(t)\| \leq H(t)$

d) Let $m(t)$ be the maximal solution of the scalar functional integral equation

$$r(t) = \gamma(\sigma,u) + G(t,r_t) + \int_\sigma^t K(t,s)W(s,r_s)ds + \int_\sigma^t H(s)ds \tag{8}$$

with the initial function $r_\sigma = u$, $\sigma > 0$.

Then the solution $x(t)$ and $y(t)$ of the functional integral equations (6) and (7) satisfy the inequality

$$\|x(t) - y(t)\| \leq m(t), \qquad t \in [\sigma-r,\infty) \tag{9}$$

Note. The existence of the maximal solution for the scalar functional equation has been proved by Lakshmikantham.

Proof of Theorem 2 is similar to the corresponding one given by Nohel [6].

Remarks. 1. Many other results obtained by Nohel in [6] can as well be developed for the functional integral equations (6) and (7).

2. Exploiting the inequality (9), we can get many interesting results, like the results on perturbation problems by Corduneanu [2], by Miller, Nohel and Wong [5] etc.

Similarly it is possible to develop some useful relationships between the solutions of functional integro-differential equations

$$x'(t) = g_1(t,x_t) + \int_\sigma^t k(t,s)f_1(s,x_s)ds + \int_\sigma^t h(s)ds \tag{10}$$

and

$$y'(t) = g_2(t,y) + \int_\sigma^t k(t,s)f_2(s,y_s)ds + \int_\sigma^t h(s)ds \tag{11}$$

On differentiating equation (11), we get functional differential equation of second order

$$z''(t) = F(t, z(t), z_t) \tag{12}$$

which yields its own interesting properties like oscillatory and periodic solutions which are analogues of corresponding types of ordinary differential equations of second order - see Norkin's work in [7].

ACKNOWLEDGEMENT

I am deeply indebted to Professor S. K. Singh for his most invaluable help and kind encouragement in my work.

REFERENCES

1. Corduneanu, C., On a class of functional-integral equations. *Bulletin Math. de la Soc. Sci. Math. de la R.S. de Roumanie 12(60)*, nr. 4 (1968).
2. Corduneanu, C., Some Perturbation problems in the theory of integral equations. *Math. Systems Theory 1*, nr. 2 (1967).
3. Hale, J. K., and Meyer, K., A class of functional differential equations of neutral type. Memoirs of the American Mathematical Society No. 76 (1967).
4. Lakshmikantham, V., Mitchel, A. R. and Mitchel, R. W., On the existence of solutions of differential equations of retarded type in a Banach Space. Tech. Rep. No. 28, Dept. of Math. Univ. of Texas at Arlington, June 1975.
5. Miller, R. K., Nohel J. A., and Wong, J. A. W., Perturbations of Volterra Integral Equations. *J. Math. Anal. Appl. 25*, nr. 3 (1969).
6. Nohel, J. A., Some problems in nonlinear Volterra integral equations. *Bull. Amer. Math. Soc. 68*, nr. 4 (1962).
7. Norkin, S. B., Differential Equations of the Second Order with Deviating Argument. Translated from Russian under the series "Translations of Mathematical Monographs", vol. 31, American Mathematical Society, Providence, R.I. (1972).

TURBULENCE AND HIGHER ORDER BIFURCATIONS

George R. Sell[1,2]

School of Mathematics
University of Minnesota
Minneapolis, Minnesota

I. SPECULATIONS ABOUT TURBULENCE

Turbulence is a phenomenon which can be observed in various hydrodynamical systems which depend on a parameter ω, e.g. an external force, or the viscosity or Reynolds number. As this parameter crosses some critical value ω_T, one notices a very complicated and chaotic behavior in the fluid flow. This behavior is oftentimes referred to as turbulence.

The central mathematical problem here is to give a mathematical description of not only the chaotic behavior but also the behavior of the flow before the onset of turbulence. At this time, a precise and widely accepted definition of turbulence does not exist, even for the simplest of hydrodynamical systems. However it is clear that a study of the changes in the underlying flow as the parameter ω varies is crucial for

[1]Supported in part by NSF Grants MCS 76-06003 and MCS 78-00907.

[2]Present address: School of Mathematics, University of Minnesota, Minneapolis, Minnesota 55455.

ISBN 0-12-186280-1

an understanding of turbulence. As this parameter varies, one does observe changes in the flow which can be described by bifurcation theory. Therefore it is expected that by a study of bifurcation theory one can find the key for an understanding of turbulence. Let us now look at a specific example.

Picture a viscous fluid between two coaxial rotating cylinders. For simplicity let us assume that the outer cylinder is fixed and that the inner cylinder rotates with an angular velocity ω. Elementary experiments [5] show that there are three critical values of the parameter $0 < \omega_1 < \omega_2 < \omega_3$ with the following properties:

For $0 < \omega < \omega_1$, one has the classical Couette flow where the fluid travels in circular streamlines about the inner cylinder. The Couette flow is represented as a stationary solution of the associated Navier-Stokes equations.

For $\omega_1 < \omega < \omega_2$, Taylor cells or Taylor vortices occur. A radial component of the velocity of the fluid flow develops and the fluid itself decomposes into a series of cells, or stacked solid tori about the axis of the cylinders. These tori, the Taylor cells, are invariant sets for the flow. The Taylor cells also are stationary solutions of the Navier-Stokes equations.

For $\omega_2 < \omega < \omega_3$, the Taylor cells develop waves which precess around the inner cylinder. These wavy Taylor vortices appear to be time-periodic solutions of the Navier-Stokes equations.

For $\omega_3 < \omega$, the behavior of the fluid flow becomes very complicated, and a complete understanding of what happens as ω crosses ω_3 is still unavailable. However a recent experiment of Gollub and Swinney [2] suggests that the time-periodic solutions, which represent the wavy Taylor vortices, develop another fundamental frequency and change to a quasi-periodic motion. In other words, it appears that for ω close to ω_3, and $\omega_3 < \omega$, there is an invariant two-dimensional torus T^2 in the associated flow. The Gollub-Swinney experiment suggests that the invariant torus T^2 persists for a range of values ω with $\omega_3 < \omega$, after which the flow breaks down into some chaotic behavior. Thus the critical value ω_T, which describes the onset of turbulence, satisfies $\omega_3 < \omega_T$ for their flow.

One of the earliest conjectures on the nature of turbulence was offered by Landau [6] and Hopf [4] about thirty years ago. They speculated that turbulence developed as a series of bifurcations. First a stationery solution (T^0) would bifurcate into a time-periodic solution (T^1). Next the time-periodic solution would bifurcate into a quasi-periodic solution with two basic frequencies (T^2), and this quasi-periodic solution would bifurcate into a quasi-periodic solution with three basic frequencies (T^3), etc. A quasi-periodic solution with k-fundamental frequencies defines an invariant k-dimensional torus T^k in the underlying flow. Thus the Hopf-Landau conjecture can be described schematically by the series of bifurcation illustrated in Figure 1.

$$T^0 \rightarrow T^1 \rightarrow T^2 \rightarrow T^3 \rightarrow T^4 \rightarrow T^5 \rightarrow \ldots$$

FIGURE 1. The Hopf-Landau Series

The basic work of Hopf in 1942 on bifurcation theory [3] gives a satisfactory mathematical theory which explains the first step $T^0 \rightarrow T^1$ in the Hopf-Landau series. Eventhough Hopf's bifurcation theory is formulated for systems of ordinary differential equations, he was convinced that the insight one gains in studying ordinary differential equations is applicable to certain partial differential equations such as the Navier-Stokes equations. This convinction has certainly been justified during recent years, as can be seen in Marsden and McCracken [7] and elsewhere.

A satisfying theory for the secondary bifurcation $T^1 \rightarrow T^2$ was developed in 1964 by Sacker [9]. However a mathematical theory for the higher order - bifurcations $T^k \rightarrow T^{k+1}$, $k \geq 2$, remained unresolved.

In 1971, Ruelle and Takens [8] reexamined the conjecture of Hopf and Landau. They did not resolve this conjecture. Instead, by using generic theories developed by Smale [13] and others in their study of structural stability, they argued that the Hopf-Landau conjecture is "unlikely". In addition

Ruelle and Takens construct an example which suggests that the Hopf-Landau series may stop at T^4. The example is somewhat complicated, but it does show that a simple rotational flow on T^4 can bifurcate into a flow with a "strange" attractor. Because of this, Ruelle and Takens offer a new explanation of turbulence. They speculate that turbulence occurs at a bifurcation point where the underlying flow develops a strange attractor.

The Notion of a strange attractor has appeared recently in a number of papers. As with turbulence, the concept of a strange attractor does not have a universally accepted definition. A nonperiodic attractor is a starting point, but this would not exclude a stable invariant torus. Ruelle and Takens require that a strange attractor have a suitable hyperbolic structure, however the Lorenz attractor [7], which does arise in bifurcation theory, does not have this hyperbolic property.

Clearly one needs a better understanding of bifurcation phenomena before one can hope to understand something as complicated as turbulence. In this note we wish to report on a recent development in the study of bifurcations [1,12]. This work does not in itself resolve the controversy over the nature of turbulence. However it is a first step in this direction. Specifically we shall outline a theory which describes when the higher order bifurcations $T^k \to T^{k+1}$, $k \geq 2$, can occur. With this theory of higher-order bifurcations, it is now expected that one can perform a critical analysis of not only the Hopf-Landau conjecture but also the Ruelle-Takens conjecture.

In the next section we shall describe the Hopf theory of first order bifurcations $T^0 \to T^1$. Next we shall consider the second-order bifurcation $T^1 \to T^2$ and then in Section IV we shall describe both the Chenciner-Iooss theory [1] and the Sell theory [12] of higher-order bifurcations. We refer the reader to the references cited below for the proofs of these bifurcation theories.

II. HOPF BIFURCATION. (FIRST-ORDER BIFURCATION)

Consider a family

$$X' = f(X,\alpha) \tag{1}$$

of ordinary differential equations on R^n where f depends on a real parameter α. Assume that $f(0,\alpha) \equiv \alpha$ for all α in some interval I. Let $A(\alpha)$ denote the linear part of f at $X = 0$. Assume that a parametric value (say $\alpha = 0$) is "critical" in the sense that the following four conditions are satisfied.

(H1). A pair of complex conjugate eigenvalues $\lambda(\alpha)$ and $\bar{\lambda}(\alpha)$ cross the imaginary axis at $\alpha = 0$.

(H2). All the remaining eigenvalues of $A(\alpha)$ stay away from the imaginary axis for α near 0.

(H3). $\mathrm{Re}\lambda(0) = 0$ and $\mathrm{Im}\lambda(0) \neq 0$.

(H4). $\mathrm{Re}\lambda'(0) \neq 0$ where $\lambda'(\alpha) = d\lambda/d\alpha$.

Hopf showed that if $\alpha = 0$ is critical in the above sense, then the system (1) admits a family of periodic orbits in the vicinity of $(X,\alpha) = (0,0)$. In addition, if a certain constant K, which depends on the nonlinear terms in f, in nonzero, then these periodic orbits $\hat{\tau}(\alpha)$ can be parameterized in terms of α (for either $\alpha > 0$ or $\alpha < 0$). Also one has $\hat{\tau}(\alpha) \to 0$ as $\alpha \to 0$.

III. SECOND-ORDER BIFURCATION

Now assume that the differential system (1) has a family of periodic orbits $\tau(\alpha)$ which vary smoothly in α. The Floquet multipliers of $\tau(\alpha)$ are the eigenvalues of the linearized Poincaré map associated with $\tau(\alpha)$. Assume that a parametric value $\alpha = 0$ is "critical" in the sense that the following five conditions are satisfied:

(J1). A complex conjugate pair of Floquet multipliers $\lambda(\alpha)$ and $\bar{\lambda}(\alpha)$ cross the unit circle at $\alpha = 0$.

(J2). All the remaining Floquet multipliers remain away from the unit circle for α near 0.

(J3). $\lambda(0) \neq 1$.

(J4). $\frac{d}{dx}|\lambda(\alpha)|\,|_{\alpha=0} \neq 0$.

These four conditions are the direct analogue of the four Hopf conditions (H1)-(H4). However for second order bifurcations the following nonresonance condition is needed:

(J5). $\lambda^m(0) \neq 1$ for $m = 1,2,3,4$.

In this case one can show that if a certain computable constant K, which depends on the nonlinear Poincaré map, is nonzero, then the system (1) admits a family of invariant two dimensional tori $\tilde{\tau}(\alpha)$ which are defined either for $\alpha > 0$ or $\alpha < 0$ (depending on the sign of K). In addition one has $\tilde{\tau}(\alpha) \to \tau(0)$ or $\alpha \to 0$.

Unlike the first-order bifurcation, the proof of the second-order bifurcation theorem does require that the nonresonance condition (J5) is satisfied and that the constant K, referred to above, is nonzero. As we now turn to the higher-order bifurcation, we shall see that the corresponding theory there is built on suitable analogues of the five conditions (J1)-(J5) together with the assumption that $K \neq 0$.

IV. HIGHER-ORDER BIFURCATION

Now assume that the differential system (1) has a family of k-dimensional invariant tori $\tau(\alpha)$, which vary smoothly for α in some open interval I, where $\alpha = 0 \in I$.

The Chenciner-Iooss theory of higher-order bifurcations is based upon a study of the associated Poincaré map which is generated by $\tau(\alpha)$. Let X be a real Banach space and let denote a neighborhood of the origin 0 in X. For $\alpha \in I$ let

$$F_\alpha : T^k \times \mathcal{V} \to T^k \times X$$

be a family of mappings of class C^k, with the property that for each α, F_α induces a C^k-diffeomorphism on $T^k \times \{0\}$. This means that F_α can be represented in the form

$$F_\alpha(\theta,x) = (\theta + \varphi_\alpha(\theta,x), \quad \Phi_\alpha(\theta,x))$$

where φ_α and Φ_α are periodic in $\theta \in R^k$ and $\Phi_\alpha(\theta,0) = 0$ and $g_\alpha(\theta) = \theta + \varphi_\alpha(\theta,0)$ is a C^k-diffeomorphism on T^k.

Chenciner and Iooss develop their theory of bifurcation by studying F_α and the associated linearized maps. We will not give a detailed description of their hypotheses here. These hypotheses are appropriate generalizations of the five conditions (J1)-(J5) referred to in the last section together with an assumption that an appropriate constant $K \neq 0$. The object of this study is to show that the fixed point for F_α, viz. $T^k \times \{0\}$, bifurcates into an invariant circle $T^k \times S^1$ in the space $T^k \times \mathcal{V}$. This then describes an invariant $(k+1)$-dimensional torus for the mapping F_α.

The approach to higher-order bifurcations used by Sell is based upon the spectral theory for invariant manifolds which is developed in Sacker and Sell [10,11]. In this framework, the appropriate generalizations of (H1) and (H2) become two geometric hypotheses (L1) and (L2) which describe the normal spectrum of $\tau(\alpha)$. A careful statement of (L1) and (L2) would be quite lenghty and we shall not present it here. However in loose terms (L1) and (L2) require that for $\alpha \neq 0$ the flow has a hyperbolic structure in the vicinity of the invariant torus $\tau(\alpha)$ and that the dimension of the unstable manifold (or the stable manifold) changes by 2 as α crosses the critical value $\alpha = 0$.

The consequence of Hypotheses (L1) and (L2) is that there is a suitable curvilinear coordinate system in the vicinity of $\tau(\alpha)$ so that the differential equation (1) becomes

$$\begin{aligned} x' &= A_{11}(y,\alpha)x + \alpha A_{12}(y,\alpha)z + F(x,y,z,\alpha) \\ y' &= G(x,y,z,\alpha) \\ z' &= \alpha A_{21}(y,\alpha)x + [B(y) + \alpha A_{22}(y,\alpha)]z + H(x,y,z,\alpha). \end{aligned} \tag{2}$$

where $x \in R^2$, $y \in T^k$, $z \in R^{n-k-2}$ and $\alpha \in I$. The terms B and A_{ij} denote matrices of the appropriate dimensions and F and H contain higher order terms in x and z. Furthermore one has $F(0,y,0,\alpha) = 0$ and $H(0,y,0,\alpha) = 0$. Also the differential equation $y' = G(0,y,0,\alpha)$ denotes the restriction of the flow generated by (1) to the torus $\tau(\alpha)$.

Let (ρ,θ_0) denote the polar coordinates in the x-plane and let $\theta = (\theta_0,y)$ denote a typical point in T^{k+1}, the $(k+1)$-dimensional torus. For any continuous function $u = u(\theta)$ defined on T^{k+1}, we let

$$M_\theta[u] = \left(\frac{1}{2\pi}\right)^{k+1} \int_0^{2\pi} \cdots \int_0^{2\pi} u(\theta_0,y)\,d\theta_0\,dy$$

denote the mean-value of u.

Now expand the (2×2) matrix $A_{11}(y,\alpha)$ in terms of α, that is, let

$$A_{11}(y,\alpha) = \Omega + \alpha W(y,\alpha)$$

where $\Omega = A_{11}(y,0)$. Let $a_{ij}(y,\alpha)$ denote the entries of $W(y,\alpha)$. Then the next three hypotheses are the following:

(L3). *The* (2×2) *matrix* Ω *satisfies*

$$\Omega = \begin{pmatrix} 0 & -\omega_0 \\ \omega_0 & 0 \end{pmatrix}$$

where ω_0 *is a nonzero constant.*

(L4). *The mean value*

$$\bar{W} = M_\theta[w_{11}(y,0)\cos^2\theta_0 + (w_{12}(y,0) + $$
$$+ w_{21}(y,0)\sin\theta_0\cos\theta_0 + w_{22}(y,0)\sin^2\theta_0]$$

is nonzero.

(L5). *There is a constant vector* $\bar{\omega} = (\omega_1,\ldots,\omega_k)$ *and a smooth function* $L(x,y,z,\alpha)$ *such that*

$$G(x,y,z,\alpha) = \bar{\omega} + \alpha L(x,y,z,\alpha) .$$

Moreover there exist positive constants c *and* γ *such that one has*

$$|n \cdot \omega| \geq c|n|^{-\gamma} \tag{3}$$

for all $(k+1)$ - *triples* $n = (n_0, n_1, \ldots, n_k) \in Z^{k+1}$ *with* $n \neq 0$. Here we adopt the usual convention that

$$n \cdot \omega = n_0\omega_0 + n_1\omega_1 + \ldots + n_k\omega_k$$

$$|n| = \max_i |n_i|$$

In the case $k = 0$, the y - variable is missing in (2). Consequently the matrix A_{11} depends only on α and one has

$$A_{11}(\alpha) = \begin{pmatrix} \mathrm{Re}\lambda(\alpha) & -\mathrm{Im}\lambda(\alpha) \\ \mathrm{Im}\lambda(\alpha) & \mathrm{Re}\lambda(\alpha) \end{pmatrix}$$

where $\lambda(\alpha)$ is given by hypothesis (H1). In this case hypothesis (L3) reduces to (H3) where $\mathrm{Im}\lambda(0) = \omega_0$. Also (L4) becomes (H4) with $\bar{W} = \mathrm{Re}\lambda'(0)$.

The nonresonance condition (J5), which first appeared in the second-order bifurcation theory, now takes the form (L5) in this higher-order theory. One should note that even for $k = 1$, our condition (L5) is much stronger than the corresponding condition (J5). Also our condition (L5) is somewhat stronger than the corresponding Chenciner-Iooss nonresonance condition.

The inequality (3) assures us that

$$n_0\omega_0 + n_1\omega_1 + \ldots + n_k\omega_k \neq 0$$

for any collection $n_0, n_1, \ldots, n_k$ of integers with $|n| \neq 0$. Also notice that the flow on the torus $\tau(0)$ is the irrational twist flow $y' = \bar{\omega}$.

In our statement of the main result, which we give below, reference is made to a constant K. This constant, which is expressed as a mean value, depends upon the low-order terms (i.e. order ≤ 3) in the Taylor series expansion of the system (2). The formula for K is too complicated to reproduce here.

Theorem. *Let* $\tau(\alpha)$ *be a family of* k-*dimensional invariant tori for Equation* (1). *Assume that hypotheses* (L1) *and* (L2) *are satisfied so that in the vicinity of* $\tau(\alpha)$ *Equation* (1) *reduces to Equation* (2). *Assume further that hypothesis* (L3), (L4) *and* (L5) *are satisfied and that the constant* K *is nonzero. Then there is a unique family of* $(k+1)$*-dimensional invariant tori* $\hat{\tau}(\alpha)$ *defined for*

$$\alpha \cdot \operatorname{sgn}(\bar{W} K) < 0$$

and one has $\hat{\tau}(\alpha) \to \tau(0)$ *as* $\alpha \to 0$. *Furthermore if* $\bar{W} > 0$, $K < 0$ *and the tori* $\tau(\alpha)$ *are asymptotically stable for* $\alpha < 0$, *then the bifurcating tori* $\hat{\tau}(\beta)$ *are asymptotically stable for* $\beta > 0$.

V. THE HOPF-LANDAU CONJECTURE

This higher-dimensional bifurcation theory now gives a framework for deciding which dynamical systems have a Hopf-Landau series of bifurcations illustrated by Figure 1. An example of a bifurcation $T^2 \to T^3$ is given in [12]. The same example can be easily modified to illustrate higher bifurcations $T^k \to T^{k+1}$, $k \geq 3$. In fact one can further modify this example to construct a series of bifurcations $T^0 \to T^1 \to T^2 \to \ldots$. Of course, for a finite-dimensional ordinary differential equation such a series is necessarily finite, since the dimension k of the torus T^k cannot exceed the dimension of the ambient space. In fact if the ambient space is the Euclidean space R^n, one must have $k \leq n-1$ for all k.

Examples constructed in this fashion need not be related to any hydrodynamical model. The existence of a higher order bifurcation $T^2 \to T^3$ remains an open question in any hydrodynamical system. This is certainly worthy of further study.

REFERENCES

1. Chenciner, A. and Iooss, G., Bifurcations de torus invariante. *Arch. Rational Mech. Anal.* (to appear).
2. Gollub, J. P. and Swinney, H. L., Onset of turbulence in a rotating fluid. *Phy. Rev. Letters 35*, 921 (1975).
3. Hopf, E., Abzweigung einer periodischen Lösung eines Differential systems. *Ber. Math. Phys. Sachsische Akad. Wiss. Leipzig 94*, 1-22 (1942). (Also see Reference 7, pp 63-193, for English translation).
4. Hopf, E., A mathematical example displaying features of turbulence. *Comm. Pure Appl. Math. 1*, 303 (1948).
5. Joseph, D. D., Stability of Fluid Motions I, II. Springer Tracts in Natural Philosophy vol. 27, 28, Springer-Verlag, New York, (1976).
6. Landau, L. D., On the problem of turbulence. *C.R. Acad. Sci. USSR 44*, 311 (1944).
7. Marsden, J. E. and McCracken, M., The Hopf Bifurcation and Its Applications. Applied Math. Science Vol. 19, Springer-Verlag, New York, (1976).
8. Ruelle, D. and Takens, F., On the nature of turbulence. *Comm. Math. Phys. 20*, 167-192 (1971), and *23*, 343-344 (1971).
9. Sacker, R. J., On invariant surfaces and bifurcation of periodic solutions of ordinary differential equations. New York University Tech. Report IMM-NYU, (1964).
10. Sacker, R. J. and Sell, G. R., A spectral theory for linear differential systems. *J. Differential Equations 27*, 320-358 (1978).
11. Sacker, R. J. and Sell, G. R., The spectrum of an invariant manifold and block triangular systems. (to appear).
12. Sell, G. R., Bifurcation of higher dimensional tori. *Arch. Rational Mech. Anal.* (to appear).
13. Smale, S., Differentiable dynamical systems. *Bull. Amer. Math. Soc. 73*, 747-817 (1967).

CONVERGENCE OF POWER SERIES SOLUTIONS OF p-ADIC NONLINEAR DIFFERENTIAL EQUATION

Yasutaka Sibuya[1,2]
Steven Sperber[3,4]

School of Mathematics
University of Minnesota
Minneapolis, Minnesota

I. INTRODUCTION

Let Q_p be the field of rational p-adic numbers, and let Ω be a complete algebraically closed extension of Q_p. We denote by ord the p-adic (additive) valuation in Ω which is normalized by the condition ord $p = 1$. Throughout this paper, "convergence" means "convergence in the p-adic sense".

In 1937, E. Lutz [5] proved the following result:

Theorem A. *Let* $F(z;y_1,\ldots,y_n)$ *be an* n-*dimensional vector whose components are convergent power series in* $(z;y)$ *with coefficients in* Ω. *Then the system* $dy/dz = F(z;y)$ *admits a unique*

[1]Partially supported by NSF Grant MCS 78-00907.

[2]Present address: School of Mathematics, University of Minnesota, Minneapolis, Minnesota 55455.

[3]Partially supported by NSF Grant MCS 78-00991.

[4]Present address: School of Mathematics, University of Minnesota, Minneapolis, Minnesota 55455.

ISBN 0-12-186280-1

formal solution of the form $y = \sum_{m=1}^{\infty} c_m z^m \ (c_m \in \Omega)$ *and this formal solution is convergent.*

Let y be a scalar, and let $a_0(z),\ldots,a_n(z)$ be convergent power series in z with coefficients in Ω. Set

$$L(y) = \sum_{j=0}^{n} a_j(z) d^j y/dz^j \quad \text{and} \quad L(z^m) = \sum_{q=q_0}^{\infty} \varphi_q(m) z^{m+q} ,$$

where m is an arbitrary integer (i.e. $m \in Z$) and $\varphi_{q_0}(m)$ is a non-zero polynomial in m. ($\varphi_{q_0}(m)$ is called the indicial polynomial of the operator L). In 1966, D. N. Clark [2] obtained the following result:

Theorem B. *Assume that every root* α *of the indicial polynomial* $\varphi_{q_0}(m)$ *satisfies the condition*

$$\operatorname{ord}(m+\alpha) = O(\log m) \quad \text{as} \quad m \to +\infty \quad \text{in } Z. \tag{C}$$

Then, for any convergent power series $f(z)$ *, every formal power series solution of the differential equation* $L(y) = f(z)$ *is also convergent.*

An element α of Ω is called a p-*adically non-Liouville number* if Condition (C) is satisfied (cf. Clark [2]). Clark's theorem, to be precise, is stronger than Theorem B, applying to the case in which y and f are Laurent series, and furthermore relating the radii of convergence of y and f. The generalization of Theorem B to a system of linear differential equations is also immediate.

Note that the case when $z = 0$ is a singular point is *not* excluded from Theorem B as long as Condition (C) is satisfied. Recently, F. Baldassarri [1] proved, by utilizing Clark's theorem, that components of the coefficient matrix of the linear transformation which reduces a given p-adic linear system to its Hukuhara-Turrittin's canonical form are convergent power series provided certain conditions corresponding to Condition (C) are satisfied.

A primary purpose of this paper is to generalize Theorem B to a nonlinear differential equation. We shall prove the following two theorems:

Theorem 1. *Let* $f(z) = \sum_{m=1}^{\infty} f_m z^m$, $a(z) = \sum_{m=1}^{\infty} a_m z^m$ *be convergent power series in* z *with* $f_m \varepsilon \Omega$ *and* $a_m \varepsilon \Omega$, *and let* $g(z,y) = \sum_{\ell=2}^{\infty} \sum_{m=0}^{\infty} g_{\ell m} z^m y^\ell$ *be a convergent power series in two variables* (z,y) *with* $g_{\ell m} \varepsilon \Omega$. *Let* $\alpha \varepsilon \Omega$ *and* $\alpha \neq 0$, *and let* r *be a positive rational integer. Then, the differential equation*

$$z^{r+1} dy/dz + \alpha y = f(z) + a(z)y + g(z,y) \tag{1.1}$$

has a unique formal power series solution of the form $y = \sum_{m=1}^{\infty} c_m z^m$ $(c_m \varepsilon \Omega)$ *and this formal solution is convergent.*

Theorem 2. *Let* f, a *and* g *be the same as in Theorem* 1. *Let* $\alpha \varepsilon \Omega$ *and satisfy a condition*

$$\lim_{m \to +\infty} \frac{\operatorname{ord}(m+\alpha)}{m^{(1-\delta)}} = 0 , \tag{1.2}$$

where δ *is a real number, independent of* m, *such that* $0 < \delta < 1$. *Assume that*

$$y = \sum_{m=1}^{\infty} c_m z^m \quad (c_m \varepsilon \Omega) \tag{1.3}$$

is a formal power series solution of the differential equation

$$z\, dy/dz + \alpha y = f(z) + a(z)y + g(z,y) . \tag{1.4}$$

Then, the formal solution (1.3) *is convergent.*

Our proof of Theorem 1 is similar to the proof of Theorem A which is due to E. Lutz. Theorem 1 may be generalized to a system of differential equations immediately. Such a generalization of Theorem 1 will allow us to establish a block-diagonalization theorem for a linear system. Theorem 2 poses for us a problem of small divisors. We utilize, in our proof, "Newton's method" which is well established in the celestial mechanics (cf. S. Sternberg [6]). We still do not know how to generalize Theorem 2 to a system. The difficulty only arises if "a root of the indicial polynomial" lies in Z_p, the p-adic integers.

One can hardly ignore the important number-theoretic consequences of results concerning p-adic differential equations. While many of the important applications have concerned linear differential equations, Lutz' early work [5] required knowledge of nonlinear equations (albeit without singularity). Of course, many nice properties of linear equations are carried over, as expected, to certain closely-related special nonlinear equations such as the Riccati equation. More generally, the underlying commutativity of the differentiation and Frobenius operators which has played such an important role in the theory of B. Dwork [3] and has prompted his conjecture concerning the widespread existence of Frobenius structures of linear differential equations [4], raises the possibility that such structures exist even in nonlinear equations. However, the evidence is not sufficient for us to formulate such a conjecture at this time.

II. PROOF OF THEOREM 1

Let us write Equation (1.1) in the form

$$\alpha^{-1} z^{r+1} dy/dz + y = \alpha^{-1} f(z) + \alpha^{-1} a(z)y + \alpha^{-1} g(z,y). \tag{2.1}$$

Set $z = c\zeta$ and $y = \gamma\eta$ where ζ and η are new variables, and c and γ are constants in Ω. Then Equation (2.1) becomes

$$c^r \alpha^{-1} \zeta^{r+1} d\eta/d\zeta + \eta = \alpha^{-1}\gamma^{-1} f(c\zeta) + \alpha^{-1} a(c\zeta)\eta + \alpha^{-1}\gamma^{-1} g(c\zeta,\gamma\eta) . \tag{2.2}$$

Let $\mathcal{O}$ be the ring of integers of Ω. We can choose c and γ so that $c^r\alpha^{-1}$ and the coefficients of the power series $\alpha^{-1}\gamma^{-1}f(c\zeta)$, $\alpha^{-1} a(c\zeta)$ and $\alpha^{-1}\gamma^{-1} g(c\zeta,\gamma\eta)$ are all in $\mathcal{O}$

Set $\eta = \sum_{m=1}^{\infty} \gamma_m\zeta^m$ $(\gamma_m \in \Omega)$ and determine coefficients γ_m through Equation (2.2), successively. Then, it is easy to prove that $\gamma_m \in \mathcal{O}$ for all m. Hence $\sum_{m=1}^{\infty} \gamma_m \zeta^m$ converges for $\operatorname{ord} \zeta > 0$. This completes the proof of Theorem 1.

III. PROOF OF THEOREM 2: PART I

To start with, we shall simplify Equation (1.4). Let us fix a positive integer k_0 once for all so that

$$\begin{aligned} &\frac{\operatorname{ord}\,(m+\alpha)}{m^{(1-\delta)}} < 1 \,, \\ & \qquad\qquad\qquad\qquad \text{for } m \in Z,\ m \geq 2^{k_0} . \\ &\frac{\operatorname{ord}\, m}{m^{(1-\delta)}} \quad 1 \,, \end{aligned} \tag{3.1}$$

Condition (1.2) and $\operatorname{ord} m = 0(\log m)$ guarantee the existence of such an integer k_0.

Set $\varphi_0(z) = \sum_{m=1}^{2^{k_0}-1} c_m z^m$ and $y = u + \varphi_0(z)$. The equation (1.4) becomes

$$z\, du/dz + \alpha u = \hat{f}(z) + \hat{a}(z)u + \hat{g}(z,u) \,, \tag{3.2}$$

where

$$\hat{f}(z) = f(z) + a(z)\,\varphi_0(z) + g(z, \varphi_0(z)) - z d\,\varphi_0(z)/dz - \alpha\varphi_0(z) \,,$$

$$\hat{a}(z) = a(z) + \frac{\partial g}{\partial y}\,(z, \varphi_0(z)) \,,$$

$$\hat{g}(z,u) = g(z, u+\varphi_0(z)) - g(z, \varphi_0(z)) - \frac{\partial g}{\partial y}\,(z, \varphi_0(z))u \,.$$

Since (1.3) is a formal solution of Equation (1.4), we have $\hat{f}(z) = \sum_{m=2^{k_0}}^{\infty} \hat{f}_m z^m$ $(\hat{f}_m \in \Omega)$. The power series $\hat{a}(z)$ and $\hat{g}(z,u)$ can be written as $\hat{a}(z) = \sum_{m=1}^{\infty} \hat{a}_m z^m$ and $\hat{g}(z,u) = \sum_{\ell=2}^{\infty} \hat{g}_\ell(z) u^\ell$ respectively, where the coefficients $\hat{g}_\ell(z)$ are convergent power series in z. Evidently, $\hat{f}$, $\hat{a}$ and $\hat{g}$ are convergent.

Note that Condition (3.1) implies that $m + \alpha \neq 0$ for $m \in Z$, $m \geq 2^{k_0}$. Therefore, considering the recursive formula the coefficients of a solution satisfy, we see that Equation (3.2) has a unique formal power series solution of the form

$$u = \sum_{m=2^{k_0}}^{\infty} c_m z^m . \tag{3.3}$$

This means that, if we prove the existence of a convergent power series solution of Equation (3.2) of the form (3.3), the proof of Theorem 2 will be completed.

Hence we may assume that $f(z)$ of Theorem 2 has a form

$$f(z) = \sum_{m=2^{k_0}}^{\infty} f_m z^m \qquad (f_m \in \Omega) . \tag{3.4}$$

We assume further that $a(z) = 0$ in Theorem 2. This assumption is justified since the transformation

$$y = u \exp \{ \int_0^z a(t) \frac{dt}{t} \}$$

is well defined, where

$$\int_0^z a(t) \frac{dt}{t} = \sum_{m=1}^{\infty} \frac{a_m}{m} z^m ,$$

a series which converges on any open disk, ord $z > r$, on which $a(z)$ converges. In fact, we can choose a real number $r_0 \geq r$ such that

$$\text{ord } (\int_0^z a(t) \frac{dt}{t}) > \frac{1}{p-1} \qquad \text{for} \qquad \text{ord } z \geqq r_0 .$$

Thus, we shall consider a differential equation

$$z\, dy/dz + \alpha y = f(z) = g(z,y) , \tag{3.5}$$

where

$$f(z) = \sum_{m=2^{k_0}}^{\infty} f_m z^m \qquad (f_m \in \Omega) ,$$

and

$$g(z,y) = \sum_{\ell=2}^{\infty} g_\ell(z)y^\ell ,$$

$$g_\ell(z) = \sum_{m=0}^{\infty} g_{\ell m} z^m \qquad (g_{\ell m} \in \Omega) .$$

The final remark is that, by utilizing a method similar to the proof of Theorem 1 of Section 2, we can reduce Equation (3.5) to a case where the following conditions are satisfied:

$$\text{ord } f(z) \geqq \rho_0 \quad \text{for} \quad \text{ord } z \geqq r_0 , \tag{3.6}$$

and

$$\text{ord } g_\ell(z) \geqq \frac{\mu}{p-1} \quad \text{for} \quad \text{ord } z \geqq r_0, \quad \ell = 2,3,\ldots , \tag{3.7}$$

where ρ_0 and μ are real numbers such that

$$\rho_0 > 0 , \quad \mu > 1 , \tag{3.8}$$

and r_0 is a suitable real number.

IV. PROOF OF THEOREM 2: PART II

In this section, we shall prove two lemmas on which our method of successive approximations is based. Note that we fixed the integer k_0 so that (3.1) is valid.

Lemma 1. *Suppose that* k *is a positive integer such that* $k \geqq k_0$, *and that a power series* $f(z) = \sum_{m=2^k}^{\infty} f_m z^m$ $(f_m \in \Omega)$ *converges for* $\text{ord } z \geqq r$, *where* r *is a real number. Suppose also that* $\text{ord } f(z) \geqq \rho$ *for* $\text{ord } z \geqq r$, *where* ρ *is another real number. Set*

$$F(z) = \sum_{m=2^k}^{\infty} \frac{f_m}{m+\alpha} z^m .$$

Then, the power series $F(z)$ *converges and satisfies the condition* $\text{ord } F(z) \geqq \rho$ *for* $\text{ord } z \geqq r + (2^{-\delta})^k$.

Proof. Our assumptions imply that $\operatorname{ord} f_m \geqq -mr + \rho$ and $m \geqq 2^k$. Hence

$$\operatorname{ord}(f_m/(m+\alpha)) \geqq -mr + \rho - \operatorname{ord}(m+\alpha)$$
$$= -m(r + (2^{-\delta})^k) + \rho + m(2^{-\delta})^k\left[1 - \frac{\operatorname{ord}(m+\alpha)}{m(2^{-\delta})^k}\right].$$

Since $m \geqq 2^k \geqq 2^{k_0}$, we have

$$1 - \frac{\operatorname{ord}(m+\alpha)}{m(2^{-\delta})^k} = 1 - \left(\frac{2^k}{m}\right)^{\delta} \frac{\operatorname{ord}(m+\alpha)}{m^{(1-\delta)}} > 1 - \left(\frac{2^k}{m}\right)^{\delta} \geqq 0 .$$

Hence

$$\lim_{m\to+\infty} \left[\operatorname{ord}(f_m/(m+\alpha)) + m(r + (2^{-\delta})^k)\right] = +\infty$$

and

$$\operatorname{ord}(f_m/(m+\alpha)) + m(r + (2^{-\delta})^k) \geqq \rho \quad \text{for } m \geqq 2^k .$$

This completes the proof of the lemma.

Lemma 2. *Suppose that* k *is a positive integer such that* $k \geqq k_0$ *and that the power series* $b(z) = \sum_{m=2^k}^{\infty} b_m z^m$ $(b_m \in \Omega)$ *converges for* $\operatorname{ord} z \geqq r$, *where* r *is a real number. Suppose also that* $\operatorname{ord} b(z) \geqq \mu/(p-1)$ *for* $\operatorname{ord} z \geqq r$, *where* μ *is a real number such that* $\mu > 1$. *Then,* $\exp\{\int_0^z b(t)\frac{dt}{t}\}$ *is well defined and satisfies the condition*

$$\operatorname{ord}\left(\exp\left\{\int_0^z b(t)\,\frac{dt}{t}\right\}\right) = 0$$

for $\operatorname{ord} z \geqq r + (2^{-\delta})^k$.

Proof. Note that

$$\int_0^z b(t)\,\frac{dt}{t} = \sum_{m=2^k}^{\infty} \frac{b_m}{m} z^m .$$

Hence, by an argument similar to the proof of Lemma 1, (but here using the second inequality of (3.1)), we can prove that

$$\operatorname{ord}\left(\int_0^z b(t)\,\frac{dt}{t}\right) \geqq \mu/(p-1) \quad \text{for} \quad \operatorname{ord} z \geqq r + (2^{-\delta})^k .$$

This proves the lemma.

V. PROOF OF THEOREM 2: PART III

As we summarized it at the end of Section 3, we consider a differential equation

$$z\,dy/dz + \alpha y = f(z) + g(z,y) , \tag{5.1}$$

where

$$f(z) = \sum_{m=2^{k_0}}^{\infty} f_m z^m \qquad (f_m \in \Omega) , \tag{5.2}$$

$$g(z,y) = \sum_{\ell=2}^{\infty} g_\ell(z) y^\ell, \quad g_\ell(z) = \sum_{m=0}^{\infty} g_{\ell m} z^m \qquad (g_{\ell m} \in \Omega) , \tag{5.3}$$

$$\operatorname{ord} f(z) \geqq \rho_0 \quad \text{for} \quad \operatorname{ord} z \geqq r_0 , \tag{5.4}$$

$$\operatorname{ord} g_\ell(z) \geqq \mu/(p-1) \quad \text{for} \quad \operatorname{ord} z \geqq r_0 , \quad \ell = 2,3,\ldots . \tag{5.5}$$

The two real numbers ρ_0 and μ satisfy the condition

$$\rho_0 > 0 , \qquad \mu > 1 . \tag{5.6}$$

In order to prove Theorem 2, it is sufficient to prove the existence of a convergent power series solution of Equation (5.1) of the form $y = \sum_{m=2^{k_0}}^{\infty} c_m z^m$ (cf. Section III). We construct a series of functions $\psi(z) = \sum_{j=0}^{\infty} \varphi_j(z)$ in such a way that the partial sums

$$\psi_j(z) = \sum_{i=0}^{j} \varphi_i(z) \tag{5.7}$$

converge to a solution ψ of our given nonlinear differential equation (5.1). Specifically we write $\psi_j(z) = \psi_{j-1}(z) + \varphi_j(z)$ and we seek $\psi_j(z)$ (or $\varphi_j(z)$) satisfying

$$z d\psi_j(z)/dz + \alpha \psi_j(z) = f(z) + g(z, \psi_{j-1}(z)) + \frac{\partial g}{\partial y}(z, \psi_{j-1}(z)) \varphi_j(z) . \tag{5.8-j}$$

Clearly, we are retaining on the right only the linear terms from the Taylor expansion of $g(z, \psi_{j-1}(z) + \varphi_j(z))$. In terms of φ_j, we derive from (5.8-j) (using (5.8-(j-1)) the differential equation

$$z d\varphi_j(z)/dz + \alpha \varphi_j(z) = h_j(z) + b_j(z) \varphi_j(z) . \tag{5.9-j}$$

where

$$h_j(z) = g(z, \psi_{j-1}(z)) - g(z, \psi_{j-2}(z)) - \frac{\partial g}{\partial y}(z, \psi_{j-2}(z)) \varphi_{j-1}(z) \tag{5.10-j}$$

and

$$b_j(z) = \frac{\partial g}{\partial y}(z, \psi_{j-1}(z)) . \tag{5.11-j}$$

Precisely speaking, Equation (5.9-j) is for $j \geqq 2$. We shall construct $\varphi_0(z)$ and $\varphi_1(z)$ respectively as solutions of the equations

$$z d\varphi_0(z)/dz + \alpha \varphi_0(z) = f(z) \tag{5.9-0}$$

and

$$z d\varphi_1(z)/dz + \alpha \varphi_1(z) = g(z, \varphi_0(z)) + b_1(z) \varphi_1(z) . \tag{5.9-1}$$

Note that, if we use Notation (5.10-j), Equation (5.8-j) is written as

$$z d\psi_j(z)/dz + \alpha \psi_j(z) = f(z) + g(z, \psi_j(z)) - h_{j+1}(z) . \tag{5.8'-j}$$

We shall construct $\varphi_0(z)$, $\varphi_1(z)$, ..., $\varphi_j(z)$, ... successively through Equations (5.9-0), (5.9-1) and (5.9-j) for $j \geqq 2$.

By virtue of Lemma 1, we get

$$\varphi_0(z) = \sum_{m=2}^{\infty} {}_{k_0} \gamma_{om} z^m \qquad (\gamma_{om} \in \Omega) \tag{5.12-0}$$

and this power series converges and satisfies the condition

$$\text{ord } \varphi_0(z) \geqq \rho_0 \tag{5.13-0}$$

for

$$\text{ord } z \geqq r_0 + c^{k_0} , \tag{5.14-0}$$

where

$$c = 2^{-\delta} . \tag{5.15}$$

As to $\varphi_1(z)$, we derive, first of all,

$$b_1(z) = \sum_{m=2}^{\infty} {}_{k_0} b_{1m} z^m \qquad (b_{1m} \in \Omega) \tag{5.15-1}$$

and this power series converges and satisfies the condition

$$\text{ord } b_1(z) \geqq \mu/(p-1) \tag{5.16-1}$$

for (5.14-0) (cf. (5.5)). Therefore, we can define $\exp\{\int_0^z b_1(t) \frac{dt}{t}\}$ by utilizing Lemma 2, and we get

$$\text{ord}(\exp\{\int_0^z b_1(t) \frac{dt}{t}\}) = 0 \tag{5.17-1}$$

for

$$\text{ord } z \geqq r_0 + 2\, c^{k_0} . \tag{5.18.1}$$

Set $h_1 = g(z, \varphi_0(z))$. Then we have

$$h_1(z) = \sum_{m=2}^{\infty} {}_{k_0+1} h_{1m} z^m \qquad (h_{1m} \in \Omega) \tag{5.19-1}$$

and

$$\text{ord } h_1(z) \geqq 2\rho_0 \tag{5.20-1}$$

for (5.14-0). Now, change $\varphi_1(z)$ by $\varphi_1(z) = \tilde{\varphi}_1(z)\exp\{\int_0^z b_1(t)\frac{dt}{t}\}$, and apply Lemma 1 to $\tilde{\varphi}_1(z)$. Then, we obtain

$$\varphi_1(z) = \sum_{m=2}^{\infty} {}_{k_0+1}\gamma_{1m}z^m \qquad ({}_{k_0+1}\gamma_{1m} \in \Omega) \tag{5.12-1}$$

and

$$\text{ord } \varphi_1(z) \geqq 2\rho_0 \tag{5.13-1}$$

for

$$\text{ord } z \geqq r_0 + 2c^{k_0} + c^{k_0+1} . \tag{5.14-1}$$

We set up the following inductive assumptions:

$$\varphi_j(z) = \sum_{m=2} {}_{k_0+1}\gamma_{jm}z^m \qquad ({}_{k_0+1}\gamma_{jm} \in \Omega) ; \tag{5.12-j}$$

$$\text{ord } \varphi_j(z) \geqq 2^j\rho_0 \tag{5.13-j}$$

for

$$\text{ord } z \geqq r_0 + 2\sum_{q=0}^{j-1} c^{k_0+q} + c^{k_0+j} ; \tag{5.14-j}$$

$$b_j(z) = \sum_{m=2}^{\infty} {}_{k_0}b_{jm}z^m \qquad ({}_{k_0}b_{jm} \in \Omega) ; \tag{5.15-j}$$

$$\text{ord } b_j(z) \geqq \mu/(p-1) \quad \text{for (5.14-(j-1))}; \tag{5.16-j}$$

$$\text{ord } (\exp\{\int_0^z b_j(t)\frac{dt}{t}\}) = 0 \tag{5.17-j}$$

for

$$\text{ord } z \geqq r_0 + 2 \sum_{q=0}^{j-1} c^{k_0+q} \; ; \tag{5.18-j}$$

$$h_j(z) = \sum_{m=2}^{\infty}{}_{k_0+j} \, h_{jm} z^m \qquad (h_{jm} \in \Omega) \; ; \tag{5.19-j}$$

and

$$\text{ord } h_j(z) \geqq 2^j \rho_0 \quad \text{for } (5.14\text{-}(j-1)) \; . \tag{5.20-j}$$

We assume that these assumptions are satisfied for j such that $1 \leqq j \leqq \nu - 1$, and verify all those conditions for $j = \nu$.

By virtue of Definition (5.11-j), Conditions (5.5) and (5.13-j), $1 \leqq j \leqq \nu - 1$, we obtain immediately

$$b_\nu(z) = \frac{\partial g}{\partial y}(z, \psi_{\nu-1}(z)) = \sum_{m=2}^{\infty}{}_{k_0} \, b_{\nu m} z^m \qquad (b_{\nu m} \in \Omega) \, , \tag{5.15-ν}$$

and

$$\text{ord } b_\nu(z) \geqq \mu/(p-1) \quad \text{for } (5.14\text{-}(\nu-1)) \; . \tag{5.16-ν}$$

Since $\psi_{\nu-1}(z) = \psi_{\nu-2}(z) + \varphi_{\nu-1}(z)$, we get

$$b_\nu(z) - b_{\nu-1}(z) = \frac{\partial g}{\partial y}(z, \psi_{\nu-2}(z) + \varphi_{\nu-1}(z)) - \frac{\partial g}{\partial y}(z, \psi_{\nu-2}(z))$$

$$= \sum_{m=2}^{\infty}{}_{k_0+\nu-1} \, \beta_m z^m \qquad (\beta_m \in \Omega)$$

and

$$\text{ord}(b_\nu(z) - b_{\nu-1}(z)) \geqq \mu/(p-1) \quad \text{for } (5.14\text{-}(\nu-1)) \; .$$

Then, applying Lemma 2 to $b_\nu(z) - b_{\nu-1}(z)$, we obtain

$$\text{ord}(\exp\{\int_0^z [b_\nu(t) - b_{\nu-1}(t)] \frac{dt}{t}\}) = 0$$

for

$$\text{ord } z \geqq r_0 + 2 \sum_{q=0}^{\nu-1} c^{k_0+q} \; . \tag{5.18-ν}$$

Since

$$\exp\{\int_0^z b_\nu(t)\frac{dt}{t}\} = \exp\{\int_0^z b_{\nu-1}(t)\frac{dt}{t}\}\exp\{\int_0^z [b_\nu(t)-b_{\nu-1}(t)]\frac{dt}{t}\},$$

it follows that

$$\mathrm{ord}(\exp\{\int_0^z b_\nu(t)\ \frac{dt}{t}\}) = 0 \quad \text{for } (5.18\text{-}\nu)\ . \tag{5.17-ν}$$

To derive (5.19-ν) and (5.20-ν), set

$$g(z,u+\psi_{\nu-2}(z)) = g(z,\psi_{\nu-2}(z)) + \frac{\partial g}{\partial y}(z,\psi_{\nu-2}(z))u + \sum_{\ell=2}^{\infty} \tilde{g}_\ell(z)u^\ell .$$

Then,

$$h_\nu(z) = \sum_{\ell=2}^{\infty} \tilde{g}_\ell(z)\ \varphi_{\nu-1}(z)^\ell \quad . \tag{5.21}$$

We can derive (5.19-ν) and (5.20-ν) from (5.21) immediately.

Finally, change $\varphi_\nu(z)$ by $\varphi_\nu(z) = \tilde{\varphi}_\nu(z)\exp\{\int_0^z b_\nu(t)\frac{dt}{t}\}$ and apply Lemma 1 to $\tilde{\varphi}_\nu(z)$. Then, (5.12-ν) and (5.13-ν) for (5.14-ν) follow immediately. This completes the construction of the sequence $\{\varphi_j(z);\ j = 0,1,2,\ldots\}$.

VI. PROOF OF THEOREM 2: PART IV

Set

$$\psi(z) = \sum_{j=0}^{\infty} \varphi_j(z)\ . \tag{6.1}$$

This series converges for

$$\mathrm{ord}\ z \geqq r_0 + r'\ , \tag{6.2}$$

where

$$r' = 2\sum_{q=0}^{\infty} c^{k_0+q}\ . \tag{6.3}$$

Now let us verify (5.8-j) or (5.8'-j). To do this, look at, first, Equations (5.9.0) and (5.9.1), and derive

$$z d\,\psi_1(z)/dz + \alpha\psi_1(z) = f(z) + g(z,\psi_0(z)) + b_1(z)\varphi_1(z) \quad . \qquad (6.4)$$

This is exactly Equation (5.8-1). Then, (5.8-j) follows inductively from (5.8-(j-1)) and (5.9-j).

Finally, (5.8'-j) and (5.20-(j+1)) allow us to take the limit as $j \to +\infty$ in the disk (6.2). Thus we get $zd\psi(z)/dz + \alpha\psi(z) = f(z) + g(z,\psi(z))$. This completes the proof of Theorem 2.

REFERENCES

1. Baldassarri, F., Differential modules and singular points of p-adic differential equations, (preprint), Princeton Univ., (1977).
2. Clark, D. N., A note on the p-adic convergence of solutions of linear differential equations, *Proc. A.M.S. 17*, 262-269 (1966).
3. Dwork, B., On the zeta function of a hypersurface, *Pub. Math. (I.H.E.S.), Paris 12*, 5-68 (1962).
4. Dwork, B., On p-adic differential equations I, *Bull. Soc. Math. France, Mémoire 39-40*, 27-37 (1974).
5. Lutz, E., Sur l'équations $y^2 = x^3 - Ax - B$ dans les corps p-adiques, *J. Reine Angew. Math. 177*, 238-247 (1937).
6. Sternberg, S., Celestial Mechanics, II. Benjamin, (1969).

UNIQUENESS OF PERIODIC SOLUTIONS OF THE LIENARD EQUATION

Ulrich Staude[1]

Mathematisches Institut
Universität Mainz, Mainz, BRD

I. INTRODUCTION

I shall give a short report concerning the uniqueness theorems for periodic solutions of the Liénard equation

$$\begin{aligned} \dot{x} &= y - F(x) \\ \dot{y} &= -x \end{aligned} \quad , \qquad (1)$$

$F(0) = 0$, $F(x) \in \mathrm{Lip}(\mathbb{R})$.

The apparently more general equation

$$\begin{aligned} \dot{x} &= y - F(x) \\ \dot{y} &= -g(x) \end{aligned} \quad , \qquad (2)$$

$g(x) \in \mathrm{Lip}(\mathbb{R})$, $x \cdot g(x) > 0$ for $x \neq 0$, the generalized Liénard equation, can be transformed to the Liénard equation by the

[1]Present address: Mathematisches Institut der Universität Mainz, Saarstrasse 21, D-6500 Mainz.

ISBN 0-12-186280-1

Conti transformation $z = \sqrt{2\,G(x)} \cdot \operatorname{sgn} x$, with $G(x) = \int_0^x g(u)\,du$. Letting $x(z)$ be the inverse function and $H(z) = F(x(z))$ we find

$$\dot{z} = y - H(z)$$
$$\dot{y} = -z$$

If $F(x) \in C^1(\mathbb{R})$ we write $f(x) = F'(x)$ and then we obtain for the systems (1), (2) the equations:

$$\ddot{x} + f(x)\,\dot{x} + x = 0\,,$$
$$\ddot{x} + f(x)\,\dot{x} + g(x) = 0\,. \qquad (3.1)$$

The Rayleigh equation

$$\ddot{v} + F(\dot{v}) + v = 0 \qquad (4)$$

can be transformed to the Liénard equation by $x = \dot{v}$, $y = -v$.

II. EXISTENCE OF PERIODIC SOLUTIONS

The origin is the only stationary point of the system (1) Therefore all nontrivial periodic solutions must circle around the origin.

The existence of at least one periodic solution is proved by constructing a Poincaré-Bendixson domain.

The most general existence theorem is due to FILIPPOV [25] (see REISSIG, SANSONE, CONTI [3], p. 156):

Theorem: Suppose

i) $\frac{F(x)}{x} < k$, $0 < k < 2$, for $0 < |x| < a$,

ii) $F(x) \leq F(-x)$, but $F(x) \neq F(-x)$ for $0 < x < a$,

iii) $\frac{F(x)}{x} > -k$ for $|x| > b$,

iv) $F(x) > F(-x)$ for $x > b$,

v) $\int_{0}^{b} \{F(x) - F(-x)\}x \, dx > 0$.

Then the Liénard equation possesses at least one nontrivial periodic solution.

There are some other existence theorems with more complicated assumptions dealing with cases which are not covered by the theorem of FILIPPOV. Such theorems have been proved by BARBALAT, HALANAY [24] and more recently by UTZ [26], who considered the Rayleigh equation (4).

III. UNIQUENESS OF PERIODIC SOLUTIONS

In order to prove uniqueness of periodic solutions one needs additional assumptions. In the literature there are numerous uniqueness results. In the list below I shall give only conditions needed by various authors to ensure the existence of at most one periodic solution. In most cases additional conditions must be assumed to obtain the existence of at least one periodic solution. This list improves an earlier list given by COPPEL [2].

For the proof of many of the uniqueness results one needs certain vortex theorems which guarantee that all cycles intersect the lines $x = -a$ and $x = b$ $(0 < a,\ 0 < b)$. If this conditions is fulfilled we write $F \in S(-a,b)$. Moreover $F \in S^*(-a,b)$ will mean that all cycles circle around the interval $[-a,b]$ of the x-axis. In most cases $F \in S^*(-a,a)$ is deduced from the assumptions $F(x) < 0 < F(-x)$ for $0 < x < a$, $F(a) = F(-a) = 0$, $F(x) > 0 > F(-x)$ for $x > a$. Clearly $F \in S^*(-a,b)$ implies $F \in S(-a,b)$. Sometimes one of these vortex theorems follows from other conditions in some other cases it must be supposed.

A)

i) $F \in S^*(-a,a)$ $\quad f(x) > 0, \ |x| > a$

LIENARD [14] (1928) : $f(x) = f(-x)$

SANSONE [19] (1949), p.174

BROWN [6] (1964)

ii)	$f(x)<0,\ \lvert x\rvert<a;\quad f(x)>0,\ \lvert x\rvert>a$
LEVINSON, SMITH [13] (1942) :	$F \in C^2(\mathbb{R})$
SANSONE [19] (1949), p.179	
iii)	$F(x)$ increasing for $\lvert x\rvert > a$; additional assumptions
YU SHU-HSIANG [23] (1964)	
iv)	$f(x) > 0,\ \lvert x\rvert > a$; additional assumptions
RYCHKOV [17] (1969)	
v)	$(F(x)-F(-x))\cdot(x-a)>0,\ 0<x,\ x\neq a$; $F(x)$ increasing for $\lvert x\rvert > a$
STAUDE [21] (1977)	

B)

i)	$f(x)$ increasing for $x > 0$, decreasing for $x < 0$
SANSONE [20] (1951) :	$\lvert f(x)\rvert < 2$
MASSERA [16] (1952)	
HUDAI-VERENOV [11] (1957)	
ii)	$f(x)<0,\ x\in(-a,b),\ f(x)>0,\ x\notin[-a,b]$; if $b\leq a$: $f(x)$ increasing for $x>b$, decreasing for $x<0$; if $a\leq b$: $f(x)$ increasing for $x>0$, decreasing for $x<-a$
CHAN CHI-FEN [7] (1958), Thm.3	
iii)	$\frac{F(x)}{x}$ increasing for $x > 0$, decreasing for $x < 0$
CONTI [9] (1952)	$\left\lvert\frac{F(x)}{x}\right\rvert < 2$
BARBALAT [5] (1954)	$F \in S^*(-a,a)$; $\left\lvert\frac{F(x)}{x}\right\rvert < 2$
SANSONE, CONTI [4] (1956), p.405; (1964), p.329/330	
CHAN CHI-FEN [7] (1958), Thm.2	

iv) DE FIGUEIREDO [10] (1960)	$\frac{F(x)}{x}$ increasing for $x > a$, decreasing for $x < 0$ $F(x) < 0 < F(-x)$, $0 < x < a$; $F(a) = 0$; $F(x) > 0$, $x > a$
v) STAUDE [22] (1978)	$(F(x)-F(-x))\cdot(x-a)>0$, $0<x$, $x\neq a$; $e \in [0,a]$, $F(e) < E < F(-e)$ $\frac{F(x) - E}{\sqrt{x^2-e^2}}$ increasing for $\|x\|>e$
C) i) CHAN CHI-FEN [7] (1958), Thm.1	$\frac{f(x)}{x}$ increasing for $x > 0$ and $x < 0$
ii) LINS, DE MELO, PUGH [15] (1977)	$F_1(x) = F(x) - cx^2$ odd; $f_1(x) = F_1'(x)$; $f_1(x) < 0$, $\|x\|<a$ $f_1(x) > 0$, $\|x\| > a$; $\frac{f_1(x)}{x}$ increasing for $x \notin [-a,a]$
iii) $F \in S^*(-a,b)$ CHERKAS, ZHILEVICH [8] (1970)	$f_1(x) = f(x)+x(c+dF(x))$, $c,d\in\mathbb{R}$ $f_1(x)<0$, $x\in(-a,b)$; $f_1(-a) = f_1(b) = 0$; $f_1(x)>0$, $x \notin [-a,b]$; $\frac{f_1(x)}{x}$ increasing for $x \notin [-a,b]$
D) RYCHKOV [18] (1970)	complicated assumptions
E) $F \in S(-a,b)$ KNOBLOCH [12] (1975)	$f(x)<0$, $x\in(-a,b)$; $f(x)>0$, $x \notin [-a,b]$; $((F(x) + c_1)\frac{f(x)}{x})'>0$, $x > b$; $((F(x) + c_2)\frac{f(x)}{x})'<0$, $x < -a$; $(\frac{f(x)}{x})'>0$, $x \in (b,b+\varepsilon)$, $x \in (-a-\varepsilon,-a)$

The following diagram shows the interdependence of the various theorems. The fact that theorem B can be derived from the theorem A is indicated by A ⟶ B.

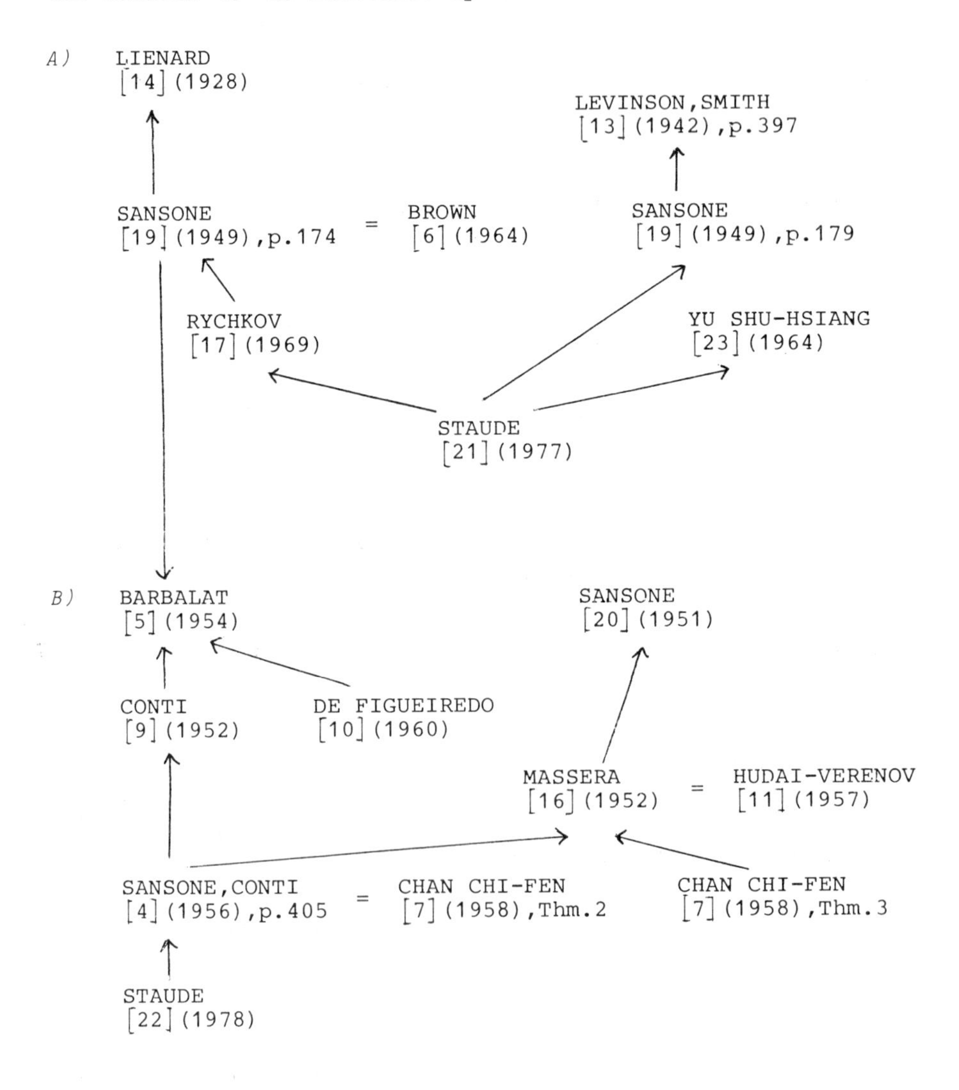

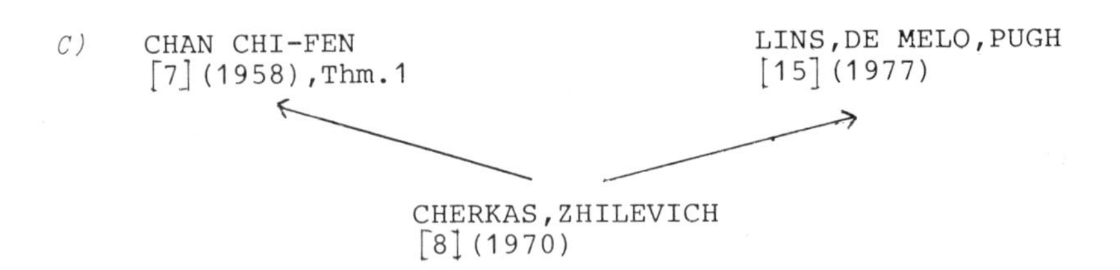

Remarks:

i) Let $F(x)$ be the simplest polynomial such that there exists at least one stable cycle. Then we have

$$F(x) = a_3x^3 + a_2x^2 + a_1x^1 \quad \text{with} \quad a_3 > 0, \quad a_1 < 0.$$

Denote by x_{max} the x value for which $F(x)$ attains its relative maximum and define x_{min} analogously. Let us denote $2D = x_{min} - x_{max}$. Then it follows that there is at least one periodic solution for $x_{min} \in (0,2D)$. The uniqueness of the periodic solution follows from the theorems of

SANSONE, CONTI [4]	provided	$x_{min} = D$,
STAUDE [21]	"	$x_{min} \in \left[\frac{D}{2}, \frac{3D}{2}\right]$,
CHERKAS, ZHILEVICH [8]	"	$x_{min} \in (0,2D)$.

The last result is sharp. It was at the first time explicitly proved by LINS, DE MELO, PUGH [15]. They were obviously unaware of the work of CHERKAS, ZHILEVICH [8]. CHAN CHI-FEN [7] proved in his theorem 1 only that in this case there is at most one stable cycle.

ii) For the most general results in *A)* and *B)* it is not necessary that $F(x) \in C^1(\mathbb{R})$. I feel that in most cases one can find better results with easier proofs by considering the system (1) instead of the equation (3.1)

REFERENCES

A) Books and review articles

1. Cherkas, L.A., Estimation of the number of limit cycles of autonomous systems, *Differential Equations 13*, 529-547 (1977).
2. Coppel, W.A., Limit cycles and rotated vector fields, MRC Technical Summary Report # 583 (1965).
3. Reissig, R., Sansone, G., Conti, R., Qualitative Theorie nichtlinearer Differentialgleichungen, Rom (1963).

4. Sansone, G., Conti, R., Equazioni differenziali non lineari, Rom,(1956).
Sansone, G., Conti, R., Nonlinear differential equations, revised edition, New York,(1964).

B) Articles with uniqueness theorems

5. Barbalat, I., Une proprieté globale des trajéctoires d'un système d'équations différentielles équivalent à l'equation des oscillations non linéaires de Liénard, *Acad. Popul Romîne, Bul. Sti., Sect. Sti. Mat. Fiz. 6*, 853-860 (1954).
6. Brown, T.A., A uniqueness condition for nontrivial periodic solutions to the Liénard equation, *J. Math. Anal. Appl. 8*, 387-391 (1964).
7. Chan Chi-Fen, On the uniqueness of the limit cycles of some nonlinear oscillation equations, *Dokl. Akad. Nauk SSR (NS) 119*, 659-662 (1958) (Russian).
8. Cherkas, L. A., Zhilevich, L. I., Some criteria for the absence of limit cycles and for the existence of a single limit cycle, *Differential Equations 6*, 891-897 (1970).
9. Conti, R., Soluzioni periodiche dell'equazione di Liénard generalizzata. Esistenza ed unicità, *Boll. Un. Mat. Ital., 3. Ser. 7*, 111-118 (1952).
10. De Figueiredo, R. P., Existence and uniqueness of the periodic solution of an equation for autonomous oscillations. Contrib. theory nonlin. oscill., vol. 5, *Ann. Math. Stud. 45*, 269-284 (1960).
11. Hudai-Verenov, M. G., Some theorems on limit cycles for the equation of Liénard, *Uspehi Mat. Nauk (N.S.) 12*, 389--396 (1957), *Amer. Math. Soc. Transl. (2) 18*, 163-171 (1961).
12. Knobloch, H. W., Die Eindeutigkeit des Grenzzyklus bei der Liénardschen Differentialgleichung, *J. Reine Angew. Math. 274/275*, 175-186 (1975).
13. Levinson, N. and Smith, O.K., A general equation for relaxation oscillations, *Duke Math. J. 9*, 382-403 (1942).
14. Liénard, A., Etude des oscillations entretenues, *Rev. Gén. Elec. 23*, 901-912, 946-954 (1928).
15. Lins, A., De Melo, W., Pugh, C. C., On Liénard's equation. Geometry and Topology, Rio de Janeiro, July 1976. Lecture Notes in Mathematics 597, 335-357 (1977).

16. Massera, J. L., Sur un théorème de G. Sansone sur l'équation de Liénard, *Boll. Un. Mat. Ital.*, *3.Ser. 9*, 367-369 (1954).

17. Rychkov, G. S., The uniqueness of the limit cycle of the system $\dot{y} = -g(x)$, $\dot{x} = y - f(x)$, *Differential Equations 5*, 563-564 (1969).

18. Rychkov, G. S., A complete investigation of the number of limit cycles of the equation $(b_{10}x + y)dy = \sum_{i+j\geq 1}^{2} a_{ij}x^i y^j dx$, *Differential Equations 6*, 2193-2199 (1970).

19. Sansone, G., Sopra l'equazione di A. Liénard per le oscillazioni di rilassamento, *Ann. Mat. Pura Appl.*, *4.Ser. 28*, 153-181 (1949).

20. Sansone, G., Soluzioni periodiche dell'equazione di Liénard. Calcolo del periodo, *Rend. Sem. Mat. Univ. e Politec. Torino 10*, 155-171 (1951).

21. Staude, U., Ein Eindeutigkeitssatz für periodische Lösungen der Liénard-Gleichung, VII Internationale Konferenz über nichtlineare Schwingungen, Berlin 975, Bd. I,2.

22. Staude, U., to appear (1978).

23. Yu Shu-Hsiang, On Filippov's method of proving existence and uniqueness of limit cycles, *Chinese Math. -Acta 5*, 496-505 (1964).

C) Further cited articles

24. Barbalat, A., Halanay, A., Un critère d'existence d'un cycle limite stable pour l'équation des oscillations non linéaires, *Stud. Cerc. Mat. 7*, 81-94 (1956).

25. Filippov, A. F., Sufficient conditions for the existence of stable limit cycles of second order equations, *Mat. Sb. (72) 30*, 172-180 (1952) (Russian).

26. Utz, W. R., Periodic solutions of second order differential equations with nonlinear nondifferentiable damping, *SIAM J. Appl. Math. 31*, 504-510 (1976).

BOUNDARY STABILIZABILITY FOR DIFFUSION PROCESSES

Roberto Triggiani[1]

Mathematics Department
Iowa State University
Ames, Iowa

In [N1] T. Nambu studied the problem described below of boundary stabilizability for a particular class of diffusion processes (Eq. (1.2)). The aim of the present paper is to show that a different approach - more on the soft analysis side than in Nambu's work - yields for particularly convenient choices of the sought after vectors g_k, much stronger results (in fact, the best results one can hope for) in a quicker way and for the most general class of parabolic systems. Nambu confined himself to the canonical self adjoint case involving the Laplacian and made use of the natural basis of eigenvectors; here we shall consider instead the general case involving uniformly strongly elliptic operators, where a basis of eigenvectors need not exist.

To facilitate the reading of the present paper in relation to Nambu's, we shall *generally* adopt his notation when convenient.

The boundary stabilizability problem studied here differs from the ones recently considered in [S2] and [Z1].

---

[1]Present address: Mathematics Department, Iowa State University, Ames, Iowa 55001.

ISBN 0-12-186280-1

I. THE STABILIZABILITY PROBLEM

Let Ω be a bounded open domain in R^n. Let $A(x,D)$ be a uniformly strongly elliptic operator in Ω in the form

$$A(x,D) = \sum_{|\alpha| \leq 2m} a_\alpha(x)\, D^\alpha$$

with real coefficients a_α. Finally, let B_i, $i = 1,\ldots,m$ be m differential boundary operators of respective orders m_i given by

$$B_i\,(x,D) = \sum_{|\alpha| \leq m_i} b^i_\alpha(x)\, D^\alpha\ .$$

The boundary $S = \partial\Omega$ of Ω is assumed of class C^{2m}. As we shall make use of the estimate known as Agmon-Douglis-Nirenberg inequality, we shall throughout assume the conditions - specified in [F1, p. 74] - under which it applies. To write it down explicitly, let A be the following operator in $L_2(\Omega)$. The domain $\mathcal{D}(A)$ of A consists of the closure in $H^{2m}(\Omega)$ of the set of functions f in $C^{2m}(\bar{\Omega})$ that satisfy the boundary conditions $B_i(x,D)f = 0$ on S $(1 \leq i \leq m)$. For every $f \in \mathcal{D}(A)$, A is defined as

$$(Af)(x) = -A(x,D)f(x)\ .$$

Then the A-D-N inequality reads (for $p = 2$)

$$\|u\|_{H^{2m}(\Omega)} \leq C(\|Au\|_{L_2(\Omega)} + \|u\|_{L_2(\Omega)}) = C\,\|u\|_G \qquad u \in \mathcal{D}(A) \tag{1.1}$$

where C is a constant independent of u [F1, p. 75] and $\|u\|_G$ indicates the graph norm on $\mathcal{D}(A)$. The operator A is closed [F1, p. 75]. Under the additional assumptions that the system $\{B_i\}$ be normal [F1, p. 76] and that the strong complementary condition [F1, p. 77] hold, the operator A *generates an analytic (holomorphic) semigroup* [F1, p. 101, Example] on $L_2(\Omega)$ which we shall conveniently indicate by e^{At}. Therefore the Cauchy problem

$$\dot{u} = Au \qquad u(0) = u_o$$

on $L_2(\Omega)$ has the unique solution $u(t) = e^{At}u_o$, $t > 0$.

Remark 1.1. The above model includes the diffusion process treated by Nambu [N1]:

$$\frac{\partial u(t,x)}{\partial t} = \Delta\, u(t,x) - q(x)u(t,x) \qquad t > 0, \quad x \in \Omega$$

$$u(0,x) = u_o(x), \qquad x \in \Omega \tag{1.2}$$

$$\alpha(\xi)\; u(t,\xi) + (1 - \alpha(\xi))\, \frac{\partial u(t,\xi)}{\partial n} = 0 \qquad t > 0, \quad \xi \in S$$

where $q(\cdot)$ is Hölder continuous in $\bar{\Omega}$, $\alpha(\xi) \in C^2(S)$, $0 \leq \alpha(\xi) \leq 1$, and $\partial/\partial n$ is the outward normal derivative at the point ξ of S: in this case the corresponding operator A is self-adjoint.

Since Ω is bounded, the resolvent $R(\lambda,A)$ is compact. Hence the spectrum $\sigma(A)$ of A is only point spectrum and consists of a sequence of eigenvalues $\{\lambda_i\}$, $i = 1,2,\ldots$ with corresponding normalised linearly independent eigenvectors Φ_{ij}, $j = 1,\ldots,l_i$, l_i being the multiplicity of λ_i. As is well known, the λ_i's are contained in a triangular sector Σ delimited by the rays $a + \rho e^{\pm i\theta}$, $0 \leq \rho < \infty$, $\pi/2 < \theta$, with no finite accumulation point. Therefore, at the right of any vertical line in the complex plane, there are only at most finitely many of them.

A standing *assumption* that we make - for the stabilizability problem described below to be significant - is that: there are $(M-1)$ eigenvalues: $\lambda_1,\ldots,\lambda_{M-1}$ at the right of the imaginary axis, ordered, say, for decreasing real part:

$$\mathrm{Re}\,\lambda_M < 0 \leq \mathrm{Re}\,\lambda_{M-1} \leq \cdots \leq \mathrm{Re}\,\lambda_1 \tag{1.3}$$

Let now γ be any continuous operator from $H^{2m}(\Omega)$ into $L_2(S)$. In applications, γ may be the operator that assigns to $f \in H^1(\Omega)$ its boundary value $f|_S \in L_2(S)$. (This special case in the one considered in Nambu's work [N1]); or the operator that assigns to $f \in H^2(\Omega)$ the boundary value of its normal derivative $\partial f/\partial n|_S \in L_2(S)$.

Define the operator $B : \mathcal{D}(B) \subset L_2(\Omega) \to L_2(\Omega)$ by

$$Bu = \sum_{k=1}^{N} (\gamma u, w_k)\, g_k \qquad u \in \mathcal{D}(B) = H^{2m}(\Omega) \tag{1.4}$$

Here the w_k's and the g_k's are fixed vectors in $L_2(S)$ and $L_2(\Omega)$ respectively, while $(.,.)$ is the inner product in $L_2(S)$. The operator B is bounded on $L_2(\Omega)$ and it is even unclosable [K1, p. 166] (take $\Omega = (0,1)$, $N = 1$, and let $u_n(x)$ be a sequence of smooth functions with compact support in, say, $(1-1/n,1)$, and satisfying $u_n(1) \equiv 1$; take $w = 2$ and $g(x) \equiv 1$. Then $u_n \to 0$ and $Bu_n \to 2g \neq 0$ in $L_2(\Omega)$). To avoid the case $Bu \equiv 0$, we assume $\gamma u \not\equiv 0$ (with reference to the diffusion process (1.2) and γ the operator $f \in H^1(\Omega) \to f|_S \in L_2(S)$, this corresponds to assuming $\alpha(\xi) \not\equiv 1$, which is Nambu's case).

The system that we wish to study and stabilize is[(1)]

$$\dot{u} = Au + \sum_{k=1}^{N} (\gamma u, w_k)\, g_k, \qquad u(0) = u_o \tag{1.5}$$

A qualitative statement of the boundary stabilizability problem, which is the main object of the present paper, is as follows: find - if possible - functions $g_k \in L_2(\Omega)$, $k = 1,\dots,N$ and conditions on the functions $w_k \in L_2(S)$ as to guarantee that the solutions of (1.5) corresponding to the largest possible class of initial conditions u_o, tend to zero, preferably in an exponential manner, as $t \to +\infty$ in the strongest possible norm.

Remark 1.2. The above stabilizability problem for the abstract system (1.5) corresponds - say for $N = 1$ - to the following output stabilizability problem in the classical finite dimensional theory given the output system

---

(1) Our entire procedure leading to the exponential decay of the solutions in the graph norm applies verbatim by simply assuming that A generate a differentiable semigroup (i.e. $\exp(At)L_2(\Omega) \in \mathcal{D}(A)$, $t > 0$) and that it have compact resolvent.

$$\dot{x} = Ax + gf \qquad x, g \in R^n$$

$$y = Hx \qquad y \in R^m$$

find the scalar input f as a feedback of the output, i.e. of the form $f = (y,k)_{R^m}$, $k \in R^m$, in such a way that the resulting feedback system:

$$\dot{x} = Ax + (Hx,k)\, g$$

be globally asymptotically stable.

The first question to settle is of course the well-posedness of eq. (1.5). This is the object of the next section. In preparation, and following the procedure originated in [T1], let $L_2(\Omega)$ be decomposed into two orthogonal subspaces E_1 and E_2, corresponding to, respectively, the subsets $\{\lambda_1, \ldots, \lambda_{M-1}\}$ and $\{\lambda_i,\ i \geq M\}$ of the spectrum $\sigma(A)$ of A satisfying (1.3). Here we appeal to the standard decomposition theorem as in [K1, p. 178] . With P denoting the orthogonal projection of $L_2(\Omega)$ onto E_1, then $(I-P)\,\mathcal{D}(A) \subset \mathcal{D}(A)$, E_1 and E_2 are invariant under A and hence under the semigroup e^{At}. Also, $\sigma(A_1) = \{\lambda_1, \ldots, \lambda_{M-1}\}$, $\sigma(A_2) = \{\lambda_i,\ i \geq M\}$, where A_j is the restriction of A on E_j, and A_1 is bounded. Finally P and $(I-P)$ commute with A, hence with the semigroup e^{At}. We shall henceforth use the notation $Pu = u_1$ and $(I-P)\,u = u_2$. Moreover, to avoid cumbersome notation, the norm in $L_2(\Omega)$ and in $L_2(S)$ will be simply denoted by $\|\ \|$, while other norms will be specified by an appropriate subscript. The norm of from $H^{2m}(\Omega) \to L_2(S)$ will be instead denoted by $|||\gamma|||$.

II. WELL-POSEDNESS OF EQ. (1.5)

For the particular case of eq. (1.5) corresponding to the diffusion process (1.2), Theorem 3.1 of [N1] claims the following: "The differential equation has a real-valued solution $u(t)$ for the real-valued initial value u_o satisfying $(I-P)\, u_o \in \mathcal{D}(A_2^{\beta})$, where $\beta > 1/2$. The solution $u(t)$ such that $Bu(t)$ has at most a summable singularity at $t = 0$ is unique". By solution it is meant a function $u(.)$ of class

$$C([0,\infty); L_2(\Omega)) \cap C^1((0,\infty); L_2(\Omega)),$$
$$\text{satisfying } Bu(.) \in C((0,\infty); L_2(\Omega)).$$

The proof of this statement given in [N1] is a rather lenghty computation. Our first theorem provides a much stronger conclusion for the well-posedness of eq. (1.5): the unique solution of eq. (1.5) corresponding to *any* initial condition $u_o \in L_2(\Omega)$ is *analytic* for $t > 0$. Our short proof is radically different from Nambu's, being based on viewing the operator B as a perturbation of the generator A.

Theorem 2.1. The operator $A + B$ with domain $\mathcal{D}(A+B) = \mathcal{D}(A)$ generates an analytic semigroup $e^{(A+B)t}$ on $L_2(\Omega)$, which gives the solution of (1.5): $u(t,u_o) = e^{(A+B)t}u_o$, $t > 0$.

Proof. First we observe that the operator B has finite dimensional range (of dimension in fact at most N). Therefore the desired conclusion follows from a recent perturbation theorem of Zabczyk [Z1, Proposition 1] - which relies on the standard perturbation result [K1, Thm 2.4, p. 497] - as soon as we prove that B is bounded with respect to A [K1, p.190]. To this end, definition (1.4) and the continuity of γ imply

$$\|Bu\| \leq c \|u\|_{H^{2m}(\Omega)}, \quad u \in H^{2m}(\Omega), \quad \dot{c} = |||\gamma||| \sum_{k=1}^{N} \|w_k\| \|g_k\|$$

and we only need to invoke the A-D-N inequality (1.1) to conclude. Q.E.D.

Remark 2.1. As Zabczyk has shown [Z1, Remark 3], the A-bound of the operator B with finite dimensional range is actually zero, i.e. we have

$$\|Bu\| \leq a \|Au\| + b \|u\| \qquad u \in \mathcal{D}(A) \subset \mathcal{D}(B) \tag{2.1}$$

where the greatest lower bound of all possible constants a in (2.1) is zero (b will generally increase as a is chosen close to zero).

III. STABILIZABILITY

In order to formulate our stabilizability result, let W_i be the $N \times l_i$ matrix defined by

$$W_i = \begin{vmatrix} (w_1, \gamma\Phi_{i1}), & (w_1, \gamma\Phi_{i2}), \ldots, & (w_1, \gamma\Phi_{il_i}) \\ (w_2, \gamma\Phi_{i1}), & (w_2, \gamma\Phi_{i2}), \ldots, & (w_2, \gamma\Phi_{il_i}) \\ \vdots & & \vdots \\ (w_N, \gamma\Phi_{i1}), & (w_N, \gamma\Phi_{i2}), \ldots, & (w_N, \gamma\Phi_{il_i}) \end{vmatrix}$$

associated with each eigenvalue λ_i of A, with multiplicity l_i and associated normalized eigenvectors $\Phi_{i1}, \ldots, \Phi_{il_i}$.

Theorem 3.1. Let A_1 be diagonalizable. Also assume the condition

$$\text{rank } W_i = l_i, \qquad i = 1,\ldots,M-1 \tag{3.1}$$

which implies $N \geq \max\{l_i, \ i = 1,\ldots,M-1\}$. Then, for any ε, $0 < \varepsilon \leq -\text{Re}\lambda_M$, there exist vectors g_k in E_1, $k = 1,\ldots,N$ (to be specified in the proof of Lemma 3.2 below) such that the solution $u(t,u_o) = e^{(A+B)t}u_o$ of the corresponding eq. (1.5) due to any initial condition $u_o \in L_2(\Omega)$, satisfies for any preassigned positive number h:

$$\|u(t,u_o)\|_{H^{2m}(\Omega)} \leq C\,\|u(t,u_o)\|_G \leq C_{\varepsilon,u_o,h}e^{-\varepsilon t} \tag{3.2}$$

$$t \geq h > 0$$

where $\|\ \|_G$ is the graph norm and $C_{\varepsilon,u_o,h}$ a constant depending on ε, on u_o and h. Hence, by the Principle of Uniform Boundedness, it follows that for the corresponding operator B one has

$$|e^{(A+B)t}| \leq C_{\varepsilon,h}e^{-\varepsilon t} \qquad t \geq h > 0 \tag{3.3}$$

where $|\ \ |$ is the corresponding operator norm. Actually a slight variation of the same proof for initial conditions

$u_o \in \mathcal{D}(A)$ shows

$$\|e^{(A+B)t}\|_{\mathcal{D}(A)} \leq C_\varepsilon e^{-\varepsilon t} \qquad t \geq 0$$

where $\| \quad \|_{\mathcal{D}(A)}$ is the operator norm corresponding to the graph norm on $\mathcal{D}(A)$.

Remark 3.1. The minimum number N os such functions g_k is equal to the largest multiplicity of the eigenvalues $\lambda_1,\dots,\lambda_{M-1}$.

Remark 3.2. The same proof will show that if one assumes rank $W_i = l_i$ true for $i = 1,\dots,I-1$ with $M \leq I$ and A restricted on the subspace corresponding to $\lambda_1,\dots,\lambda_{I-1}$ diagonalizable, then in the conclusion of the theorem one can take any ε with $0 < \varepsilon \leq -\text{Re}\,\lambda_I$ while the g_k's are taken in such subspace. In particular, if rank $W_i = l_i$ holds for all i and A is normal, then the exponential decay of the solution can be made arbitrarily fast.

Remark 3.3. Even in the special case studied by Nambu regarding the diffusion process (1.2), where $m = 2$ and γ only continuous from $H^1(\Omega) \to L_2(S)$, our Theorem 3.1 - as well as our Theorem 3.2 below - are much stronger - than his Theorem 4.2 in [N1]: in fact Nambu's Theorem 4.2 only gives an exponential upperbound in the weaker $H^1(\Omega)$ - norm and only for initial data u_o with projection $u_{20} = (I-P)u_o \in \mathcal{D}(A_2)$, $\beta > 1/2$. His g_k are not taken in E_1, but 'close' to it (i.e. $\|(I-P)g_k\|$ 'small').

Proof. In (3.2) the inequality on the left is the A-D-N inequality (1.1). To prove the right hand side of (3.2), we select preliminarily the vectors g_k to be in E_1, so that

$$PBu = \sum_{k=1}^{N} (\gamma u, w_k)\, g_k \in E_1, \quad \text{while} \quad (I-P)\,Bu \equiv 0 .$$

The projections of eq. (1.5) onto E_1 and E_2 are

$$\dot{u}_1 = A_1 u_1 + \sum_{k=1}^{N} (\gamma u_1, w_k)\, g_k + \sum_{k=1}^{N} (\gamma u_2, w_k)\, g_k \tag{3.4}$$

and

$$\dot{u}_2 = A_2\, u_2 \tag{3.5}$$

respectively. Since A_2 generates an analytic semigroup on E_2, it satisfies the spectrum determined growth condition [T1, § 2] and hence

$$\|u_2(t,u_{20})\| = \|e^{A_2 t}\, u_{20}\| \leq e^{-\varepsilon_2 t}\, \|u_{20}\| \tag{3.6}$$

for all $u_{20} \in E_2$ and any ε_2, $0 < \varepsilon_2 \leq -\text{Re}\lambda_M$. Due to the analyticity of $e^{A_2 t}$ we have

$$\begin{aligned} \|A_2\, e^{A_2 t}\, u_{20}\| &= \|A_2\, e^{A_2 (t-h)}\, e^{A_2 h}\, u_{20}\| \\ &= \|e^{A_2 (t-h)}\, A_2\, e^{A_2 h}\, u_{20}\| \\ &\leq e^{-\varepsilon_2 t}\, e^{-\varepsilon_2 h}\, \|A_2\, e^{A_2 h}\, u_{20}\|, \quad t \geq h > 0 \end{aligned} \tag{3.6'}$$

and hence (3.6) and (3.6') imply

$$\|u_2\, (t,u_{20})\|_G = \|e^{A_2 t}\, u_{20}\|_G \leq C_{u_{20},\varepsilon_2,h}\, e^{-\varepsilon_2 t}, \quad t \geq h > 0 . \tag{3.7}$$

The unperturbed part of eq. (3.4) is

$$\dot{z} = A_1\, z + \sum_{k=1}^{N} (\gamma z,\, w_k)\, g_k, \qquad z \in E_1 \tag{3.8}$$

and can be rewritten in matrix form as

$$\dot{z} = A_{g,w}\, z \tag{3.8'}$$

where $A_{g,w}$ is a square matrix of size equal to $\dim E_1$, depending on A_1, the g_k's and the w_k's. This can be seen by using in E_1 the (non necessarily orthogonal) basis of normalised eigenvectors Φ_{ij}, $i = 1,\ldots,M-1$, which make the matrix corresponding to the operator A_1 diagonal. The exponential decay of (3.8') for a suitable choice of the g_k's is handled by the following lemma.

Lemma 3.2. Assume condition (3.1). Then for any $\varepsilon_1 > 0$, there exist vectors $g_k \in E_1$, $k = 1,\dots,N$, such that the solution $z(t,z_o)$ due to the initial datum z_o of the corresponding equation (3.8') satisfies

$$||z(t,z_o)|| = ||e^{A_{g,w}t} z_o|| \leq \bar{C}_{z_o,\varepsilon_1} e^{-\varepsilon_1 t}, \quad t > 0 \tag{3.9}$$

in the norm of E_1 inherited from $L_2(\Omega)$. The minimum number N of such g_k's is equal to $\max\{l_i, i = 1,\dots,M-1\}$.

Proof of Lemma 3.2. See appendix for a constructive proof.

It remains to show exponential decay of the perturbed equation (3.4). The analyticity of the semigroup implies $e^{A_2 t} u_{20} \in \mathcal{D}(A_2)$ for all $t > 0$ and all $u_{20} \in E_2$. The A-D-N inequality (1.1) and the inequality (3.7) give

$$\begin{aligned} ||u_2(t,u_{20})||_{H^{2m}(\Omega)} &= ||e^{A_2 t} u_{20}||_{H^{2m}(\Omega)} \\ &\leq C\, ||e^{A_2 t} u_{20}||_G \leq C_{u_{20},\varepsilon_2,h}\, e^{-\varepsilon_2 t}, \quad t \geq h > 0 \end{aligned} \tag{3.10}$$

for any ε_2, $0 < \varepsilon_2 \leq -\mathrm{Re}\,\lambda_M$. From now on let the vectors g_k be the ones of Lemma 3.2. Starting from (3.9), one easily obtains

$$||z(t,z_o)||_G = ||e^{A_{g,w}t} z_o||_G \leq C_{\varepsilon_1,z_o}\, e^{-\varepsilon_1 t}, \quad t > 0. \tag{3.11}$$

Finally, we write the variation of parameter formula for the perturbed system (3.4):

$$\begin{aligned} u_1(t,u_o) &= e^{A_{g,w}t}(u_{10}+v_h) \\ &+ \int_h^t e^{A_{g,w}(t-\tau)} \sum_{k=1}^{N} (\gamma u_2(\tau),w_k) g_k \, d\tau , \end{aligned} \tag{3.12}$$

where

$$v_h = \int_0^h e^{-A_{g,w}\tau} \sum_{k=1}^{N} (\gamma u_2(\tau),w_k)\, g_k \, d\tau .$$

As the unperturbed system (3.8') satisfies the exponential bound (3.11), while the perturbing term of (3.4) satisfies a bound related to (3.10), we finally obtain from (3.12):

$$\|u_1(t,u_o)\|_G \leq C_{\varepsilon_1,u_{10},v_h} e^{-\varepsilon_1 t} + K\int_h^t e^{-\varepsilon_1(t-\tau)} e^{-\varepsilon_2\tau} d\tau$$

$$\leq C_{\varepsilon_1,u_{10},v_h} e^{-\varepsilon_1 t} + K \frac{e^{-\varepsilon_2 t}}{\varepsilon_1 - \varepsilon_2}$$

$$\leq \text{const}_{\varepsilon_2,u_o,h} e^{-\varepsilon_2 t}, \qquad t \geq h > 0 \tag{3.13}$$

where

$$K = C_{u_{20},\varepsilon_2,h} \; |||\gamma||| \sum_{k=1}^{N} \|w_k\| \; C_{\varepsilon_1,g_k}$$

and where ε_1 is now chosen larger than the preassigned $\varepsilon_2 \in (0, -\text{Re}\,\lambda_M]$, say $\varepsilon_1 = 2\varepsilon_2$.

The desired right side of inequality (3.2) then follows from (3.7) and (3.13). Q.E.D.

Remark 3.4. As noticed in [N1] on the basis of results of [S1], condition (3.1) is also necessary for choices of g_k restricted to E_1. In fact, in this case, failure of (3.1) at some λ_i makes λ_i an eigenvalue of $(A+B)$.

If one insists on selecting stabilising vectors g_k *not* in E_1 [N1, Remark in § 4], the following theorem, whose proof is more elaborate, serves the purpose.

Theorem 3.2. Under the same assumptions as in Theorem 3.1, given any ε, $0 < \varepsilon < -\text{Re}\,\lambda_M$, one can select suitable vectors g_k, with $0 \neq Qg_k \in \mathcal{D}(A_2)$ such that for the solutions of the corresponding eq. (1.5) the same conclusion as in Theorem 3.1 holds. Here $Q = I - P$.

Proof. For simplicity of notation we write the proof only when $N = 1$, only trivial changes being needed when $N > 1$. The projections of the solution $u(t,u_o) = e^{(A+B)t} u_o$ onto E_1 and E_2 are:

$$u_1(t,u_o) = e^{A_{Pg,w}t} u_{10} + \int_o^t e^{A_{Pg,w}(t-\tau)} Pg\ (\gamma u_2(\tau),w)\ d\tau$$

$$u_2(t,u_o) = e^{A_2 t} u_{20} + \int_o^t e^{A_2(t-\tau)} Qg\left[(\gamma u_2(\tau),w)+(\gamma u_1(\tau),w)\right] d\tau .$$

For any $h > 0$ and $t \geq h$, these can be rewritten as

$$u_1(t,u_o) = e^{A_{Pg,w}t} (u_{10} + r_h) + \int_h^t e^{A_{Pg,w}(t-\tau)} Pg(\gamma u_2(\tau),w)\ d\tau \tag{3.14}$$

$$u_2(t,u_o) = e^{A_2 t} u_{20} + e^{A_2(t-h)} v_h + \int_h^t e^{A_2(t-\tau)} Qg\left[(\gamma u_2(\tau),w) + (\gamma u_1(\tau),w)\right] d\tau \tag{3.15}$$

where:

$$r_h = r_h(g) = \int_o^h e^{-A_{Pg,w}\tau} Pg(\gamma u_2(\tau),w)\ d\tau$$

$$v_h = v_h(g) = \int_o^h e^{A_2(h-\tau)} Qg\left[(\gamma u_2(\tau),w) + (\gamma u_1(\tau),w)\right] d\tau \ \varepsilon\ \mathcal{D}(A_2) .$$

Notice that $u_2(t,u_o) = Q\ e^{(A+B)t} u_o\ \varepsilon\ \mathcal{D}(A) \cap E_2 = \mathcal{D}(A_2)$ for $t > 0$. We seek a suitable stabilising g with $0 \neq Qg\ \varepsilon\ \mathcal{D}(A_2)$. For $t \geq h > 0$, (3.15) yields

$$A_2 u_2(t,u_o) = e^{A_2(t-h)} \left[A_2\ e^{A_2 h} u_{20} + A_2 v_h\right] + \int_h^t e^{A_2(t-\tau)} A_2\ Qg\left[(\gamma u_2(\tau),w) + (\gamma u_1(\tau),w)\right] d\tau . \tag{3.16}$$

Therefore, for a suitable choice of the projection Pg in E_1 as dictated by Lemma 3.2, eqs. (3.14), (3.15), (3.16) and the A-D-N inequality (1.1) yield for $t \geq h > 0$:

$$\|u_1(t,u_o)\|_G \leq C_{\varepsilon_1,u_{10}+r_h} e^{-\varepsilon_1 t}$$

$$+ \int_h^t C_{\varepsilon_1,Pg} e^{-\varepsilon_1(t-\tau)} |||\gamma||| \; \|w\| \; \|u_2(\tau)\|_G \, d\tau \tag{3.17}$$

$$\|u_2(t,u_o)\|_G \leq \mu_1 e^{-\varepsilon_2 t}$$

$$+ \int_h^t \mu_2 e^{-\varepsilon_2(t-\tau)} [\, \|u_2(\tau)\|_G + \|u_1(\tau)\|_G] \, d\tau \tag{3.18}$$

where ε_1 is an arbitrary positive constant and ε_2 is constant satisfying $\varepsilon < \varepsilon_2 < -\mathrm{Re}\,\lambda_M$

$$\text{(i)} \quad \mu_1 = \mu_1(g) = \max\{\|u_{20}\| + e^{\varepsilon_2 h}\|v_h\| , \; e^{\varepsilon_2 h}\|A_2(e^{A_2 h} u_{20} + v_h)\|\} \tag{3.19}$$

$$\text{(ii)} \quad \mu_2 = \mu_2(Qg) = C \, |||\gamma||| \; \|w\| \max\{\|Qg\| \; , \|A_2 Qg\|\}$$

Here we choose to indicate for μ_i only the dependence on the projections of g. By means of (3.17) we then compute:

$$\int_h^t e^{\varepsilon_2 \tau}\|u_1(\tau)\|_G \, d\tau \leq C_{\varepsilon_1,u_{10}+r_h} \frac{1 - e^{(\varepsilon_2-\varepsilon_1)t}}{\varepsilon_1 - \varepsilon_2} \tag{3.20}$$

$$+ C_{\varepsilon_1,Pg} |||\gamma||| \; \|w\| \int_h^t \frac{e^{(\varepsilon_2-\varepsilon_1)} - e^{(\varepsilon_2-\varepsilon_1)t}}{\varepsilon_1 - \varepsilon_2} e^{\varepsilon_1 s} \|u_2(s)\|_G \, ds$$

where the second term on the right side was obtained after a change in the order of integration. Hence selecting $\varepsilon_1 > \varepsilon_2$ yields

$$\int_h^t e^{-\varepsilon_2(t-\tau)} \|u_1(\tau)\|_G \, d\tau \leq C_{\varepsilon_1,u_{10}+r_h} \frac{e^{-\varepsilon_2 t}}{\varepsilon_1 - \varepsilon_2}$$

$$+ C_{\varepsilon_1,Pg} \frac{|||\gamma||| \; \|w\|}{\varepsilon_1 - \varepsilon_2} \int_h^t e^{-\varepsilon_2(t-s)} \|u_2(s)\|_G \, ds \, . \tag{3.21}$$

Finally, we plug (3.21) into (3.18) to get

$$\|u_2(t,u_o)\|_G \leq M_1 e^{-\varepsilon_2 t} + \int_h^t M_2 e^{-\varepsilon_2(t-\tau)} \|u_2(\tau)\|_G d\tau \qquad (3.22)$$

where

$$\text{(i)} \quad M_1 = M_1(g) = \mu_1 + C_{\varepsilon_1, u_{10}+r_h} \frac{\mu_2}{\varepsilon_1 - \varepsilon_2}$$

$$\text{(ii)} \quad M_2 = M_2(g) = \mu_2 + C_{\varepsilon_1, Pg} \frac{|||\gamma||| \; \|w\| \mu_2}{\varepsilon_1 - \varepsilon_2} \qquad (3.23)$$

We now need to invoke a standard result [L1, Corollary 1.9.1, p. 38] with

$$m(t) = e^{\varepsilon_2 t} \|u_2(t,u_o)\|_G , \quad n(t) \equiv M_1 , \quad v(t) \equiv M_2$$

to get

$$\|u_2(t,u_o)\|_G \leq M_1 e^{-M_2 h} e^{-(\varepsilon_2 - M_2)t} , \quad t \geq h > 0 \qquad (3.24)$$

Analyzing (3.19) (ii) and (3.23) (ii), we see that μ_2, hence M_2, can be made as small as we please by suitable selecting Qg. In fact, the range $\mathcal{R}(A_2)$ of A_2 being dense in E_2, we can take $y \in \mathcal{R}(A_2)$ with $\|y\|$ small and define $Qg = A_2^{-1} y$ so that $\|Qg\|$ and $\|A_2 Qg\|$ are so small as to make

$$-\text{Re}\, \lambda_M < -(\varepsilon_2 - M_2) < -\varepsilon$$

where ε is the preassigned constant in the statement of the theorem. Hence

$$\|u_2(t,u_o)\|_G \leq K_{\varepsilon,u_o,h} e^{-\varepsilon t} \qquad t \geq h > 0 . \qquad (3.25)$$

Plugging (3.25) into (3.17) finally yields

$$\|u_1(t,u_o)\|_G \leq \bar{K}_{\varepsilon,u_o,h} e^{-\varepsilon t} \qquad t \geq h > 0 \qquad (3.26)$$

where ε_1 is selected greater than ε. Eqs (3.25) and (3.26) provide the desired conclusion.

APPENDIX. A *constructive* proof of Lemma 3.2 is sketched here. In [N1] instead, a well known existence result on pole assignment, essentially due to Wonham, is invoked from [S1] for its proof. First, consider an arbitrary eigenspace S_i of dimension l_i, corresponding to the eigenvalue λ_i, $1 \leq i \leq (M-1)$. Using the (non necessarily orthogonal) basis, $\Phi_{i1}, \ldots, \Phi_{il_i}$, one can show by direct computations that the restriction of the matrix $A_{g,w}$ over S_i is given by the following $l_i \times l_i$ matrix:

$$\lambda_i I_{l_i} + [W_i G_i]^T, \qquad G_i = \begin{vmatrix} g^1_{i,1}, & \ldots, & g^1_{i,l_i} \\ \vdots & & \vdots \\ g^{l_i}_{i,1}, & \ldots, & g^{l_i}_{i,l_i} \end{vmatrix}$$

where the rows of G_i are, respectively, the coordinates of $g_1, \ldots, g_{l_i}$ restricted over S_i with respect to the chosen basis. Since W_i is of full rank, there is a matrix G_i in S_i such that $W_i G_i = -\alpha_\varepsilon I_{l_i}$ with

$$\alpha_\varepsilon > \max\{\operatorname{Re} \lambda_i, \quad i = 1, \ldots, M-1\} + \varepsilon,$$

and ε is an arbitrarily preassigned positive number. Therefore for such a choice of the G_i's, we have

$$\|e^{(\lambda_i I_{l_i} + [W_i G_i]^T) t}\| \leq C_\varepsilon e^{-\varepsilon t}, \qquad t > 0 \tag{A.1}$$

with ε and C_ε independent on i, $1 \leq i \leq (M-1)$. Next, construct vectors $g_1, \ldots, g_N$ in E_1 by setting:

$$g_1 = [g^1_{1,1}, \ldots, g^1_{1,l_i}, \ldots, g^1_{M-1,1}, \ldots, g^1_{M-1,l_{M-1}}]$$

$$\ldots\ldots\ldots\ldots\ldots\ldots\ldots\ldots \qquad (A.2)$$

$$g_N = [g^N_{1,1}, \ldots, g^N_{A,l_i}, \ldots, g^N_{M-1,1}, \ldots, g^N_{M-1,l_{M-1}}]$$

where one sets $g^k_{ij} = 0$ if $k > l_i$. Then $N = \max\{ l_i, i = 1,\ldots,M-1 \}$.

Finally, since each S_i is invariant under the motion, the desired exponential bound

$$\|e^{A_{g,w}t}\| \leq \bar{C}_\varepsilon \, e^{-\varepsilon t} \qquad t > 0$$

for such g_i's as in (A.2) is obtained from (A.1) plus finitely many applications of the law of cosines. Q.E.D.

REFERENCES

F1 Friedman, A., Partial Differential Equations, reprinted by Robert E. Krieger Publishing Company, Huntington, New York, (1976).

K1 Kato, T., Perturbation theory for linear operators, Springer Verlag, New York, (1966).

L1 Lakshmikantham, V. and Leela, S., Differential and Integral Inequalities, Vol. I, Academic Press, New York, (1969).

N1 Nambu, T., Feedback stabilization for distributed parameter systems of parabolic type, manuscript, revised.

S1 Sakawa, Y. and Matsushita, T., Feedback stabilization of a class of distributed systems and construction of a state estimator. *IEEE Trans. Autom. Contr. AC-20*, 748-753 (1975).

S2 Slemrod, M., Stabilization of boundary control systems, *J. Diff. Equat. 22*, 420-425 (1976).

T1 Triggiani, R., On the stabilizability problem in Banach space, *J. Math. Anal. Appl. 52*, 383-403 (1975); Addendum, Ibid., *56* (1976).

Z1 Zabczyk, J. On decomposition of generators, *SIAM J. Control*, to appear in 1978.

Z2 Zabczyk, J., On stabilizability of boundary control system, Université de Montreal, Centre de Recherches Mathématiques, Report CRM - 785, March 1978.